The Historical Development of Quantum Theory

Jagdish Mehra
Helmut Rechenberg

The Historical Development of Quantum Theory

VOLUME 4

Part 1

The Fundamental Equations of Quantum Mechanics
1925–1926

Part 2

The Reception of the New Quantum Mechanics
1925–1926

Springer-Verlag

New York Heidelberg Berlin

Library of Congress Cataloging in Publication Data
Mehra, Jagdish.
The fundamental equations of quantum mechanics, 1925–1926.
The reception of the new quantum mechanics, 1925–1926.
(The historical development of quantum theory; v. 4)
Bibliography: p.
Includes index.
1. Quantum theory—History. 2. Dirac, Paul
Adrien Maurice, 1902– . 3. Physicists—Great
Britain—Biography. I. Rechenberg, Helmut.
II. Title. III. Title: The reception of the new
quantum mechanics. IV. Mehra, Jagdish. The historical
development of quantum theory; v. 4.
QC173.98.M44 vol. 4 530.1′s [530.1′2] 82-3162
AACR2

Printed in the United States of America.

9 8 7 6 5 4 3 2 1

ISBN 0-387-90680-0 Springer-Verlag New York Heidelberg Berlin
ISBN 3-540-90680-0 Springer-Verlag Berlin Heidelberg New York

Contents

Foreword

This volume deals with Paul Adrien Maurice Dirac's intellectual development, his student years in Bristol and Cambridge, his preparation for entry into research on problems of quantum theory, and finally his formulation of the fundamental equations of quantum mechanics and a consistent mathematical scheme (the algebra of q-numbers) to treat atomic problems.

I first became acquainted with Paul Dirac in April 1955, but it was only from summer 1965 that our contacts began to flourish. I have derived enormous pleasure from meetings and conversations with Paul Dirac at various places in Europe and America, from which I have learned much about his own work on the development of quantum mechanics and recent physics. Dirac has kindly allowed me to use quotations in this work from my conversations and other interviews with him, and I wish to express my gratitude to him. I am also indebted to Mrs. Margit (Mancie) Dirac for many conversations about Dirac and their life in Cambridge.

Subrahmanyan Chandrasekhar has told me about the years he spent in Cambridge. We occasionally discussed the Cambridge scene, the presence there at the time of personalities like J. J. Thomson, Ernest Rutherford, Joseph Larmor, A. S. Eddington, R. H. Fowler, E. A. Milne and, of course, the scientific work and habits of Paul Dirac. I am grateful to Chandrasekhar for these conversations. I also wish to thank Peter Kapitza for a conversation about his years in Cambridge.

I am indebted to Eugene Wigner who has given me over the years every opportunity of pursuing with him scientific, historical and philosophical questions, including the work and personality of Dirac, his 'famous brother-in-law.'

This volume also contains a part on the Reception of the New Quantum Mechanics, i.e., the reception by the physics community of the schemes formulated by Born, Heisenberg and Jordan, and by Dirac. A number of physicists, apart from the principal architects themselves, both in Europe and America—Hendrik Kramers, Lucy Mensing, Otto Halpern, David M. Dennison, J. Stuart Foster, Ralph Kronig, Frank C. Hoyt, John Slater, John H. Van Vleck, Igor Tamm and Lev Landau—applied the matrix scheme to treat atomic problems before the conceptually more intuitive and mathematically more straightforward wave mechanical scheme took over. The new theory was propagated by the principal authors themselves: Born gave lectures at M.I.T., Heisenberg in Göttingen, and Dirac in Cambridge. Soon after the appearance of the work of

Schrödinger, the equivalence of matrix mechanics and wave mechanics was established by Schrödinger himself, by Pauli, and by Carl Eckart.

I am grateful to Aage Bohr for giving me complete access to the *Bohr Archives* and the *Archives for the History of Quantum Physics* (the latter deposited, among other places, at the Niels Bohr Institute in Copenhagen). These rich source materials have been most useful.*

JAGDISH MEHRA

*An inventory of these sources is contained in: T. S. Kuhn, J. L. Heilbron, P. Forman and L. Allen, *Sources for History of Quantum Physics*, The American Philosophical Society, Philadelphia, 1967.

Part 1

The Fundamental Equations of Quantum Mechanics 1925-1926

Introduction

On 28 July 1925 Werner Heisenberg delivered a talk at the Kapitza Club before young research students, associates and friends of Peter Kapitza from the Cavendish Laboratory, who used to gather for informal discussions of recent experimental and theoretical results on physics. Heisenberg spoke on '*Termzoologie und Zeemanbotanik*' ('Term Zoology and Zeeman Botany'), that is, the theory of the observed multiplet structure of spectral lines, a topic which had been reviewed several times in earlier meetings of the Kapitza Club.[1] Ralph Howard Fowler, then Stokes Lecturer at Trinity College, Cambridge, with whom Heisenberg had become acquainted the previous spring in Copenhagen, and who now had invited him to stay with him, was very interested in Heisenberg's work, including his most recent attempts to find a consistent quantum formulation of mechanical laws describing atomic phenomena. As Heisenberg recalled later, 'Fowler was certainly the one who was deeply involved in these things. He knew about all the troubles [in quantum theory]. But I do not know whether there was anybody else who really was interested in these things. I have tried to remember the names of those who [were] at the Kapitza Club at that time, but I just don't know' (Heisenberg, Conversations, p. 275).[2] However, Fowler became excited

[1] For example, D. R. Hartree had discussed Niels Bohr's long paper on line spectra and atomic structure (Bohr, 1923c) at the 40th Meeting of the Kapitza Club on 22 January 1924; P. M. S. Blackett spoke on multiplets and their Zeeman effects on 4 February 1924 (42nd Meeting); H. W. B. Skinner discussed the extension of the laws of X-ray doublets to optical spectra (65th Meeting on 21 October 1924); and L. H. Thomas reviewed Heisenberg's 1924 paper on the anomalous Zeeman effect (Heisenberg, 1924e) at the 73rd Meeting on 20 January 1925.

[2] Heisenberg later recalled that in his talk at Cambridge he had also discussed his most recent work on the quantum-theoretical re-interpretation of kinematics and mechanics (Heisenberg, 1925c). He said: 'Fowler took me out when he played golf and we went for walks. Then he took me to Kapitza Club on time . . . I was asked to speak about this paper of mine [i.e., Heisenberg, 1925c], and I explained all of the details of the paper' (Heisenberg, Conversations, pp. 272–273). However, from the title of his talk, 'Term Zoology and Zeeman Botany,' it is evident that he discussed his earlier theoretical work dealing with the multiplet structure of spectral lines and their associated anomalous Zeeman effects (Heisenberg, 1925b). On being questioned about this particular point, Heisenberg remarked: 'I would think that I talked about all those things in which I was interested at that time, and since these things were for myself a continuation of all the problems on these two subjects [i.e., the theory of multiplet structure, on the one hand, and the quantum-theoretical re-interpretation of kinematics and mechanics, on the other] I just [treated] them under this title . . . I was so much engaged with the whole problem that I could scarcely avoid talking about it' (Heisenberg, Conversations, pp. 487–488).

Although Heisenberg rightly considered his new approach of summer 1925 as 'just a continuation of the older attempts [such as the dispersion-theoretic approach of Kramers and himself to spectroscopic problems, and his own work on multiplet theory of spring 1925], but a more systematic

about the ideas which Heisenberg discussed with him; he told Heisenberg, 'As soon as you have something ready in print, could you send me proofs of it?' (Heisenberg, Conversations, p. 273). Heisenberg did so; as soon as he received the proof-sheets of his paper on the quantum-theoretical re-interpretation of kinematics and mechanics from the *Zeitschrift für Physik* (Heisenberg, 1925c)—and this happened around mid-August, while he spent his summer vacation in Munich and the Bavarian Alps—he immediately forwarded one copy to Fowler in Cambridge.[3] Fowler found the paper interesting, but was uncertain about it and handed it over to Paul Dirac, his research student, for a closer inspection of its contents.

Although Dirac had been a member of the Kapitza Club since fall 1924, and himself spoke on 'Bose's and Louis de Broglie's Deduction of Planck's Law' at the 95th Meeting of the Kapitza Club on 4 August 1925, one week after Heisenberg's talk, neither he nor Heisenberg retained any impression of their mutual encounter on the occasion of Heisenberg's visit.[4] 'It seems that I did not meet Dirac [on that first visit],' Heisenberg has said, 'at least I don't remember it'

---

continuation of which one could hope, but not know, that it would be a consistent scheme' (Heisenberg, Conversations, p. 488), there are good reasons why his talk at the Kapitza Club did not impress the audience as an important breakthrough in atomic theory. First, by the end of July 1925 Heisenberg had not been able to work out a detailed theory of the multiplet structure on the basis of his new ideas; therefore he could not possibly have presented this approach to the members of the Kapitza Club, most of whom were not experts in the atomic spectroscopy and quantum theory as it had developed lately in Copenhagen and Göttingen. Second, Heisenberg's speaking time was limited, for in the same 94th Meeting on 28 July 1925 another guest, the astrophysicist Harold Delos Babcock of Mt. Wilson Observatory, addressed the Club on 'Systematic Errors in Measurements of Close Pairs of Spectral Lines,' a topic which had some connection with Heisenberg's own (see *Minute Book of the Kapitza Club*: For the 94th Meeting, Babcock's entry with signature follows Heisenberg's). Evidently, the very fact that a second talk was given at the meeting of 28 July did not enhance the opportunity for a detailed discussion of Heisenberg's lecture. As a result, Heisenberg did not discover whether 'there was anybody else who was really interested in these things,' as he remarked later.

[3] Max Born had submitted Heisenberg's paper (1925c) to the *Zeitschrift für Physik*, where it was received on 29 July 1925. In the middle of August Born asked Heisenberg (in a letter to Jordan) to send Pascual Jordan a copy of the proof-sheets. In response to this request Heisenberg informed Jordan: 'Unfortunately I have no proof-sheets anymore' ('*Korrekturen hab ich leider keine mehr*,' Heisenberg to Jordan, 20 August 1925), and sent him his original manuscript. It means that Heisenberg had received the proofs before Born's request and had sent one set to Fowler; he returned the second set with the necessary corrections to the editor.

[4] Dirac's name first occurred on the list of members of the Kapitza Club in fall 1924, together with the names of N. Ahmad, P. M. S. Blackett (who spent some time in Göttingen in 1924 and 1925, and was known to Heisenberg), J. C. Boyce, J. A. Carroll, J. D. Cockcroft, R. W. Ditchburn, R. W. Gurney, D. R. Hartree, J. E. Jones, P. Kapitza, L. H. Martin, M. Rogers, H. W. B. Skinner, L. H. Thomas, W. L. Webster and W. A. Wooster. This list of members was entered into the *Minute Book of the Kapitza Club* following the 60th Meeting on 19 August 1924. The 61st Meeting took place on 7 October 1924.

Dirac signed up for his first talk at the 95th Meeting on 4 August 1925. He spoke on several occasions during the following months: on 'Direction and Ejection of Photo-Electrons,' 99th Meeting, 13 October 1925; on Debye's 'Note on the Scattering of X-rays by Liquids' (Debye, 1925), 100th Meeting, 6 November 1925; on 'The Light Quantum Theory of Diffraction,' 107th Meeting, 15 December 1925; on 'Fundamental Equations of Quantum Mechanics,' 109th Meeting, 19 January 1926.

(Heisenberg, Conversations, p. 273). However, he recalled that Fowler had told him at the time about 'an extremely good mathematician,' who had studied electrical engineering and who might 'perhaps be interested' in his ideas on the quantum-theoretical re-interpretation of kinematics. Dirac did not participate in the private discussions between Fowler and Heisenberg. Confessed Dirac, 'I did not hear anything about Heisenberg's "matrices" until I got a copy of the proofs from Fowler' (Dirac, Conversations, p. 94).[5] And even when he received and read the paper about a month after Heisenberg's visit, he was not particularly impressed by it. Said Dirac: 'I have often tried to recall my early reaction [to Heisenberg's paper], but I cannot remember what it was. I suppose it was just some disparity between that and the Hamiltonian formalism. I was so impressed then with the Hamiltonian formalism as the basis of atomic physics that I thought anything not connected with it would not be much good. I thought there was not much in it [Heisenberg's paper] and I put it aside for a week or so' (Dirac, Conversations, p. 58; see Mehra, 1972, p. 31).

During the previous year Dirac had become thoroughly acquainted with the problems of atomic theory and started to make his own contributions to it, especially in two papers dealing with the adiabatic invariance of quantum systems (Dirac, 1925a, c). He had also become convinced that the future quantum theory, which would account for the entire bulk of empirical data collected by the spectroscopists, had to be formulated in terms of a Hamiltonian scheme. He, therefore, found it difficult to understand Heisenberg's formulation of a quantum-theoretical kinematics, arising from calculations of the anharmonic oscillator, which seemed to have nothing to do with a Hamiltonian theory. It took him more than a week until he discovered that Heisenberg's principal step had been the introduction of a fundamentally new idea into quantum theory: namely, the noncommutativity of physical quantities. As Dirac recalled later: 'It was already in September of 1925 that I realized that this paper would mean a breakthrough. In fact, it seemed pretty obvious then, and I really find it more difficult to understand why I did not see it the first time I looked at this paper. But then I worked quite hard to try to connect these "matrices" [i.e., Heisenberg's dynamical variables] with the action and angle variables' (Dirac, Conversations, p. 58).

After several weeks of intensive search, Dirac got the decisive idea. 'During a long walk on a Sunday it occurred to me,' said Dirac, 'that the commutator might be the analogue of the Poisson bracket, but I did not know very well what a Poisson bracket was then. I had just read a bit about it, and forgotten most of what I had read. I wanted to check up on this idea, but I could not do so because I did not have any book at home which gave Poisson brackets, and all the libraries were closed. So, I had to wait impatiently until Monday morning when

[5]To the best of his recollection Dirac attended Heisenberg's talk on 28 July 1925. Even if Heisenberg did refer to his work on the quantum-theoretical re-interpretation of kinematics and mechanics, it was still too early for him to talk about 'matrices.'

the libraries were open and check on what Poisson bracket really was. Then I found that they would fit, but I had one impatient night of waiting' (Dirac, Conversations, p. 54; see Mehra, 1972, p. 31).

With the help of the analogy between the commutator and the Poisson bracket Dirac was back on familiar ground; with Heisenberg's noncommuting quantum-theoretical variables he could build a new kind of Hamiltonian dynamics which would replace classical Hamiltonian mechanics for describing atomic phenomena. After he had obtained some definite results towards this new formulation, he showed them to his supervisor. Fowler was very impressed with what Dirac had extracted from Heisenberg's work, and he saw the need to publish the new findings without delay. He persuaded Dirac to write up a paper which he communicated to the *Proceedings of the Royal Society*. 'He told the Royal Society,' Dirac remembered, 'to give priority to my paper and they published it extremely quickly. I can thank Fowler for that, because he appreciated that it was urgent, and I suppose that he was thinking that there might be competition from other places [especially from Göttingen and Copenhagen]' (Dirac, Conversations, p. 59). Dirac's paper, entitled 'The Fundamental Equations of Quantum Mechanics' was received on 7 November 1925 and appeared in the 1 December issue of the *Proceedings of the Royal Society* (Dirac, 1925d).[6] With this paper, and the other which followed it in quick succession, Dirac joined the ranks of the principal architects of quantum mechanics.

[6]R. H. Fowler had been elected a Fellow of the Royal Society of London in 1925. James Hopwood Jeans, a former Cambridge scientist, was Secretary of the Royal Society from 1919 to 1929 and thus the editor responsible for the *Proceedings of the Royal Society*, Series *A*. Jeans was helpful in the rapid publication of Dirac's early papers on quantum mechanics.

Chapter I
Paul Dirac's Intellectual Development

I.1 Growing Up in Bristol

Paul Adrien Maurice Dirac was born on 8 August 1902 in Bristol, England, the son of Charles Adrien Ladislas Dirac and his wife Florence Hannah Holten. He was the second of three children, having an older brother and a younger sister. His father, who was Swiss by birth, had left home for England as a young man, married in Bristol and settled there.[7] Charles Dirac was a teacher of French in the Merchant Venturer's Secondary School in Bristol, the school in which Paul also received his early education. He was a retiring man. As Dirac recalled, 'Nobody ever came to our house, except a few pupils of my father, who came from lectures. No one ever came for social purposes' (Dirac, Conversations, p. 5). The need for social contacts was not emphasized; even the meals were not taken together in the Dirac family. As a rule Paul would eat with his father in the dining room, while his brother and sister sat in the kitchen with their mother. 'I would have preferred to be with the others,' said Dirac, 'but there were not enough chairs in the kitchen' (Dirac, Conversations, p. 6). At home, young Dirac followed the rule, which his father had set, of talking to him only in French in order to learn the language. Whenever Paul found that he could not express himself well in French, he would stay silent. This arrangement rather early led to a habit of reticence. Paul was an introvert and, being often silent and alone, he devoted himself to 'thinking about problems of nature.' Thus, he early acquired two habits: the ability to think for himself and a love for the quiet contemplation of nature.[8]

[7]Dirac heard from his mother that his father had had an unhappy childhood. 'He ran away from home without telling his parents and went to England. His parents never knew that he was married until some years later when he visited them with a couple of children' (Dirac, Conversations, p. 6).

Charles Dirac's family came from Monthey in Canton Valais, Switzerland. When Paul Dirac, as a child of two, first visited Switzerland his father's family lived at *3 Rue Winkelried* in Geneva. Paul's own relationship with his father had not been a happy one and as a result of his memories of his childhood he did not visit his father's 'home,' i.e., Geneva, again for seventy years. In summer 1973 I was in Geneva; Paul and Margit Dirac came to see me there and stayed from 1 to 4 July. Dirac remembered numerous sights from his visit to Geneva as a child: e.g., *Rue Winkelried, L'Isle Rousseau*, and the little fishing village of *La Bellotte* on Lake Geneva. (J. Mehra)

[8]Some of the most intensely personal discussions I have had with Dirac took place in Geneva (see footnote 7). In one of these, at dinner in the *Pavillon de Ruth* at Lake Geneva, 2 July 1973, Dirac remarked: 'Things contrived early in such a way that I should become an introvert.' (J. Mehra)

Dirac's father appreciated the importance of a good education and encouraged him to study mathematics. Merchant Venturer's School in Bristol was a very good school, concentrating on science (mathematics, physics, and chemistry, but no biology), modern languages and a little history and geography. In contrast to most secondary schools, Latin and the classics were not included in the curriculum; students intending to go to Oxford or Cambridge, where Latin was required for admission, would learn it as a separate subject. Most of the students going on to college from Dirac's school would go to Bristol University and pursue studies in science or engineering.[9] The school shared its buildings with the engineering college of Bristol University, so that it was a secondary school in the daytime and a technical college in the evening. As a result the laboratory facilities of the college were available to the school, and Dirac even obtained some practical experience of metal work.

The Merchant Venturer's School left a deeper impression on young Paul Dirac than on other boys of his age, who participated in a richer family life. He considered himself lucky to have gone to this school. Certainly it was rather specialized and emphasized science over arts, but this agreed with Paul's personal tastes as he did not like the arts side of the curriculum anyway. For instance, he read only a few novels, mostly in connection with his study of English. As he confessed later, 'I was always a slow reader and did not like poetry. In fact, I never understood good poetry. My whole interest was on the scientific side and I was really ignorant about matters outside my school work' (Dirac, Conversations, p. 18).[10] At school he concentrated mainly on the science courses. In physics there were three hours a week of lectures with one afternoon of practical work, and the course covered mechanics, heat, light and sound. Dirac's chemistry teacher believed in teaching chemistry in the modern way and introduced atoms and chemical equations very early in the course, using atomic weights rather than equivalent weights right away. In most of the subjects, which also included courses in English, French and German, Paul did not go much beyond the class, with the exception of mathematics.

Mathematics, which interested Dirac most, was divided into algebra, geometry and trigonometry. He improved and extended his knowledge of mathematics by going thoroughly through the books which he had and others which he obtained from his teachers; thus he did a lot of mathematical reading on his own, at a

[9] Bristol University was established in 1909; it grew out of Bristol University College, which was founded in 1876. The University grew rapidly thanks to generous endowments, especially from the Wills family. The main building opened in 1925.

[10] Dirac once remarked to J. Robert Oppenheimer: 'How can you do physics and poetry at the same time? The aim of science is to make difficult things understandable in a simple way; the aim of poetry is to state simple things in an incomprehensible way. The two are incompatible' (reported to J. Mehra by Archibald MacLeish in conversations, 6 April 1966). Dirac's attitude towards literature comes through in a story told by George Gamow. Dirac had read through an English translation of Dostoevsky's *Crime and Punishment*, and on being asked how he liked it, he answered: 'It is nice, but in one of the chapters the author made a mistake. He describes the Sun as rising twice on the same day.' This was Dirac's only comment on Dostoevsky's novel (Gamow, 1966, pp. 121–122).

more advanced level than the rest of the class. He worked rather early through books on calculus (Edwards) and geometry (Hall and Knight).[11] Dirac also became acquainted with the methods of non-Euclidean geometry, but it did not attract him. As he recalled later: 'When I was at school, one of my mathematics masters had told me that I would probably be very interested in non-Euclidean geometry, and suggested books which I should read on the subject. However, I was not interested in it. The reason for that is that I was interested in the real physical world, and it seemed to me to be obvious that the real physical world was based on Euclidean geometry. There was therefore no need to consider any other kind of geometry' (Dirac, 1977, p. 112).

Besides mathematics Dirac developed an early interest in the mathematical formulation of fundamental physical questions. 'I would spend much time just thinking about them,' he recalled. 'For instance, I thought out for myself that there might be a connection between space and time coordinates such that when one changes one's time axis, one would also rotate one's space axis' (Dirac, Conversations, p. 3). However, he did not make much headway with this problem. 'I knew nothing about hyperbolic geometry at that time. I could [only] see that it wouldn't work just with Euclidean geometry' (Dirac, Conversations, p. 3). He learned later that the specific problem of simultaneous rotation of space and time axes had already been solved; however, he would continue to think about the fundamental problems of space and time and make profound new contributions to them.

Dirac's interest in mathematics, and independent studies in it, helped him to develop a personal method of learning a new subject that he described as follows: 'I like to jump forward and back again, [and do so] continually. One cannot very well do that if one is listening to an ordered presentation like a lecture. When I go to lectures I usually just get stimulated to think on certain lines, and then I think along those lines instead of listening to every word the lecturer says. I perhaps miss a good deal of the lecture for that reason and have to make it up later in my own reading. But all my learning in mathematics has been rather along these lines' (Dirac, Conversations, p. 3).

Dirac progressed through the Merchant Venturer's School quite rapidly, the reason being that the older students were sent off to do war work (World War I), leaving empty the higher classes; Dirac had the advantage of being pushed into a class higher than would correspond to his age throughout his secondary school, thereby obtaining the opportunity of learning things, especially the sciences, quite early.

On completing school at the age of sixteen, Paul Dirac followed in the footsteps of his elder brother who had studied engineering. 'I did not have much initiative of my own,' Dirac said. 'This path was rather set out for me, and I did

[11] Dirac did not need either too many or any special books to learn a subject. As he remarked: 'I don't think that it matters very much which books one reads if one has reasonably good books' (Dirac, Conversations, p. 2).

not know very well what I wanted' (Dirac, Conversations, p. 3). Thus it came about that he entered the University of Bristol at the age of sixteen as a student of electrical engineering in fall 1918.

I.2 The Education of an Electrical Engineer

The passage from high school to the engineering college was very smooth for young Dirac. His studies were carried on in the same building as before, and even some of the teachers were the same.[12] One of them, David Robertson, who had taught him physics at school was now his professor of electrical engineering. Robertson taught both theoretical and practical subjects. Under his influence Dirac chose to specialize in electrical engineering, but he continued to nourish his interest in physics and mathematics. Robertson was paralyzed in his legs from polio and had to get around in a wheel chair; he had organized his life very methodically and he impressed upon his students the same need for organization. Dirac took over some of Robertson's attitude in organizing his daily life.[13]

In Robertson's courses on electrical engineering Dirac learned all about electrical circuits and how to design them, as well as the rules for winding electrical motors and dynamos. He found the mathematics in the equations and rules used for these purposes quite interesting. 'I am grateful to David Robertson for explaining these things in a way that would show mathematical beauty in the calculations,' Dirac said later (Dirac, Conversations, p. 18). However, Robertson did not teach Maxwell's electromagnetic theory in a systematic manner. The curriculum was concentrated on power engineering; electromagnetic waves, wireless, and other electrotechnical topics were not included. Dirac did become familiar with the Heaviside calculus while learning to solve linear differential equations with constant coefficients, which occurred in the problems of electromagnetic circuits. In engineering courses the emphasis was on mathematical rules with the help of which problems could be solved, without giving strict proofs or

[12] The mathematicians and physicists at Bristol University were housed in a different building about half a mile away. During his engineering studies Dirac never went to their building and had little or no contact with them. The possibility of staying on in familiar surroundings contributed to Dirac's decision to study engineering.

[13] Dirac's habit of organizing things is illustrated by a story which Heisenberg has recalled:

> Toward the end of my stay in America [in 1929], I made arrangements to return home with Paul Dirac. We would meet in Yellowstone Park, go walking for a few days, then sail across the Pacific to Japan and return to Europe via Asia. Our meeting place was the hotel in front of Old Faithful [Geyser], and since I got there a day earlier, I did some mountaineering on my own . . . . In my letter to Paul Dirac I had mentioned that we might perhaps look at some of the other geysers, and that with luck we might see one or two in action. It was characteristic of Paul's careful and systematic ways that, when we met, he had already worked out a precise itinerary in which he had not only marked the times of activity of these natural fountains, but had mapped out precise routes that would bring us, in the course of the afternoon, to the greatest possible number of geysers just in time to watch them spring into action (Heisenberg, 1971, p. 100).

asking how or why the rules worked. Dirac remembered that there always remained a kind of magic about these rules, and that frequently he had a strange feeling about how he ever got answers out of them.

Dirac also learned certain aspects of general engineering. For instance, he obtained some experience of testing materials by pulling metal wires and other objects until they would break, calculating the breaking stress in individual cases. Indeed, the calculation of stresses in solid structures, say the stress on a whirling shaft, played an important role in his training. Again, in all these courses in the engineering college, whether they were on dynamics or general engineering, the emphasis was always on learning the equations that would give the right answer so that one could then calculate the result with the help of a slide rule. Although, from the physicist's point of view, this procedure was quite unsatisfactory, Dirac derived some benefit from it. 'I think it was probably [this] sort of training that first gave me the idea of a delta function. Because when you think of loads of engineering structures, sometimes you have a distributed load and sometimes you have a concentrated load [at a point]. In the two cases you have somewhat different equations. Essentially, the attempt to unify these two things leads to the delta function' (Dirac, Conversations, p. 19). At that time Dirac did not manage to discover such a unified treatment, but he strongly felt the urge for doing so.[14]

While Dirac enjoyed the theoretical part of his engineering studies, he did little experimental work. On one occasion, during the long university vacation in 1920, Dirac worked at the British Thompson–Houston Works in Rugby. There he obtained some practical experience, but it was not satisfactory; he failed to please his employers and they sent an unfavourable report about his work to David Robertson.[15] As time went on, he never had to do any practical work in engineering to earn his living. The immediate reason for this was less Dirac's dislike of such work than the fact that, when he graduated from Bristol University in 1921 with a bachelor's degree, Great Britain was in the midst of an economic depression; Paul could not find a position in engineering. Robertson suggested to him that instead of waiting around he should do some research work. He gave him a problem on stroboscopes, but Dirac did not get very far with it, for soon he was offered the opportunity of continuing his studies in mathematics which suited his talents better.

His father Charles Dirac had hoped that Paul would continue his education at

[14]When Dirac finally introduced the symbolic delta function in fall 1926, he did not refer to the earlier uses of a similar idea by Oliver Heaviside (1893a, b); he had come upon this concept independently on the basis of his studies in electrical engineering.

[15]At the British Thompson–Houston Works in Rugby Dirac learned to turn things on lathes and did metal work like filing and drilling. Robertson showed Dirac the report from Rugby about his work; it complained that he lacked keenness and was slovenly.

The two months at the factory in Rugby was the first time that Dirac lived away from home. His brother was working in the same factory, but the two never took the opportunity to come into contact. 'If we passed each other in the street, we didn't exchange a word' (Dirac, Conversations, p. 7).

an appropriate place upon taking his engineering degree from Bristol. To this end he sent Paul in summer 1921 to Cambridge to try for a scholarship at the university. The examinations for the open scholarships usually took place in December, and Paul was too late for them; but St. John's College offered some 1851 Exhibition Studentships for which the examination was held in June.[16] Paul took the examination and succeeded in getting one of the Studentships, worth £70 per annum. This amount was not enough to support him in Cambridge, and he was not able to raise any other funds to supplement it. He, therefore, remained in Bristol and continued to stay with his parents.

The mathematicians at Bristol had learned about Dirac's mathematical ability from the record of his examinations at school. They had hoped that he would specialize in mathematics, and were quite disappointed when he decided to go into engineering. Although he liked mathematics very much and would have liked to pursue it, he did not know that one could earn a living from pure science. The thought of pursuing an education in mathematics and becoming a school teacher did not appeal to him. He had studied engineering in the belief that it would lead to a satisfactory career; but in fall 1921, having completed his engineering education, he found himself without a job. The mathematics department at Bristol University offered Dirac free tuition to study mathematics; since he had no definite plans for the future, he accepted.[17]

Dirac never regretted the time he had spent on engineering. 'I owe a lot to my engineering training,' he said, 'because it did teach me to tolerate approximations. Previously to that I thought that any kind of approximation was really intolerable and one should just concentrate on exact equations all the time. Then I got the idea that in the actual world all our equations are only approximate. We must just *tend* to greater and greater accuracy. In spite of the equations being approximate they can be beautiful' (Dirac, Conversations, p. 20; see Mehra, 1972, p. 19). Dirac thus learned the important lesson, which he later applied successfully time and again, that in constructing physical theories one has first to neglect a lot of factors, as the actual situation to be described is usually much too complicated. The real task consists in finding suitable approximations. Dirac became convinced that to have a general feeling about what is important and what can be neglected would not only make a good engineer but a good physicist as well. As a result of this conviction he has 'never been much interested in questions of mathematical logic or in any form of absolute measure or accuracy, an absolute standard of reasoning.' He always felt that 'these things are just not

[16]The Great Exhibition of 1851 in London realized a considerable profit and Albert, the Prince-Consort, decided to establish a fund to help the education of deserving young scientists by the award of scholarships.

[17]The mathematician whom Dirac knew best in the mathematics department at Bristol University was Peter Fraser, and it is very likely that it was he who informed Dirac about the possibility of studying mathematics with free tuition. Dirac, being very retiring, was not personally known to mathematicians other than his school teachers; nor did he approach them himself in such matters.

important and that the study of nature through getting [successive] improving approximations is the profitable line of procedure' (Dirac, Conversations, p. 1).[18]

I.3 Studying Applied Mathematics

The course of mathematics at Bristol University normally lasted three years, but because of Dirac's previous training they let him off one year. During the first year Dirac studied both pure and applied mathematics, but in the second year he had to specialize in one of the two. When it was time to decide, he did not know which one to choose, as he liked both. However, Dirac and a Miss Beryl M. Dent were the only two students in the honours course in mathematics. The professor wanted both of them to take the same option for specialization, as they would then not have to give two sets of courses. Since Miss Dent was quite determined to pursue applied mathematics, Dirac also decided upon it.[19] This was his way back to science; the choice had been made for him.

In the course of his mathematical studies at Bristol University Dirac came under the influence of his teachers Henry Ronald Hassé and Peter Fraser. Hassé taught applied mathematics, and Fraser lectured on pure mathematics. Both of them had come from Cambridge. Hassé, the Head of the Mathematics Department, had been appointed after World War I in 1919.[20] Fraser, however, had already joined the staff of Bristol University College, which soon thereafter became Bristol University, in 1906; he kept the mathematics department going after its first head, F. R. Barrell, died in 1915 until Hassé arrived. Fraser was a man whose 'interests ranged over the whole of the pure mathematics of his day, but his approach was always from the geometrical rather than from the analytical side' (Hodge, 1959, p. 111). Dirac thought that Peter Fraser was the best

[18]Dirac's approach to physical problems differed strikingly from that of other physicists. For instance, Heisenberg expressed his belief that 'you can never solve only one difficulty at a time, you have to solve quite a lot of difficulties at the same time' (Heisenberg, 1969a, p. 46). Dirac's response: 'A good many people have Heisenberg's view, but it is just too difficult to solve a whole lot of problems together. I rather early (already in Bristol) got the idea that everything in Nature was only approximate, and that science would develop through getting continually more and more accurate approximations but would never attain complete exactness. I got that point of view through my engineering training, which I think has influenced me very much' (see Mehra, 1972, p. 51).

[19]Miss Dent graduated with Dirac in applied mathematics. Later, she worked in theoretical physics. When John Edward Lennard-Jones from Cambridge became a Reader in Mathematical Physics at Bristol University in 1925, Miss Dent became one of his first collaborators. They published joint papers dealing with the forces between atoms and ions and on an application of the theory developed there to carbonate crystals (Lennard-Jones and Dent, 1926, 1927a, b). The collaboration of Lennard-Jones and Miss Dent on problems of molecular physics continued at Bristol (see, e.g., Lennard-Jones and Dent, 1928).

[20]H. R. Hassé was in Cambridge and knew E. Cunningham before World War I. Cunningham thanked him in the preface to his book *The Principle of Relativity*, signed June 1914 (Cunningham, 1914, p. vi). During the war Hassé collaborated with J. B. Henderson on the application of thermodynamics to develop a consistent theory of the explosions in a gun. They published their results later on (Henderson and Hassé, 1922). He retired from his Bristol chair in 1949.

teacher he ever had.[21] Fraser introduced him to ideas of mathematical rigour and made pure mathematics really attractive. He gave rigorous proofs in differential and integral calculus. This was a new experience for Dirac; previously he had just learned the rules well enough for engineering purposes and, therefore, found it hard to appreciate that rigour was needed at all. He felt then, and continued to believe later, that when one is confident that a certain method gives the right answer, one does not have to bother with mathematical rigour.

Fraser emphasized the geometrical approach to mathematical thinking and one course of Fraser's, which Dirac found most interesting, was on projective geometry. He was fascinated by this topic and recalled later: 'One could get quite powerful results, like theorems about straight lines and conics intersecting each other, just from elementary arguments about one-to-one correspondence. That appealed to me very much. All my later work since has been very much of a geometrical nature rather than of an algebraic nature' (Dirac, Conversations, p. 4). The last statement is surprising in view of the fact that Dirac's main contributions to theoretical physics rest firmly on algebraic methods. Nevertheless, Dirac always maintained that his own thinking about physical problems was largely 'geometrical' and that he could not carry through algebraic operations without picturing what the equations meant.[22] Fraser's geometrical view of mathematics had left a decisive impression on Dirac.

[21] Some information about Peter Fraser is contained in an obituary notice written by William Vallance Douglas Hodge who, as a lecturer in mathematics at Bristol University from 1926 to 1931, had been Fraser's colleague (Hodge, 1959). Fraser was born on 7 October 1880 in Inverness-shire, the only son of a farmer. He was educated at the Inverness Royal Academy and Aberdeen University; he graduated B.Sc. with special distinction in 1902, and also obtained his M.A. with first class honours in mathematics. He went up to Cambridge and devoted himself mainly to geometry. He won the Ferguson scholarship in mathematics in 1904 and was 10th Wrangler in 1905. In Part II of the Mathematical Tripos he was placed in Division 2 of the First Class. From 1906 to 1946 Fraser served on the staff of Bristol University, having been appointed Reader in Geometry in 1926. On retirement he lived near Torquay, where he died on 18 June 1958.

Fraser never published any mathematical paper. However, as his biographer Hodge observed that he kept himself informed about the development of mathematics up to the day of his retirement. 'He always seemed to have things in perspective. It was this grasp of the essentials in mathematics that made him a real power in the mathematical life of the community. His pupils who went on to make a name for themselves in mathematics owed nearly everything to him; for he sought them out at their most formative stages and talked to them, over coffee, on country walks, or on the golf course, and shared his wisdom with them. He was at the service of all who needed his help' (Hodge, 1959, p. 111). Hodge recalled that often, when someone ran into difficulties with his research work, he would suggest a game of golf to Fraser. 'When the match was under way he would propound his difficulties, and in between incredibly bad shots on both sides Peter, as he was known to everyone, would give his views or expound some theory in simple, lucid terms, and the difficulties would melt away' (Hodge, 1959, p. 112).

Fraser was accessible not only to his students and younger colleagues, but also older and renowned mathematicians such as his teacher Henry Frederick Baker from Cambridge, benefited by his wisdom on occasion. 'It was his grasp of mathematics as a whole and his preference for the perfect edifice that the really important things in mathematics formed that made him [Fraser] reluctant to add anything that seemed a useless embellishment.' However, Fraser was not unproductive. To visitors who told him about some results of their current research he would show 'a bundle of notes, yellow with age in which, years earlier, he had obtained the same result in a more elegant form' (Hodge, 1959, p. 112).

[22] For example, Dirac found it difficult to follow some of the post-World War II developments in theoretical physics, especially in elementary particle physics—such as dispersion relations and

Of course, Dirac learned much algebra too, though mainly on his own. In view of his later work on gamma matrices in connection with the relativistic electron equation, it is of interest to note that he also read an old-fashioned book about quaternions.[23] He found this book very difficult to work through since its terminology, in Dirac's opinion, was very clumsy: for example, there was too much emphasis on separating the scalar and vector parts of quaternions instead of treating them as one entity. Dirac felt that he could not do much with the quaternions described in this book, nor did he apply the results of the theory of quaternions in his papers later on. He always thought that he would have appreciated the quaternions more if at that time he had come across a book better suited to his understanding.

The two years of mathematical study at Bristol helped Dirac in becoming a more mature person, who knew better where his interests lay. Not that he had changed his shy and retiring character: he still had little contact with people, though David Robertson and Peter Fraser, who appreciated his gifts, invited him to their homes now and then. But he had been introduced to pure science, a field which he would now pursue. He planned to go up to Cambridge University. Both of his mathematics teachers, Fraser and Hassé, knew people in Cambridge to whom they recommended Dirac: Fraser had contacts with Henry Frederick Baker, Lowndean Professor of Astronomy and Geometry in Cambridge, and his geometry circle, and Hassé knew Ebenezer Cunningham. Hassé had helped Cunningham in the preparation of his book on *The Principle of Relativity* (Cunningham, 1914). In fact, it was Cunningham who had examined Dirac in summer 1921 for the 1851 Exhibition Studentship. For a long time Dirac had been fascinated by the theory of space and time; he looked forward to meeting Cunningham again in Cambridge and to working with him in electromagnetic theory and relativity.

Regge-pole theory—which employed mainly algebraic methods. He also never cared for the renormalization theory in quantum electrodynamics for the simple reason that he had 'not found the corresponding picture for the renormalization of charge' (see Mehra, 1972, p. 50).

[23] Dirac did not remember exactly which book on quaternions he had read in Bristol. From his description it appears likely that it was one of the two books on that topic by Peter Guthrie Tait: either *An Elementary Treatise on Quaternions* (Tait, 1867) or *Introduction to Quaternions* (Kelland and Tait, 1873). Many old treatments of quaternions, including the books of P. G. Tait, dealt not only with quaternions as such, but also discussed their connection with vectors and their operations. For example, one could transform a vector into another vector by multiplying it with a quaternion, the latter being composed of a scalar and a vector part. At the time when vector calculus was developed such relations were very important. Later, however, the vector methods became dominant in the language of physical theories, and quaternions disappeared. Dirac always felt that quaternions ought to have played a larger role, especially in quantum theory, than they did. The reason for this belief was the fact that quaternions provided the most general algebra containing the operations of division and associative multiplication, and that was what was needed for quantum theory. However, although Dirac attempted to make use of quaternions in quantum theory and also searched for a relation between q-numbers and quaternions, he did not get very far with these efforts. After Dirac had invented the algebra of γ-matrices for the description of the relativistic electron, Sommerfeld immediately recognized (as early as August 1928) their connection with Hamilton's quaternions (see Sommerfeld, 1929, Chapter II, Section 10E).

I.4 Fascination of Relativity Theory

Early in 1919, 'a wonderful thing happened. Relativity burst upon the world, with a tremendous impact. Suddenly everyone was talking about relativity. The newspapers were full of it. The magazines also contained articles written by various people on relativity, not always for relativity but sometimes against it. Relativity was understood in a very wide sense, and was taken up by philosophers and by people in all walks of life' (Dirac, 1977, p. 110).

What were the reasons for such a wide interest in a scientific theory? The First World War had just ended. It had been a long and difficult war, which had given rise to innumerable casualties and vast tragedy. Everybody was sick and tired of the war; there was hunger for new thoughts and a completely different outlook. Exactly at that moment Einstein's general relativity theory was available. 'It was an escape from the war,' said Dirac. 'The impact that relativity produced I think has never been equalled either before or since by any scientific idea catching the public mind' (Dirac, 1977, p. 110).

In Great Britain two facts contributed to the rapid propagation of relativity theory: first, a great authority, the astronomer Arthur Stanley Eddington, was the champion of relativity at that time; second, the results of the British Solar Eclipse Expedition of 1919. Before these events occurred, little had been known about relativity theory, though certain British physicists had contributed to the foundation of relativity and electron theory. Einstein's work of 1905 had received the attention of only a few specialists outside the Continent and his name was not generally known. To pragmatic minds special relativity theory did not seem to be particularly interesting, as it added only minor details to the understanding of the behaviour of electrons. However, the situation changed with the advent of general relativity, which claimed to improve upon and replace Newton's theory of gravitation. It was also in this context that Eddington, then Plumian Professor of Astronomy and Director of the Cambridge Observatory, became involved, and it came about as follows.

Willem de Sitter, Professor of Astronomy at the University of Leyden and later Director of the Leyden Observatory, had received a reprint of Einstein's long memoir on general relativity—published in issue No. 7 of Volume 49 of *Annalen der Physik*, which was distributed on 11 May 1916 (Einstein, 1916c)—and he passed it on to the Royal Astronomical Society of London, of which he had been an Associate Member since 1909. Eddington, then Secretary of the Society, was the official recipient of this paper, which was then not otherwise available in England because of the war. He recognized its importance and asked de Sitter to write for the *Monthly Notices* two articles explaining general relativity and its consequences in astronomy (de Sitter, 1916a, b). Eddington was impressed by the new development, especially by Einstein's calculation of the perihelion motion of Mercury, and he brought it to the attention of the public already in his report on 'Gravitation' at the 1916 meeting of the British Association (Cunningham and Eddington, 1916). He very soon mastered the necessary mathematical tools and started to contribute actively to general relativity. He also gave popular lectures and wrote articles on this subject. For example: 'On 1 February 1918 he delivered a Friday lecture at the Royal Institution, and, being thereby tempted to go deeper into the subject, he wrote out a systematic treatment of the theory, with special reference to the astronomical tests, which was published as a Report to the Physical Society in the summer of 1918 [(Eddington, 1918)]' (Dreyer, 1924, p. 556). Eddington's talk at the Royal Institution and the published *Report* called public attention to the possibility of testing the theory by observing the deflection of light from stars passing close to the sun during the total eclipse of the sun.

The early publications on general relativity in the *Monthly Notices* had been noticed by the Astronomer Royal, Sir Frank Watson Dyson. In a note to the same journal, dated 2 March 1917, he drew attention to the unique opportunity of verifying or disproving Einstein's prediction concerning the deflection of starlight in the sun's gravitational field on 29 May 1919 when many bright stars would be observed in the vicinity of the sun (Dyson, 1917). A total solar eclipse was then expected in some regions of South America and Africa, for which astronomical expeditions had to be prepared. In spite of the difficult wartime conditions—the U-boat blockade was tightening on Great Britain—Dyson was given £1000 by the Government and a Joint Permanent Eclipse Committee of the Royal Society and the Royal Astronomical Society was set up under his chairmanship. 'In 1918, in the darkest days of the war, two expeditions were planned, one by Greenwich Observatory and one by Cambridge, to observe, if the state of civilisation should permit when the time came, the eclipse of May 1919 with a view to a crucial test of Einstein's generalized relativity' (Jeans, 1925a, p. 677).

Fortunately for the astronomers the Armistice was signed in November 1918. In March 1919 the scientists and leaders of the two teams, Andrew Claude de la Cherois Crommelin and C. R. Davidson from the Greenwich Observatory—who were supposed to go to Sobral, Brazil—and Eddington and E. T. Cottingham from Cambridge—who would go to Principe Island on the Gulf of Guinea—met in Flamsteed House, Greenwich. Eddington was most optimistic about the fact that the outcome would prove general relativity. Thus, when his collaborator asked in Dyson's study: 'What will it mean if we get double the Einstein deflection?,' Dyson said: 'Then Eddington will go mad and you will have to come home alone' (Douglas, 1956, p. 40). On 29 May, during the solar eclipse at Principe Island, the weather conditions nearly prevented Eddington from obtaining the necessary photographs, and only on 3 June did he find out that on one plate he had a result and that one agreed with Einstein's prediction. Turning to his companion he said: 'Cottingham, you won't have to go home alone' (Douglas, 1956, p. 40). The Crommelin team in Sobral, on the other hand, brought seven excellent photographs confirming the general relativistic deflection of light.

It was not until 6 November 1919 that the official results of the two eclipse expeditions were reported to the Fellows of the Royal Society and the Royal Astronomical Society who met in Burlington House, London. After Dyson had read the reports, Joseph John Thomson, then President of the Royal Society, hailed Einstein's general relativity theory as 'the greatest discovery in connection with gravitation since Newton enunciated his principles.' This remark was reported in London's daily press, which gave enormous publicity to Einstein's general theory by devoting leading articles to explain its consequences.[24] A few weeks later Eddington informed Einstein that 'all England has been talking about your theory' (Eddington to Einstein, 1 December 1919).

For Eddington the confirmation of Einstein's theory by the British Solar Eclipse Expeditions of 1919 turned out to be a personal triumph.[25] He gave brilliant lectures on the theory that attracted large audiences. For example, on 2 December 1919 he talked in the hall of Trinity College, Cambridge, and the correspondent of *Nature* gave the following report:

> Fifteen minutes before the lecture began there was a queue half-way across the Great Court of men anxious to obtain admittance, and during the lecture the hall was entirely filled with dons and students listening breathlessly to hear an intelligible account, if one could be given, of the new theory. The keen interest was due, no doubt, largely to the curiosity stimulated by the newspaper accounts of the subject, but also partly to the feeling, to which at least some hope of satisfaction can be given, that a further great unifying principle is needed in natural philosophy. Whatever be the reason, however, the size and appreciation of the audience were not less extraordinary than the subject of the lecture and the brilliance of its exposition. (*Nature* **104**, p. 385, issue of 11 December 1919)

Some months after the release of the solar eclipse data, he gave a popular account of relativity theory in his book on *Space, Time and Gravitation* (Eddington, 1920). No doubt Eddington 'was the great authority whom everyone listened to with the greatest respect, and he was rather regarded as the chief exponent of relativity. Einstein was in the remote background' (Dirac, 1977, p. 110). And his fame in connection with relativity also reached Bristol and the engineering student Paul Dirac.

Since his early, rather lonely youth Dirac had thought about the problems of nature and had continued to develop this interest throughout his studies at the

[24] The London *Times* reported about this meeting the following day (7 November) and presented a leading article entitled 'The Fabric of the Universe' on 8 November, in which Thomson's remark was quoted. On 15 November, the *Times* printed another leading article on 'The Revolution in Science,' and two weeks later it published an article by Einstein outlining the meaning of general relativity in popular language. (See Clark, 1971, Section 10.)

[25] Einstein had his own characteristic way of regarding this success. When, on 27 September 1919, a cable from H. A. Lorentz informed him about Eddington's preliminary result confirming the existence of the bending of light in the gravitational field, he showed it to Ilse Rosenthal-Schneider, a research student of his and Planck's. She recalled later: 'When I gave expression to my joy that the result agreed with his calculation, he remarked totally without emotion, "I knew that the theory was correct." And when I told him what if there had been no verification of his prediction he replied, "I would have felt sorry for the dear Lord! The theory *is*, of course, right" ' (Rosenthal-Schneider, 1957).

Despite his conviction that general relativity was right, Einstein did not forget that there was still a third prediction of the theory, the red-shift of the spectral lines in gravitational fields, that had to be verified. 'If it were proved,' he wrote to Eddington, 'that this effect does not exist in nature, then the whole theory would have to be abandoned' (Einstein to Eddington, 15 December 1919, quoted in Douglas, 1956, p. 41).

secondary school and at Bristol University. Dirac had not heard anything about relativity theory during his school education. But after the war he was drawn into the general excitement about the new ideas on space and time and he tried to learn as much about them as he could. He recalled: 'I first got some accurate information about relativity through attending a course of lectures by [C. D.] Broad. . . . He gave a course of about ten or twelve lectures on relativity, discussing it from the point of view of philosophy. Several of the engineering students attended his course at the beginning, but they rather dropped out. I stayed on till the end' (Dirac, 1977, p. 111).[26]

Broad's course, in spite of its general philosophical outlook, provided some exact knowledge about special and general relativity. 'I remember (I think it was in the second or third lecture),' Dirac recalled many years later, 'he [Broad] wrote a formula on the blackboard.' This was the equation,

$$ds^2 = dx^2 + dy^2 + dz^2 - c^2dt^2, \tag{1}$$

for ds, the infinitesimal line element, in terms of infinitesimal changes in the Cartesian coordinates, x, y, z, and the time t, c being the velocity of light *in vacuo*. 'Now, when I saw that minus sign,' Dirac said, 'it produced a tremendous effect on me' (Dirac, 1977, p. 111). Although he had pondered earlier about space and time, the three plus and one minus signs in Eq. (1) were 'really something new which I had never thought of in my speculations' (Dirac, Conversations, p. 5). It had occurred to him earlier that time was very much like another dimension, and there was perhaps some connection between space and time which ought to be considered from a general four-dimensional point of view. But in all his thinking he had always considered Euclidean geometry; hence 'if space and time were to be coupled in any way, they would have to be

[26]The English philosopher Charlie Dunbar Broad, known particularly for his investigations on sense perception, was born on 30 December 1887 at Harlesden, Middlesex. He specialized in science subjects at Dulwich College and in 1906 entered Trinity College, Cambridge, where he first studied natural sciences and then switched to the Moral Science Tripos, graduating in 1910 with first class honours and special distinction. A year later he was elected Fellow of Trinity College, but he left for St. Andrews University, Scotland, to become an assistant to the Professor of Logic (1911–1914); afterwards he assumed a lectureship in philosophy at University College, Dundee. In 1920 he was appointed to the chair of philosophy in Bristol University. The lectures on relativity that Dirac attended were probably given in winter 1920–1921.

In 1922 Broad returned to Trinity College, Cambridge, as Fellow and Lecturer in Moral Science, and in 1933 he was appointed Knightbridge Professor of Moral Philosophy. Upon his retirement twenty years later he remained at Trinity and died in Cambridge on 11 March 1977. Broad received many professional honours; for example, he was a Fellow of the British Academy and of the American Academy of Arts and Sciences. He wrote several books. The first, on *Perception, Physics and Reality* (1914), arose from his thesis for the Trinity Fellowship; in it he analyzed, with the help of the philosophical methods he had learned from G. E. Moore and Bertrand Russell, the information which physical science provides about the reality of perceptible objects. From the title and contents of this book it seems to be natural that Broad should have become interested in relativity—which had very early attracted several people in Cambridge, including Broad's teacher Russell. The studies in relativity led to his second book, *Scientific Thought*, in which Broad dealt with the modifications of traditional scientific concepts produced by the recent scientific discoveries.

coupled with a plus sign here [i.e., only plus signs would occur in Eq. (1)], and it was very easy to see that that would not work, and led to nonsense as soon as one tried to make any big change in one's time axis' (Dirac, 1977, p. 111). Indeed, Broad's formula gave him a new insight into geometry, for it acquainted him with hyperbolic geometry. And, 'I was soon able to figure out by myself the basic relations of special relativity' (Dirac, 1977, p. 112).

From Broad, Dirac also learned about the notion of the constancy of the velocity of light and the concept of curved space, facts which again turned his thinking into completely new directions. In addition he tried to understand the philosophical point of view. Thus, he even borrowed a copy of John Stuart Mill's *System of Logic* from the library and read it through. At that time he thought that philosophy would perhaps contribute to the progress of science, but later on decided that it 'will never lead to important discoveries. It is just a way of thinking about discoveries which have already been made' (Dirac, Conversations, p. 21).[27] In this context Dirac discovered that he was 'not interested just in logical developments, or in seeing what the possibilities are when one follows from a different set of axioms' (Dirac, 1977, p. 112). His interest was in the real physical world, an interest which was buttressed by the engineering training he had received in Bristol.

Dirac obtained further information about relativity from Eddington's popular book on *Space, Time and Gravitation*. When this book appeared he obtained a copy and worked through it, learning special and general relativity simultaneously. He also became acquainted with the mathematical tools of tensor calculus, learning enough to follow the development of relativity theory.[28] He could discuss these things with a friend and fellow student in engineering, H. C. Wilshire, who later on taught at an aeronautical college near Bedford, not far from Cambridge. Together with him Dirac studied some of the literature on relativity, especially the articles and letters that appeared on this subject in the issues of *Nature*. During the years immediately following the British Solar Eclipse Expeditions of 1919 the pages of *Nature* provided ample discussions of various aspects of relativity theory, from the philosophical concepts to the experimental

[27] Unlike Niels Bohr, who constantly thought about philosophical questions and talked a good deal about them—discussing problems as far removed from physics as the question of the freedom of the will—Dirac never mentioned philosophical thoughts to other people. He believed that talking about such issues as the problem of free will would not lead to any definite conclusions. This 'nonphilosophical' attitude would also play a role in the way in which Dirac reformulated Heisenberg's ideas on quantum mechanics in fall 1925. Instead of invoking such 'philosophical' guidelines as the correspondence principle or the strict use of observable quantities, he preferred to argue on rigorously mathematical grounds.

[28] Eddington's *Space, Time and Gravitation* (1920) contained only a few mathematical discussions. Eddington also did not go through the details of special relativity, but proceeded soon to treat general relativity (already in Chapter V, after p. 76 out of 200 pages). He put emphasis on quickly getting those results that could be tested by experiment. This book thus served quite a different purpose than his later book, *The Mathematical Theory of Relativity*, in which he dealt with the formal apparatus (Eddington, 1923b).

consequences.[29] On the basis of these studies Dirac became well acquainted with the subject.

Among those who contributed to relativity theory in England, the name of Ebenezer Cunningham occurred very early and frequently. Actually, he was the first British physicist who made use of the theoretical ideas of Lorentz and Einstein in his analysis of the stability of a moving electron (Cunningham, 1907). He wrote numerous papers and two books on special relativity.[30] When Dirac went up to Cambridge in fall 1923, after completing his B.A. in applied mathematics in Bristol, he definitely hoped to do research in electromagnetic theory under Cunningham's supervision. He thought that in this field he would be able to use his previous experience in electrical engineering, applied mathematics and relativity. This seemed to be a reasonable plan, the first one which Dirac had consciously made for his own future. However, it did not materialize.

On arrival in Cambridge, Dirac would soon become engaged in completely

[29] For example, the entire issue of 17 February 1921 (No. 2677) was devoted to Relativity. Starting with a brief outline by Einstein on the development of the special and the general theory, all the different aspects were described in thirteen articles mostly by British authors, and an impressive list of references including more than eighty books and about fifty articles in *Nature* was added at the end. E. Cunningham discussed the changes which the Principle of Relativity (implying that only relative motion exists) had undergone since the time of Newton; Sir Frank Dyson, the organizer of the British Solar Eclipse Expeditions of 1919, summarized the results on the bending of light beams; A. C. D. Crommelin, a member of the team that observed the eclipse at Sobral, Brazil, described the empirical status concerning the perihelion motion of the planet Mercury, while Charles E. St. John of Mt. Wilson Solar Observatory presented an analysis of the up-to-date results concerning the red-shift of spectral lines in the gravitational field of the sun. More theoretical points of view were raised by G. B. Mathews (Non-Euclidean Geometries), J. H. Jeans (The General Physical Theory of Relativity), H. A. Lorentz (The Michelson–Morley Experiment and the Dimensions of Moving Bodies), H. Weyl (Electricity and Gravitation) and A. S. Eddington (The Relativity of Time). General and philosophical questions were treated in the contributions of Norman Campbell (Theory and Experiment of Relativity), Dorothy Wrinch and Harold Jeffreys (The Relation between Geometry and Einstein's Theory of Gravitation) and H. Wildon Carr (The Metaphysical Aspects of Relativity).

Remarkably enough the only contributor to this issue who raised rather critical objections against relativity theory was Sir Oliver Lodge (The Geometrization of Physics, and its Supposed Basis on the Michelson–Morley Experiment). He had arrived at the conclusion that 'it is not necessary to invoke a real FitzGerald contraction in order to explain the result of the Michelson [–Morley] experiment' (Lodge, 1921a, p. 796). However, Lodge admitted that 'undoubtedly general relativity, not as a philosophical theory but as a powerful and comprehensive method, is a remarkable achievement; and an ordinary physicist is full of admiration for the equations and the criteria, borrowed from hyper-Geometers, applied by the genius of Einstein, and expounded in this country with unexampled thoroughness and clearness by Eddington' (Lodge, 1921a, p. 800). In spite of this 'admiration,' Lodge continued to battle against special and general relativity in a long article which appeared in the following volume of *Nature* (Lodge, 1921b).

[30] In the list of articles on relativity theory, quoted in *Nature*'s special issue of 17 February 1921, Cunningham's name appeared most often. Eddington entered the scene only after 1916. Cunningham's original contributions had been concerned exclusively with special relativity and electron theory, which were also the subjects of his two books: *The Principle of Relativity* (1914) and *Relativity and Electron Theory* (1915).

It is possible that Dirac knew Cunningham's name before he personally met him in the examination at Cambridge, for he had followed the discussion of relativity theory in *Nature*. In the special issue devoted to relativity theory Cunningham had written a report on the development of the idea of relativity (Cunningham, 1921).

new ideas of the theory of atoms. He would learn for the first time about quantum theory and discover that it was as fundamental for the understanding of nature as the theory of space and time. In spite of his new involvement, however, he would never give up relativity theory: some of his early research papers were devoted to special problems in relativity, while in others relativistic arguments played a crucial role. Soon after formulating the fundamental laws of quantum mechanics he would start upon the central problem of his scientific career: the simultaneous application of quantum mechanics and relativity to describe the phenomena of atomic physics. Within less than five years after his arrival in Cambridge, Dirac would expound his relativistic equation for the electron, in which quantum mechanics and special relativity were united harmoniously and from which revolutionary consequences followed about the behaviour of matter. He would thereby add his name in the pantheon of his illustrious predecessors in Cambridge.

Chapter II
Student in Cambridge

Paul Dirac arrived in Cambridge in the autumn of 1923. His stay was supported by a grant from the Department of Scientific and Industrial Research for work in higher mathematics, together with the 1851 Exhibition Studentship which he had won two years earlier. Since, in addition to his engineering degree, he had also completed the course of applied mathematics and obtained a bachelor's degree in it from Bristol University, he became a research student immediately. He had done well in examinations in Bristol and many people there had been impressed with his mathematical ability; yet he was not quite sure of his standing in a university with a great reputation in science such as Cambridge. 'I was really very ignorant of the world,' he recalled about his days as a beginner in Cambridge (Dirac, Conversations, p. 31). And he set out to overcome this ignorance.

Although, on the basis of his examinations for the Studentship, Dirac had been admitted to St. John's College, he did not obtain immediate residence as there were not enough rooms in the College.[31] He was allowed to stay one out of three years in College. Thus, he lived for the first year in private lodgings or 'digs' as they were called; during the second year he was in residence in the College, and in the third year in digs again. Over the entire period, however, he could and did take his meals in the College hall every evening. As his private lodgings were often cold, Dirac preferred to work in a warm library. Several libraries were available in Cambridge for his use: the library of the Cambridge Philosophical Society, the University Library, the library of St. John's College, and the small library at the Cavendish Laboratory. In these libraries, especially at the Cavendish, it was possible to work undisturbed at a table for a whole morning. Dirac also found the libraries useful because he always relied more on his own reading for precise and detailed information than on lectures, where he just got general ideas.

The grant from the Department of Scientific and Industrial Research had been given to Dirac for doing research in an advanced subject in mathematics.

[31] A student at Cambridge University had to be a member of one of the colleges. St. John's College, with which Dirac became associated, was founded in 1511 by Lady Margaret Beaufort, mother of Henry VII, on the advice of John Fisher. (Fisher had been appointed earlier by Lady Margaret as the first occupant of the oldest professorship at Cambridge, a Professorship of Divinity, which she had established in 1502.) St. John's College is not the oldest college in Cambridge—the first to be established was Peterhouse (1284)—but it was founded before Trinity College (1546). Among the Fellows of Trinity College were Isaac Newton, James Clerk Maxwell and J. J. Thomson. Dirac would become the most distinguished physicist to have been a Fellow of St. John's.

He did not have any particular research project in mind, but he knew that it was possible to obtain help in this regard. Since he had been accepted as a research student in Cambridge, he needed a supervisor to guide him. The choice of the supervisor was important, for the supervisor determined the field and the topic of a student's research. From his examination for the Exhibition Studentship in 1921 Dirac knew Ebenezer Cunningham, who was a tutor in St. John's College. He worked on electromagnetic theory, which interested Dirac, not so much because of his previous training as an electrical engineer, but mainly because it was closely related to relativity theory. He hoped therefore that Cunningham would become his supervisor. For some reason, which Dirac was never able to find out, Cunningham could not accept him as a research student, and he was assigned to Ralph Howard Fowler of Trinity College.[32] At that time Fowler was interested in statistical mechanics and atomic theory, and a research student of his would work on these topics, too. As often before in other situations that determined the course of his life, Dirac had not decided the question of his supervisor and, together with it, the topic of his research himself. In the beginning he was not quite happy with the choice that had been made for him, for he had not really studied statistical mechanics before and 'thought it to be a relatively ugly subject' (Dirac, Conversations, p. 32). Fowler did not leave him much time to brood about it and set him immediately to work on a specific problem: the question of the dissociation of molecules under a temperature gradient. He also gave Dirac the necessary literature to read and suggested which lectures to attend.

The fact that Fowler became Dirac's supervisor not only set him up on the road to research in physics, it also opened a new world for him: the world of atoms. Cambridge was especially suitable for learning about atoms, for Ernest Rutherford was around as Cavendish Professor. Rutherford, as the leader of experimental research in atomic physics, determined an important part of the scientific atmosphere in Cambridge. Besides Rutherford there were many other famous scientists and scholars at Cambridge University, who were involved in teaching, guiding students and performing widely acknowledged research work. They upheld the traditionally high level of scholarship in physics and mathematics at this unique place in England, to which Dirac had gone to complete his studies and to do research. Whenever the needs of his work or his more general personal interests required, he would make use of the possibility of learning from the masters, a possibility which was amply provided by his new environment.

II.1 The Cambridge Environment

The University of Cambridge, whose history goes back to early thirteenth century, is one of the oldest universities in Europe and Great Britain. It has been

[32] The reason probably was that Cunningham, who was no longer active in research, did not wish to take research students anymore. (See Dirac, 1977, p. 115.)

particularly distinguished by the role that science played there since the establishment of the *Lucasian Professorship of Mathematics* in 1664. This chair was first held by Isaac Barrow, who resigned it six years later in favour of Isaac Newton.[33] Newton occupied the Lucasian Professorship for more than thirty years and gave the study of mathematics a unique position at Cambridge; this was reflected by the fact that the first honour's examination which came into being during the eighteenth century, the so-called 'Tripos,' was primarily a mathematical one.[34] After a period of stagnation the scientific reputation of the university increased greatly during the nineteenth century, three reasons being primarily responsible for this fact. First, in 1849 George Gabriel Stokes became Lucasian Professor, 'rescuing the chair from the doldrums into which it had fallen, and restoring it to the eminence it had when held by Newton' (Parkinson, 1976, p. 74).[35] Second, the *Sadleirian Professorship of Pure Mathematics* was established in 1863, and Arthur Cayley was appointed first professor. He had been Senior Wrangler of the Tripos in 1842 and made his reputation by publishing about three hundred papers on invariant theory, metrical geometry, algebra of matrices and group

[33]Isaac Barrow, born in London in 1630, entered Trinity College, Cambridge, at the age of thirteen and became a Fellow in 1649. Before accepting the Lucasian Professorship he was professor of geometry at Gresham College, London. In 1669 he became Royal Chaplain in London, then Master of Trinity College in 1672, and finally Vice-Chancellor of Cambridge University in 1675. He died in London on 4 May 1677. Barrow became one of the first Fellows of the Royal Society after its incorporation in 1662. His main scientific contributions were in geometry. However, his optical lectures inspired Isaac Newton, his successor in the Lucasian Chair, whom he helped in his career. Many years later Newton mentioned in an autobiographical note that Barrow procured for him 'a fellowship [at Trinity College] in 1667 and the Mathematick Professorship two years later' (Newton, around 1716, MS Add. 3968 §41, fol. 117, reprinted in I. B. Cohen, *Isaac Newton*, *Dictionary of Scientific Biography*, Charles Scribner's Sons, New York, 1974, Vol. X, p. 86).

[34]The name 'Tripos' stems from the three-legged stool on which the candidates used to sit to dispute. According to their performance the successful candidates were put into three classes, and those placed in the first class were known as the 'wranglers.' The Tripos was increasingly respected as a test of ability, and to be the best candidate—the so-called 'Senior Wrangler'—was considered a fine accomplishment. A number of senior wranglers later became famous mathematicians and physicists; for example, George Gabriel Stokes, Arthur Cayley and Joseph Larmor. In the 19th century, Tripos examinations were established in other fields: the Classical Tripos in 1824, the Tripos in Natural Sciences and the one in Moral Sciences in 1851. Most of them were divided, like the Mathematical Tripos, into two parts, which could be taken by the candidates in different years. A candidate for the bachelor's degree had to pass both parts of a Tripos.

[35]George Gabriel Stokes was born on 13 August 1819 in Skreen, Sligo County, Ireland. He attended school in Dublin and went to Bristol College to prepare for the university. In 1837 he entered Pembroke College, Cambridge, and studied mathematics; he graduated in 1841 as Senior Wrangler and first Smith's Prizeman, and was elected Fellow of his College. Since the Lucasian Chair was poorly endowed, Stokes had to teach at the Government School of Mines in London in the 1850s in order to supplement his income. In his distinguished career he became President of the Cambridge Philosophical Society (1859–1861), Secretary of the Royal Society (1854–1885) and President (1885–1890); he also represented the University of Cambridge in Parliament at Westminster from 1887 to 1891. He was knighted for his services in 1889, and in 1902 Pembroke College elected him Master. His ample work on theoretical and experimental physics (in the fields of hydrodynamics, elasticity, diffraction of light—the ether problem, and fluorescence) as well as on mathematics won him numerous honours: he became a Fellow of the Royal Society in 1851 and received the Copley Medal in 1903. He died on 1 February 1903, having held the Lucasian Professorship for nearly fifty-four years.

theory, creating most of the fields in which he became involved.[36] Finally, in 1871 the university created—with the help of its Chancellor, William Cavendish, seventh Duke of Devonshire and descendent of the physicist Henry Cavendish—the *Cavendish Professorship of Experimental Physics* and elected James Clerk Maxwell to this chair. The Cavendish Professor had at his disposal a laboratory, in which he was able to carry out fundamental measurements. The experimental researches performed by Maxwell and his successors at the *Cavendish Laboratory* made it famous throughout the world.[37]

In the period following World War I, when Dirac studied in Cambridge, a large number of distinguished men held positions in the University or the Colleges. Among them were the historian George Macaulay Trevelyan, the economist John Maynard Keynes, and there were scientists like Joseph John Thomson, Ernest Rutherford and Arthur Stanley Eddington. Dirac, because of his earlier interest in relativity, knew the latter well, and he saw him occasionally. But Eddington was not the only mathematical celebrity in Cambridge. There was, for example, Sir Joseph Larmor, the Lucasian Professor of Mathematics,

[36]We have given some details of Arthur Cayley's life and scientific career in Volume 3, footnote 19, for he was the founder of matrix calculus. It was customary at Cambridge at the time that the most eminent pure mathematician would occupy the Sadleirian chair. After Cayley's death on 26 January 1895, one of his students, Andrew Russell Forsyth (1858–1942), succeeded him; he, in turn, was followed by Ernest William Hobson (1856–1933); both of them were former Senior Wranglers in the Mathematical Tripos. Hobson became one of the leaders for the reform of the Tripos, and the order of merit was abolished in 1910.

[37]British universities, like most other universities, did not acquire any elaborate experimental facilities in physics until the second half of the nineteenth century when the industrial revolution created the necessity of carrying out exact measurements of mechanical and electrical quantities. The first physical laboratory at a British University, in which such measurements could be undertaken, was installed by William Thomson (later Lord Kelvin) in the late 1840s at the University of Glasgow. The construction of the Clarendon Laboratory at Oxford was begun in 1868 and completed in 1872. The history of the Cavendish Laboratory started at about the same time. In 1869 the Commission, which considered the matter, recommended the construction of a lecture room, class rooms, a large laboratory, and the acquisition of a stock of physical apparatus. The site of the new building was already available; however, the costs of construction were so enormous (the initial estimate of £6300 was eventually surpassed by the actual amount of £8450) that the Duke of Devonshire, Chancellor of the University, offered to pay them. The essential part of the Cavendish, containing the laboratory, was ready for use by the Michaelmas term of 1873.

In searching for the first Cavendish Professor one had initially thought of William Thomson, but since he did not wish to leave Glasgow, the next choice was James Clerk Maxwell. In 1865 Maxwell had retired from his former position as Professor of Natural Philosophy at King's College, London, to write his *Treatise on Electricity and Magnetism*. Maxwell already had great achievements to his credit, especially in kinetic gas theory and electromagnetism, but the publication of his most important book was still to come. In Cambridge he established courses of lectures and began to instruct students who were to carry out experiments on their own. The Cavendish flourished, and when Maxwell died in 1879 there were about twenty students working there. John William Strutt, third Baron Rayleigh, who succeeded him, improved the equipment; he started the 'Apparatus Fund,' himself contributing the first £500. He also hired an efficient mechanic, George Gordon, a Liverpool shipwright, who set up the workshop for building and maintaining instruments both for lectures and research work. Lord Rayleigh thus transformed the Cavendish into a modern laboratory for teaching and research. Upon his resignation in 1884, Lord Kelvin again declined to go to Cambridge and Joseph John Thomson was elected to the Cavendish Professorship.

holder of the chair which Dirac would inherit from him just nine years after his arrival in Cambridge. Larmor, a Fellow of St. John's, had been one of the pioneers of relativity theory. He had proved, for instance, that the transformation proposed by Hendrik Antoon Lorentz for describing the motion of a charged body through ether was correct to second order in the ratio of its velocity and the velocity of light *in vacuo*. Since all the crucial experiments, which showed the absence of an absolute frame of reference, had to do with these second-order terms, Larmor's step was extremely important. He had also given, at the same time as Lorentz, a theoretical explanation of the Zeeman effect based on the assumption of rotating charges. Larmor's theory was later taken over into atomic physics. All these results were contained in a great memoir, parts of which had been published between 1894 and 1897; Larmor recast it in an improved form and won the Adams Prize for 1898 with it. It was published two years later as *Aether and Matter* (Larmor, 1900b). 'It is a difficult book—unnecessarily difficult, because Larmor was certainly not gifted with lucidity of style,' said Eddington many years later. 'But to the student of the period 1900–1905 [as he had been] it was the one gateway to new thought, revolutionary and inspiring' (Eddington, 1943, p. 198). In succeeding years Larmor did not contribute any more to what became the theory of relativity, and soon he would be considered as old-fashioned as his book. He preferred to work on the applications of classical theories and became a leading authority on geomagnetism. When Dirac saw Joseph Larmor, and he could often be seen at St. John's, he was a retiring man in his mid-sixties. He still had students and published a paper occasionally, but he had no active interest in quantum theory or the other subjects in which Dirac worked. 'He seemed a man,' recalled Eddington, 'whose heart was in the nineteenth century, with the names of Faraday, Maxwell, Kelvin, Hamilton, Stokes ever on his lips—as though he mentally consulted their judgment on all the modern problems that arose' (Eddington, 1943, p. 205). In Joseph Larmor's opinion, true scientific progress had ceased about 1900.[38]

A man of greater scientific and public reputation was Joseph John Thomson, the former Cavendish Professor. Dirac saw him from time to time in the Cavendish Laboratory, where he kept a room, though he no longer took an active part in the colloquia. The impact of his work, both in science and administration, was very much in evidence in Cambridge, especially at the Cavendish. Thomson had begun his career in 1880 as Second Wrangler behind

[38] Even when writing about a 'modern' subject, such as the properties of atomic lattices (Larmor, 1921), Larmor would largely refer to ideas that were available at the turn of the century. The same 'classical' attitude was reflected in a paper on 'Generalized Relativity' (Larmor, 1919), in which he dealt with an extension of relativity by postulating a five-component potential in a five-dimensional Euclidean space in order to describe the laws of electrodynamics and gravitation in a unified way.

Dirac recalled that R. Schlapp, one of his fellow students at St. John's, was supervised by Joseph Larmor in research on a classical problem. On receiving his doctorate in Cambridge, Schlapp went to Edinburgh. There Charles Galton Darwin was Tait Professor of Natural Philosophy, and got him to work on problems of quantum mechanics. Thus, Schlapp contributed to one of the first applications of Dirac's equation for the electron by studying the Stark effect of hydrogen (Schlapp, 1928).

Larmor. He had then worked in electromagnetic theory, deriving, among other results, the important conclusion that radiation possesses momentum, that in any volume where electromagnetic energy is contained there is also stored an amount of mechanical momentum which, apart from a constant, is given by the vector product of the electric and magnetic field vectors. In 1884 Thomson, not yet quite 28 years of age, had been appointed to the Cavendish Chair. Soon he started a broad and detailed investigation of gaseous discharges. The study of the properties of cathode rays, a part of this programme, led him to his most famous experimental accomplishment, the discovery of the electron, which, he concluded, represented the lightest corpuscle or the basic unit constituting matter (Thomson, 1897a, b). These researches established him as one of the foremost experimental physicists of his time.

During the thirty-five years of his tenure in the Cavendish Chair, Thomson was the active head and inspirer of a great research school. The number of his students and collaborators had increased especially after 1895, when the change of university regulations permitted the admission of so-called 'advanced students' from other universities. One of the first to come under the new arrangement had been Ernest Rutherford from New Zealand, who, soon after Henri Becquerel's discovery of radioactivity in 1897, started his researches in this new field under Thomson's guidance. Other experimental investigations at the Cavendish, which contributed important results to the understanding of the structure of matter and of atoms, had been carried out under Thomson's general supervision by Charles Thomas Rees Wilson (on the creation of cloud tracks by charged particles, leading to the construction of the cloud chamber in 1911), by Owen Willans Richardson (on the emission of electricity by hot bodies), and by William Lawrence Bragg (on the X-ray interferences from crystals in 1912). Finally, Francis William Aston, since 1910 a research assistant at the Cavendish, had discovered, when collaborating with Thomson on the deflection of positively charged ions by combined electrostatic and magnetic fields, the two isotopes of neon, Ne20 and Ne22 (see Thomson, 1913a); these were the first isotopes observed outside the field of radioactivity. By improving Thomson's apparatus, Aston developed his spectrograph for isotope research, which won him the Nobel Prize in Chemistry in 1922.

While the above-mentioned researches were being performed, the situation at the Cavendish Laboratory changed considerably. This began with one of the first acts of J. J. Thomson as Director: he had provided the students with a reading room, equipped with the personal library of James Clerk Maxwell, which he obtained from the latter's widow. Later, the increase of the number of students and visitors from England and abroad had made the Cavendish an extremely crowded place, and Thomson had expressed a wish for more space.[39] In reply to

[39] At the turn of the century the average number of research students was about twelve; it rose to twenty-five between 1902 and 1908, and to thirty during the years before the First World War.

this request Lord Rayleigh, Thomson's predecessor in the Cavendish Chair, had assigned £5000 of his Nobel Prize in Physics, which he had won in 1904, for that purpose. It covered most of the expenses of a new building containing a number of rooms for research, a library, a chemical room, a demonstrator's room and a large basement. The extension to the Cavendish had come into full use in 1908, and from that time on all the physics lectures in the university were delivered there. These included the courses of general lectures on the properties of matter and on electricity and magnetism given by Thomson, the practical classes given by the demonstrators (such as C. T. R. Wilson or G. F. C. Searle), and the physics colloquia. The Cavendish courses of Cambridge University were supplemented by the courses of the college tutors, which had also been moved to the Laboratory.[40] 'With the concentration of the physics teaching and research the problems of organization were simplified,' commented a historian of the Cavendish Laboratory on the lasting achievements of the Thomson period. 'Those working in physics were all in touch with each other, developing a feeling of comradeship. This stimulated the growth of the Cavendish tradition, which was transmitted to future generations in Cambridge and elsewhere' (Crowther, 1974, p. 159). The Thomson era came to an end in March 1919 when the Professor, having been appointed Master of Trinity College the year before, resigned and his former student Ernest Rutherford succeeded him in the Cavendish Chair.

Rutherford, since 1907 Langworthy Professor of Physics and Director of the Physical Laboratories at Victoria University, Manchester, had just made one of his most important discoveries. On the basis of experiments performed in 1917 he had concluded the occurrence of artificial disintegration of an element: the transmutation of nitrogen nuclei into oxygen nuclei when bombarded by α-particles (Rutherford, 1919). In spite of that success Rutherford felt 'very rusty scientifically after the war' (Rutherford to Geiger, June 1919), and he was eager to work energetically in his new position at Cambridge. This was going to be necessary in any case, for as Cavendish Professor he had to direct and manage the biggest research laboratory and to guide the largest group of collaborators and associates he had ever had. The task involved a lot of administrative work, the teaching of hundreds of students, the selection and guidance of devoted research scholars, and the hunt for financial support of the experimental work. In the course of performing his duties in Cambridge, Rutherford received the highest honours available to a great scientist and a man of extraordinary distinction: in 1922 he was awarded the Copley Medal of the Royal Society, in 1925 King George V conferred on him the Order of Merit, and in 1931 he was made a peer and became Baron Rutherford of Nelson.

[40]The main duty of the College Tutors was to supervise the students in residence and to represent them in their dealings with the University. They could also give additional courses to supplement the material offered in university courses. There existed separate positions for University and College Lecturers, and the teaching and research staff of the University did not have to belong to any College. On the other hand, the relationship between the University and the Colleges in Cambridge was very close, and the students had to be associated with one of the Colleges.

Although Rutherford had come to a research institution which was famous throughout the world, he had to reorganize the place completely.[41] He immediately submitted a memorandum on the needs of the Cavendish Laboratory to the University in which he expressed the necessary demands: the provision of increased laboratory and lecture space for the teaching of physics; the provision of new, well-equipped laboratories for applied physics, optics and properties of matter; the provision of three additional lecturers of high standing, competent to direct advanced study and research in the new departments mentioned above; and the endowment of another chair of physics in the University. He estimated the cost of the new buildings and endowments to be £200,000, which was a very large sum then and much more than the University was prepared or able to spend in 1919. Thus, the big plans did not become a reality immediately, but still Rutherford managed to modernize the equipment of his laboratory and to adapt it to the requirements of his own experimental work. This modernization had become necessary, for the previous director had not placed any emphasis on adequate, not to say expensive, apparatus. For instance, the construction of Wilson's cloud chamber had cost only about £5. Although the equipment, which Rutherford had used earlier in his experiments on radioactivity and atomic structure, was basically simple—he had brought his personal apparatus along to Cambridge from Manchester together with the 250 mg of radium that had been lent to him by the Vienna Academy of Sciences in 1908—he did not believe in continuing the 'string-and-sealing-wax' conception of scientific apparatus in Cambridge. He was convinced that the new tasks in physics required large-scale research and equipment, even though he was not terribly fond of it himself.

Rutherford had definite plans for his own research. He wanted to study systematically the disintegration of lighter elements by α-particles in order to obtain a deeper insight into the structure and properties of atomic nuclei. One particular point to investigate in this context was the question of how electrons and hydrogen nuclei, which in those days were regarded as the basic constituents of matter, were put together to form the nuclei.[42] Thus, he drew attention very early to the nature of the lightest nuclei that might arise. 'It seems very likely,' he said in his Bakerian Lecture of 1920, 'that one electron can also bind two H nuclei and possibly one H nucleus. In the one case, this entails the possible existence of an atom of mass nearly 2 carrying one charge, which is to be regarded as an isotope of hydrogen. In the other case, it involves the idea of the possible existence of an atom of mass 1, which has zero nucleus charge' (Rutherford, 1920, p. 396). If the neutral object existed, which Rutherford thought of as a hydrogen nucleus–electron system that is much more tightly bound than a neutral hydrogen atom, and which, therefore, exhibits novel

[41] During the First World War not only had the scientific talent of the Cavendish Laboratory been dispersed, but its workshops had been used for military-related work (such as the construction of gauges).

[42] For the hydrogen nucleus Rutherford suggested the name 'proton,' i.e. the 'first,' at the Cardiff Meeting of the British Association for the Advancement of Science, 1920. (See the report in *Nature* **106**, issue of 11 November 1920, p. 357.)

properties, then it might play an important role in atomic physics. For example, it should be able to penetrate easily into the interior of atoms and might either unite with the nucleus to form a new heavier isotope or lead to the disintegration of the atom, together with the emission of an electron or of a hydrogen nucleus or of both. 'If the existence of such atoms be possible,' Rutherford further speculated about the light neutral object, 'it is to be expected that they may be produced, but probably only in very small numbers, in the electric discharge through hydrogen, where both electrons and H nuclei are present in considerable numbers. It is the intention of the writer to make experiments to test whether any indication of the production of such atoms can be obtained under these conditions' (Rutherford, 1920, p. 396).

It is probably fair to say that the search for the lightest neutral nucleus determined to a large extent the experimental work at the Cavendish in the 1920s. And yet, in spite of the fact that Rutherford and his collaborator James Chadwick tried all available means, they did not make any progress. Only after new methods had been developed and many new effects had been discovered at the Cavendish as well as other places, would Chadwick be able to confirm the existence of the so-called 'neutron' in 1932. However, it was clear from the very beginning that Rutherford had opened a new era at the Cavendish by bringing a completely new field—nuclear physics—to Cambridge, and that he transformed the Laboratory into the world's foremost place in that exciting field of physics.[43]

Rutherford not only introduced new physics to Cambridge, he also brought new people. First, he got his former Manchester pupil James Chadwick, whom he appointed Assistant Director of Research in 1923.[44] Chadwick helped Rutherford in running the Cavendish Laboratory: he trained the students and helped to select research tasks for them; he looked after the equipment they needed and he went through and criticized their papers. 'Chadwick read all the journals and knew everything which happened in branches of physics of interest in the

[43]Nuclear physics did not exhaust the activities at the Cavendish Laboratory in the Rutherford era. As under Thomson's directorship, many different projects went on. Besides the experiments on nuclear disintegration and transmutation, and on the determination of size, structure and electric fields of nuclei, in which Rutherford was personally involved, there was the work of F. W. Aston on the separation of isotopes of chemical elements. Then there were researches which had nothing to do with atomic and nuclear physics: for example, Geoffrey Taylor studied the turbulence of the atmosphere, the motion of liquids and certain mechanical properties of metals, and Edward Appleton investigated the propagation of radio waves.

[44]James Chadwick was born in Cheshire, England, on 20 October 1891. After attending Manchester High School, he entered Manchester University and graduated from the Honours School of Physics in 1911. From 1911 to 1913 he worked in the Physical Laboratory at Manchester University on various problems of radioactivity under Rutherford's supervision and obtained his M.Sc. degree in 1913. Then he went to Berlin with an Exhibition Scholarship to do research at the *Physikalisch-Technische Reichsanstalt* under Hans Geiger, Rutherford's former assistant and collaborator. When the First World War started he was interned in the civilian prison camp (*Zivilgefangenenlager*) at Ruhleben close to Berlin. Helped by German scientific colleagues and friends, including H. Rubens, W. Nernst and E. Warburg, he organized a little research laboratory, together with Charles D. Ellis, a British Army Officer. When Chadwick returned to Manchester after the war, Rutherford gave him a job as temporary demonstrator and later took him along to Cambridge. In 1919 Chadwick obtained the Wollaston Studentship at Gonville and Caius College and became a Fellow two years later.

Cavendish,' recalled Marcus Oliphant who came to Cambridge for experimental research in 1927. 'If something appeared which seemed worthwhile following up, it was not long before someone in the Laboratory was working on it' (Oliphant, 1972, p. 68). Chadwick was Rutherford's closest collaborator and colleague in Cambridge. He joined him in accomplishing the transmutation of light elements by bombardment with α-particles and in investigating the structure of nuclei. It may be considered as an act of historical justice that Chadwick would discover the neutron eventually in 1932, the lightest neutral nucleus predicted by Rutherford a dozen years earlier.

Together with Chadwick came Charles D. Ellis, his collaborator during the war. He worked as a research student and continued the earlier work done in Manchester on β- and γ-radiation, using mainly photographic methods. Another student at the Cavendish during Rutherford's early directorship was the Japanese Takeo Shimizu, who remodelled Wilson's cloud chamber to allow for more or less continuous photography of α-particle tracks. After his return to Japan in 1921, this programme was carried on by Patrick Maynard Stuart Blackett, a former lieutenant of the British Navy, who had been sent after the war to the University of Cambridge to attend 'holiday courses' and decided to stay on at the Cavendish.[45] In 1924 he succeeded in taking the first photographs showing the transmutation of nitrogen nuclei into oxygen nuclei under the bombardment of α-particles (Blackett, 1925). While this research confirmed Rutherford's earlier conclusion of 1919, based on the scintillation method, Blackett would perfect the photographic technique and later score an important triumph: together with Giuseppe Occhialini he would confirm the existence of the positron, the antiparticle of the electron, as predicted by Dirac's theory of 1928 (Blackett and Occhialini, 1933).

Like these students and collaborators of Rutherford's, Peter Leonidovich Kapitza, who arrived in 1921 as a visitor at the Cavendish, also started to work on problems of nuclear physics. However, he soon became interested in the

[45] P. M. S. Blackett was born on 18 November 1877 in Kensington, London. He entered the Osborne Naval College just before his thirteenth birthday and spent two years there; he spent another two years at Dartmouth College. When the war broke out in August 1914, he was immediately appointed midshipman on a cruiser. He took part in several battles on the sea, including the battle in the Falkland Islands and the battle of Jutland, and was finally in-charge of the fire of the destroyer H.M.S. Sturgeon which operated against U-boats. In 1918 he was promoted to Lieutenant. After the war ended he was sent to attend a six-month course of general studies at Cambridge University. He arrived at Magdelene College in January 1919. Impressed with what he saw at the Cavendish Laboratory, he decided to resign from the Navy to become an undergraduate student. In 1921 he finished Part II of the Tripos, in which he got a First; he was then elected to a Bye Fellowship of his College. He began his distinguished scientific career as a research student under Rutherford in fall 1921. From 1921 to 1933 (with the exception of the year 1924–1925, which he spent with James Franck at Göttingen) Blackett was at the Cavendish. Then he became Professor of Physics at Birkbeck College, London (1933–1937), the University of Manchester (1937–1953) and at Imperial College (1953–1965). Blackett died in London on 13 July 1974.

Blackett worked on problems of atomic, nuclear and cosmic ray physics, and on the magnetic fields of the earth and other astronomical bodies. He received many academic and public honours: Fellow of the Royal Society, 1933; President of the Royal Society, 1967; Nobel Prize in Physics, 1948; U.S. Medal of Merit, 1946.

physical phenomena occurring in the presence of strong magnetic fields, a subject which Rutherford found important enough to help Kapitza establish a laboratory of his own, the Royal Society's Mond Laboratory. The latter was built in the early 1930s beside the Cavendish, on the site of the old electrical substation; it was the first of the extensions of the Cavendish for which Rutherford had applied in 1919, and it was the only one to be achieved under his directorship.[46]

With Rutherford's new people the Cavendish quickly began to fill. For instance, in 1919 J. J. Thomson supervised four students and Rutherford fourteen, in 1921 the total number of people approached thirty and soon an average number of thirty-five was reached. Those associated with Rutherford, the 'boys' as he used to call them, formed an intimate community, one may even say a 'family.' Rutherford always retained a fatherly interest in his boys, which he continued far beyond the time of their stay at the Cavendish. He helped students, especially those coming from abroad, who found themselves in financial difficulties, and he helped them in their research. Thus, he regarded Chadwick as a member of his family, until he got married in 1925. Rutherford was a sociable man and he enjoyed the company of his fellows. They were invited every Sunday afternoon at 4:30 to have tea at Newnham Cottage, the home of the Rutherfords. In return the members of the Cavendish entertained the Professor at a dinner once a year. And each summer the 'family' was photographed. The social occasions were not the only ones when they gathered; on alternate Wednesdays throughout the year the Cavendish Physical Society met. Tea was served in the space between the blackboard and the lecture bench, at which Lady Rutherford presided.[47] At the first meeting each year in October, Rutherford would give a general talk on the work in progress at the Laboratory. This meeting and the others, at which eminent visitors gave talks, were well attended. The so-called

[46]The Mond Laboratory was built by the Royal Society with the help of a grant of £15,000 from the bequest of Ludwig Mond. Mond, the chemist and industrialist, had donated a sum of £50,000 to the Royal Society to meet the expenses of editing an International Catalogue of Scientific Literature, but that project had been halted by the First World War.

[47]Tea was also served to the members of the Cavendish every afternoon in the room next to the library, the tea being provided by Lady Rutherford and the buns by the people themselves. Marcus Oliphant has recalled that: 'This tea-break was a valuable institution at which members of the Laboratory chatted together about their work or the weather, but it wasted little time as there was nowhere to sit down! Sticky Chelsea buns were the favourites at tea. Before the meetings of the Cavendish Physical Society, the tea was provided by the Professor, and Lady Rutherford poured. At these special teas, we got only a half, or even a quarter of a Chelsea bun' (Oliphant, 1972, pp. 52–53).

The Cavendish Physical Society had been founded by J. J. Thomson in 1893. His son, George Paget Thomson, wrote:

> It was an unusual society with no rules and no subscription. It met about once a fortnight in term and one or sometimes two pieces of work, experimental or theoretical, were described each by one of the research students or, more rarely, by a member of the staff. The paper did not necessarily contain the speaker's own work, though a man would be asked to give an account of what he himself had been doing at least once in his time as a research student. There was a discussion after each paper. The Cavendish appears to have been the first laboratory in this country to adopt this custom, at least in physics, but it was then common in Germany and known as a "seminar" or "colloquium." Tea was provided by Mrs. Thomson and one or two ladies connected with the laboratory before the serious work started. (G. P. Thomson, 1964, p. 91)

Research Colloquia, which were more specialized and drew smaller audiences, met on the other Wednesdays at 5 P.M.[48]

Rutherford was completely happy in his work and he transmitted this happiness to a large extent to those who worked at the Cavendish. Still the 1920s, when the field of nuclear physics was created, were not an easy period. Not only the concepts which had to be employed were not known, but the available experimental tools were antiquated or even insufficient. For example, in Rutherford and Chadwick's researches on the disintegration of light nuclei, the same scintillation methods, with which ten years earlier the existence of the atomic nucleus had been proved, were used.[49] The techniques for counting α-particles and protons remained unaltered until 1929; only then were linear amplifiers, responding to the small current impulses produced in a shallow ionization chamber by a fast charged particle, developed at the Cavendish. This method allowed one to register a widely extended range of statistically significant results in nuclear physics. Another tool used in Cambridge since the mid-1920s was the automatically recycling cloud chamber of Shimizu and Blackett. The Geiger-counter, which had been considerably improved by that time, was not used at the Cavendish until 1930; soon afterwards, however, it would be used effectively by Blackett and Occhialini in their discovery of electron–positron pair production.[50] While one of the reasons why only limited and cumbersome methods were used in the early research on the nucleus was that better techniques were yet to come, the other reason was the small budget of Rutherford's Laboratory. For example, in 1920 the cost of apparatus and materials was £2548, and the utilities (fuel, water, gas and electricity) amounted to £328; in 1925 the expenditures had reached £9628, and five years later they were £15,274. Of the latter, £812 was spent on the equipment for Cockcroft and Walton's high voltage laboratory and a liquid air plant. In addition to the regular funds available in 1930, Kapitza

[48]There were other colloquia as well, attendance at which was confined to graduates; they provided research students the opportunity to talk about their research and to discuss the recent advances in physics. A regular enterprise of this sort was the *Kapitza Club*, of which we shall speak later. Sometimes there took place impromptu meetings at which someone would speak on a subject outside the interest of the Laboratory. For example, Oliphant recalled, 'one meeting was addressed by an expert on the control of the tsetse fly in Africa, whom Rutherford had met on a visit to South Africa' (Oliphant, 1972, p. 53).

[49]This was still being done when Marcus Oliphant joined the Cavendish in 1927. He has written:

> When I arrived in Cambridge, Rutherford and Chadwick were still working together on the disintegration of light nuclei by bombardment with alpha-particles. They used a zinc sulphide scintillation screen and microscope of high light-gathering power to detect the product protons, working with dark-adapted eyes in an underground laboratory in the older part of the Cavendish. They were helped by Rutherford's research assistant, George Crowe, who had prepared most of the radioactive sources under Chadwick's watchful eye.
>
> Scintillation counting, especially for the detection of protons as the rare product of a nuclear transformation, was a slow and tiring task, requiring great concentration by the observer. Chadwick has written to me: "The normal procedure was for an observer to count for one minute (sometimes less), being then relieved by another observer, and each observer might have up to 20 periods of one minute each during an experiment. The total duration of an experiment was limited by the decay of the active deposit source, as well as by the fatigue of the observers." (Oliphant, 1972, p. 35)

[50]G. Occhialini brought with him from Florence the technique of coincidence counting, which Bruno Rossi had developed for cosmic ray research.

received £2622 from the Department of Scientific and Industrial Research for the Magnetic Research Laboratory.

The comparatively large sums going for special apparatus at the end of the first decade of Rutherford's directorship of the Cavendish signified a change in the way in which experimental nuclear physics was beginning to be done. By that time it had become evident that, in order to extend the existing knowledge in this field, larger and more expensive machines had to be built. As a result, the research workers at the Cavendish would not continue to be able to make and put together all parts of the apparatus themselves, but they had to be helped by properly trained technicians and engineers. The first of this new category to come was John Cockcroft, who had studied electrical engineering at Manchester and proceeded to Cambridge on a scholarship to study for the Mathematical Tripos.[51] Upon graduating with distinction in 1924 he became a member of the Cavendish Laboratory, where he had already obtained some practical training, and began to work with Kapitza on producing high magnetic fields and low temperatures. In late 1928, stimulated by Rutherford's conclusions based on empirical results and by George Gamow's wave mechanical theory of the disintegration of light nuclei by α-particles, he turned to the investigation of the acceleration of protons by high voltage.[52] He was soon joined by Ernest T. S.

[51] John Douglas Cockcroft was born on 27 May 1897 at Todmorden, England, where his family had for several generations been cotton manufacturers. He was educated at Todmorden Secondary School and studied mathematics at Manchester University under Horace Lamb from 1914 to 1915. From 1915 to 1918, during World War I, he served as signaller in the Royal Field Artillery. Then he studied electrical engineering at the College of Technology in Manchester under Miles Walker, following which he became a student apprentice with Metropolitan-Vickers Electrical Company for two years. Walker nominated him for a scholarship that had been founded in memory of the victory in the war.

At Cambridge, Cockcroft joined St. John's College and took the Mathematical Tripos in 1924. He became a Fellow of St. John's in 1929 and then, successively, demonstrator, lecturer and (in 1939) Jacksonian Professor of Natural Philosophy in the University of Cambridge. During World War II he joined the Ministry of Supply (1939) and later became Head of the Air Defense Research and Development Establishment. In 1944 he went to Canada to take charge of the Canadian Atomic Energy Project, from where he returned to the United Kingdom as Director of the Atomic Energy Research Establishment, Harwell (1946–1958). In 1959 he was elected Master of Churchill College. Cockcroft received many honours, including the Nobel Prize in Physics, which he shared with E. T. S. Walton in 1951. He died in Cambridge on 18 September 1967.

[52] In his 1927 Presidential Address to the Royal Society, Rutherford had raised the question of the necessity for obtaining charged particles of high energy in nuclear research. After referring to the α-particles expelled from radium-C with an energy of about 8 MeV, he said:

> So far the α-particle has the greatest individual energy of any particle known to science, and for this reason it has been invaluable in exploring the inner structure of the atom and giving us important data on the magnitude of the deflecting field in the neighbourhood of atomic nuclei and of the dimensions of the nuclei. In case of some of the lighter atoms, the α-particle has sufficient energy to penetrate deeply into the nucleus and to cause its disintegration manifested by the liberation of swift photons.
>
> It would be of great scientific interest if it were possible in laboratory experiments to have a supply of electrons and atoms of matter in general, of which the individual energy of motion is greater even than that of the α-particle [from radium-C]. This would open up an extraordinarily interesting field of investigation which could not fail to give us information of great value, not only on the constitution and stability of atomic nuclei but in many other directions' (Rutherford, 1928, p. 310).

In a paper following his work on the explanation of α-decay of nuclei, George Gamow discussed

Walton, who had come to Cambridge in 1927 from Dublin, Ireland, on an 1851 Exhibition Research Scholarship.[53] In 1932 they were able to obtain energies of up to 700,000 eV (Cockcroft and Walton, 1932a). The disintegration of lithium and other light nuclei, which was obtained immediately afterwards, represented one of the experimental triumphs of the Cavendish Laboratory (Cockcroft and Walton, 1932b). At the same time as Cockcroft and Walton introduced the method of using artificially accelerated particles in nuclear physics, another even more powerful instrument had become available; this was the cyclotron constructed by Ernest Orlando Lawrence in Berkeley, California. The results of the Cambridge group were confirmed in the United States; moreover, since the energies of particles accelerated by cyclotrons could be increased considerably above 10^6 eV, new investigations calling for high energies could be undertaken. The cyclotron would become the primary tool of future research in nuclear physics, and Berkeley would become its outstanding centre. With the technical complexity of Cockcroft and Walton's apparatus and the expenses of building it, the limits had been reached that Rutherford found acceptable in his laboratory. Though later on, in 1936, he agreed to have a cyclotron, he did not live to see its completion.[54]

The new era of particle accelerators, which Rutherford had ushered with his physics and to which his collaborators had made important contributions,

---

the disintegration of atoms. Cockcroft, on receiving a typewritten manuscript, recognized the implication that high-energy projectiles would be extremely useful in nuclear research and proposed to build a machine that would produce protons of 300 keV. In December 1928 he persuaded Rutherford to obtain £1000 from the University to purchase a 300-keV transformer and the components for constructing rectifying and accelerating tubes.

[53] Ernest Thomas Sinton Walton, born on 6 October 1903 at Dungarvan, County Waterford, on the Southcoast of Ireland, was educated at the Methodist College, Belfast, and at Trinity College, Dublin. He took honours courses in mathematics and experimental science, specialized in physics, and graduated in 1926 with first class honours. After receiving the M.Sc. degree in Dublin, he obtained the scholarship for doing research under Rutherford. In Cambridge he obtained a senior research award of the Department of Scientific and Industrial Research and completed his doctorate in 1931; he stayed on as Clerk Maxwell Scholar until 1934 when he returned to Trinity College, Dublin. In 1946 he was appointed Erasmus Smith's Professor of Natural and Experimental Philosophy in 1946 at Trinity College, Dublin, and in 1960 he was elected Senior Fellow of the College. He shared both the Hughes Medal of the Royal Society of London in 1938 and the 1951 Nobel Prize in Physics with John Cockcroft.

At the Cavendish Cockcroft and Walton first worked in the same room as T. E. Allibone, who had come from Metropolitan-Vickers Electrical Company, the firm at which Cockcroft had worked earlier and which occasionally gave the Cavendish help in technical problems. Allibone had brought along a 500-kV Tesla Coil and he now produced electrons of up to 300 keV energy, but their flux was unsteady. Walton first tried without success to design a machine to accelerate electrons in circular orbits. Then Cockcroft became interested and collaborated with Walton on a different method; they used Allibone's technique of applying high voltages to vacuum tubes which contained some hydrogen gas. However, they wanted to have more stable voltages for the acceleration of protons than Allibone had available. Hence they employed the so-called Greinacher-circuit for doubling voltages and invented a clever design with the help of which they produced steady direct voltages from alternating currents using switches and rectifiers. After some time Cockcroft and Walton moved their experiment to a larger room, where they built a bigger apparatus and were able to reach voltages of over 300 kV.

[54] The question of whether research at the Cavendish should move in the direction of using big apparatus eventually caused a breach between Rutherford and Chadwick, his closest collaborator. In 1935 Chadwick, having tried unsuccessfully to persuade Rutherford to extend the Cavendish by

brought to a close the period of one and a half decades since 1919 during which the Cavendish Laboratory dominated research in nuclear physics. During this period, a transition took place in the style of performing experiments. While earlier, individuals had attacked fundamental problems in physics with their own heroic efforts, from the 1930s onwards the big successes would be obtained in general by teams of researchers. The type of work at the Cavendish itself represented this transition. Rutherford had used the opportunity of putting his collaborators and students into small groups, and the members of each worked devotedly on a particular problem. But there was neither enough space nor money available to form larger teams or afford really big apparatus. 'We've got no money, so we've got to think,' Rutherford used to say (Andrade, 1964, p. 188). After all, he had achieved his fundamental results with lots of imagination and hard work and little money.[55]

With all these limitations the Cavendish Laboratory under Rutherford was the most remarkable place of its kind in the world. Although the most spectacular results in nuclear physics were not achieved in the 1920s, the scientific atmosphere of the Cavendish could not be matched by any other laboratory. In addition the connection between experimental work and theory was good and fruitful. Though Rutherford—unlike his predecessors in the Cavendish Chair—never mingled in theory and he always used to tease 'our theoretical friends,' as he would call them, he respected and maintained a close personal relationship with some theoreticians, especially Niels Bohr. And this relationship with the theoreticians was certainly helped by the fact that Ralph Fowler, who represented quantum theory in Cambridge, was married to his only child Eileen. Thus, there was hardly any major quantum physicist of the day who did not pass through the Cavendish and deliver a talk in front of Rutherford.[56]

---

building a new high tension laboratory, left Cambridge. He became Lyon Jones Professor of Physics in the University of Liverpool; there he started without delay to build a cyclotron.

During World War II, Chadwick took part in the Manhattan Project. In 1948 he was elected Master of Gonville and Caius College, Cambridge (until 1959). He also served as a member of the U.K. Atomic Energy Authority (1957–1962). Chadwick had been knighted in 1945. He received numerous other honours: Fellow of the Royal Society of London, 1927; Hughes Medal, 1932; Copley Medal, 1950. For his discovery of the neutron, he was awarded the Nobel Prize in Physics in 1935. Chadwick died in Cambridge on 24 July 1974.

[55] Most of his researches using radioactive decay products, especially α-particles, had been made with the 250 mg of radium that had been lent to him in 1908 by the Vienna Academy. By the late 1920s no real change had occurred, for as M. Oliphant recalled later:

> One day I had taken R[utherford] up to the Radium Room so that he could assure himself that all was in order. We had at that time about 400 mgm. Ra in solution for the preparation of radon and active deposit sources. I remember well how, as we were coming down the stairs, I said that we did not have enough radium so that I had to allocate the sources very carefully to meet the demands; and I said what a pity it was that somebody or other had not made a gift to him of a gram of radium, as the women of the United States had made to Madame Curie. His reply astounded me. It was: "Well my boy, I am very glad that nobody did. Just think: at the end of every year I have to say what I have done with it. How on earth could I justify the use of a whole gram of radium?" (Oliphant, 1972, pp. 40–41)

[56] Mark Oliphant relates that:

> At question time, following lectures at which [Rutherford] had presided, he would probe mercilessly after the physics behind their mathematical techniques, often declaring: "I am a simple man and I want a simple answer." I once heard him say to Heisenberg, in proposing a vote of thanks for his lectures: "We are all much obliged for your exposition of a lot of

While the Cavendish, with its emphasis on atomic and nuclear physics, clearly determined the kind of research that flourished in Cambridge, astrophysics was the other field of physics in which the University was particularly distinguished. Its principal representative was Arthur Stanley Eddington, who, since 1913, held the Plumian Professorship of Astronomy. Eddington, the son of a Somerset Quaker, had obtained his education at Owens College, Manchester, and at Trinity College, Cambridge. He had been Senior Wrangler in the Mathematical Tripos in 1904 and, upon graduation in 1905, had taken up the appointment as Chief Assistant at the Royal Observatory, Greenwich.[57] In this position, Eddington obtained practical experience in great many different projects—such as the evaluation of the photographic observation of the planetoid Eros, the determination of the longitudes in Malta, to name a couple—which placed him in the forefront of astronomical research within a few years. Thus, already in 1912 he had led a solar eclipse expedition to Brazil. During his stay at Greenwich, Eddington had also occupied himself with studies in theoretical astronomy; he

---

interesting nonsense, which is most suggestive." After the Scott Lectures on the uncertainty principle, by Niels Bohr, he remarked: "You know Bohr, your conclusions seem to me to be as uncertain as the premises upon which they are built." (Oliphant, 1972, pp. 28–29)

[57] Arthur Stanley Eddington was born on 28 December 1882 at Kendal in Westmorland, where his father was proprietor and headmaster of Stramongate School, where John Dalton had taught from 1781 to 1793. Since Arthur's father died already in 1884 in a typhoid epidemic, his mother moved with him and his elder sister to Weston-super-Mare, where he entered school in 1890 after earlier training at home. Eddington was interested in numbers before he could read and from the age of six he also became attracted by astronomy. At school he obtained a deep appreciation of English literature, a keen interest in natural history and a good foundation in mathematics. Before he was sixteen he won a Somerset County Scholarship for three years, amounting to £60 per annum, and he entered Owens College, Manchester. There he studied for three years (1898–1902), with physics as his main subject and Arthur Schuster as his professor. At the same time Joseph Ernest Petavel, the physicist and engineer, was a research student there, and George Clarke Simpson and John William Nicholson were his co-students. Eddington also attended all the honours lectures in mathematics, which were taught by Horace Lamb, who exerted considerable influence on him. Financial support came from further scholarships: in the London matriculation of January 1899 he was placed twelfth and was awarded an exhibition of £15 for two years; in the Victoria University Preliminary of June 1899 he won the Gilchrist Scholarship of £50 for three years; in 1900 the Heginbottom Physical Scholarship added £15; finally, at the end of 1901 he gained a minor entrance scholarship of £75 per annum in natural science at Trinity College, Cambridge, which was converted into a Major Scholarship of £100 per annum in mathematics in March 1903. In 1903 he took the London B.Sc. with first class honours in mathematics and a third class in physics and was awarded a University Scholarship of £50 for three years. In 1902 Eddington took residence at Trinity College. After two years of intensive concentration on mathematics under the guidance of the famous coach R. A. Herman, who stressed both the logic and elegance of mathematical reasoning, he won the position of Senior Wrangler in Part I of the Mathematical Tripos in 1904; the following year he was placed in Class I, Division I, in Part II. (Herman was the successor of the great Tripos coaches Hopkins and Routh; he coached nine out of the last eleven Senior Wranglers before the examination system was changed. He was an artist in mathematics, scrupulously rigorous and a lover of elegance. Through him Eddington acquired the most typical Cambridge mathematical style and outlook.) Eddington attended courses given by E. T. Whittaker, A. N. Whitehead and E. W. Barnes. After obtaining the B.A. degree in 1905 he started some research at the Cavendish Laboratory on thermionic emission and composed his first mathematical essay on the expansion of a retarded potential in terms of simultaneous variables. The appointment of Frank Dyson as Astronomer Royal for Scotland in 1905 left a vacancy at the Royal Observatory, Greenwich, and Sir William Christie selected Eddington to be Chief Assistant the same year.

had begun his researches of the stellar movements in 1906 and made them the theme of his Smith's Prize Essay of 1907 and of his first book (Eddington, 1914).[58] He was elected to succeed Sir George Darwin, the late Plumian Professor of Astronomy, who had been distinguished for his researches on lunar theory and on cosmogony.[59] The appointment of a practical astronomer served to restore the original duties of the Plumian Professor, namely, observational astronomy, especially after he also became—following the death of Robert Ball—the Director of the Cambridge Observatory in 1914. Since Ball's successor as Lowndean Professor of Astronomy and Geometry was Henry Frederick Baker, a mathematician, Eddington was responsible for instruction in the entire field of astronomy; he carried this unusual burden successfully and, in addition, continued both his observational tasks and his theoretical investigations of stellar systems. However, his later fame would be connected mainly with two subjects: his pioneering work in astrophysics, especially on stellar structure, and his theoretical and experimental researches in relativity theory.

Since its formulation in 1905 relativity theory had not received widespread attention in England even though Joseph Larmor and Joseph John Thomson had made fundamental contributions to its foundation.[60] The first British physicist to write upon topics connected with relativity theory had been Ebenezer Cunningham, the Senior Wrangler of 1902 in Cambridge, Smith's Prizeman of 1904, and afterwards a lecturer in mathematics at the University of Liverpool. He had demonstrated that Max Abraham's argument about the instability of the deformable electron was based on incorrect assumptions derived from the model of a

[58] Eddington won the Smith's Prize; later in 1907 he was elected a Fellow of Trinity College.

[59] George Howard Darwin, brother of the botanist Francis Darwin (1848–1925) and the eugenicist Leonard Darwin (1850–1943) and father of the physicist Charles Galton Darwin, was born on 9 July 1845 at Down, Kent, as the eldest son of the great Charles Robert Darwin. He was educated at Trinity College, Cambridge and became Plumian Professor of Astronomy in 1883. He worked on the application of mathematical techniques to astronomy and geology. In particular he developed a cosmogony of the solar system and a theory of the earth–moon system based on mathematical and physical principles. George Darwin died on 7 December 1912 in Cambridge.

The Plumian Chair, founded in 1704, was originally assigned to observational astronomy. For example, it was held by George Biddell Airy from 1828 to 1835; he also directed the then new Cambridge Observatory. In 1861 John Couch Adams, since 1859 Lowndean Professor of Astronomy and Geometry, took over the Observatory. Although the arrangement was planned for some time only, it remained effective until Adams' death and was continued under his successor Robert Ball.

[60] As we have mentioned earlier, Larmor had discovered the fact that Maxwell's equations are invariant under Lorentz transformations to the second order in the ratio v/c, v being the velocity of a moving charged particle described by the equations and c the velocity of light (Larmor, 1897b, p. 229). But he did not go further for as his student Ebenezer Cunningham later remarked:

> I did not think he went any further than the second-order invariance of the transformation, nor did I notice any suggestion in the book [Larmor, 1900b] that it was possibly an exact transformation . . . I left Cambridge for Liverpool in 1904, and there I continued to read and think about the matter and quite on my own discovered the exactness of the transformation. I wrote to Larmor saying that I had found this to be so, and he replied briefly that he knew this was so, but made no reference to how long he had known it. He certainly had not referred to it in any lecture or publication at that time . . . He did not seem at all enthusiastic about the idea that an algebraic transformation happened to be exact. Nor do I recall that he took any particular interest in the speculation that the relativity suggested by the transformation might have far-reaching significance. (Cunningham to Kittel, 14 December 1963, quoted in Kittel, 1974, p. 728)

rigid electron (Cunningham, 1907). He had then continued to work on the relativistic theory of the electron (Cunningham, 1908, 1909). After his return to Cambridge and St. John's College in 1911, Cunningham gave courses of lectures on relativity theory and wrote two books on the subject (Cunningham, 1914, 1915).[61] Relativity theory was thus introduced to Cambridge in the 1910s, but more important consequences followed when Eddington became interested in it in 1916. Following the confirmation of the prediction concerning the bending of light in a strong gravitational field by the Solar Eclipse Expedition of 1919, Eddington became an enthusiastic supporter of Einstein's theory of gravitation. Eddington had himself taken part in the Expedition and photographed the bending of light; with his reports and books he secured for the general relativity theory a most successful start in England (Eddington, 1918, 1920, 1923b). He continued to work on the theory in his own peculiar way, combining elegant mathematics with bold speculation towards a fundamental theory of matter, on the one hand, and deriving astronomical consequences, on the other. Thus, he not only obtained new results from relativity theory, but also renewed popular interest in this subject again and again.

Eddington considered the relativistic theories of matter—that is, the deeper theoretical understanding of its constituents, the electron and the proton—and of the universe as the two fundamental cornerstones on which the explanation of all physical phenomena had to be founded.[62] Physical phenomena, for him, included the astronomical observations as well. Since 1915 Eddington had worked on the theory of the internal constitution of stars, building on the earlier considerations on gaseous spheres of Homer Lane in the nineteenth century and on radiative equilibrium of Karl Schwarzschild in the beginning of the twentieth century. He developed the photoelectric theory of absorption of radiation inside a star (Eddington, 1922a, b; 1923a), he discovered the relation between the masses and the luminosity of stars (Eddington, 1924), he calculated the hydrogen content of the stars (Eddington, 1932), and he developed a theory of the variation of the luminosity (pulsation) of the Cepheid stars (Eddington, 1941), just to mention a few outstanding examples of his work. Because of his powerful combination of empirical knowledge and theoretical imagination, his strength in

[61] Cunningham recalled that the first decade of the twentieth century was 'a period of transition [in Cambridge], which culminated in 1910 when the order of merit in the Tripos was abolished and teaching in all subjects took a much more theoretical turn, while the Cavendish Laboratory moved on into atomic theory . . . Modern physics was in its infancy. I had to find for myself the work of Lorentz, Planck etc. after I had graduated. Eddington, Jeans and others were still to come' (Cunningham to Goldberg, quoted in Goldberg, 1970, p. 124).

[62] Eddington always made use of the latest developments in fundamental theories. Thus, he quickly incorporated quantum theory as well as the new developments in electron theory into his astrophysical calculations. Later, he would become especially fond of Dirac's relativistic theory of the electron; he would connect it with a general relativistic cosmology in order to obtain what he would call the 'Fundamental Theory.' This was the theory from which the most fundamental structure constants of the universe were supposed to follow: the fine-structure constant, the total number of particles in the universe, the ratio of the masses of the proton and the electron, and the ratio of the gravitational force to the electromagnetic force between a proton and an electron.

finding elegant mathematical formulations and solutions of problems, and his brilliant and persuasive lectures and writings, Eddington may be considered to have been the most influential astrophysicist of his time.[62a]

Still, in Cambridge he was not the only astrophysicist. There was, for example, James Hopwood Jeans, well known for his work in gas dynamics, statistical mechanics and the theory of radiation.[63] Since 1909 he had devoted himself mainly to astrophysical problems, investigating in particular the dynamical behaviour of rotating spheres under the influence of their own gravitation. He presented the results in his Adams Prize Essay, which was published under the title *Problems of Cosmogony and Stellar Dynamics* (Jeans, 1919). In it, Jeans discussed, among other topics, a theory of the origin of stars from gaseous matter existing in stellar systems, a theory describing the time evolution of galactic systems, and a theory of the creation of the solar system, all based on the principles of Newtonian gravitation and dynamics and the theory of heat radiation. Though Jeans, unlike Eddington, was hesitant in applying recent physical ideas—for example, he included relativity only in his second book on cosmology (Jeans, 1928)—he discovered many previously unknown results, which helped to establish cosmology as a modern physical theory. Through his researches and books—the technical as well as the popular ones—Jeans exercised much influence.

But even Eddington and Jeans did not exhaust all astrophysics in Cambridge. The official man in this field was Hugh Frank Newall, Professor of Astrophysics and since 1913 Director of the Solar Physics Observatory. Newall, son of the engineer, inventor and manufacturer Robert Stirling Newall (who had constructed most of the first submarine cables), had in fact introduced astrophysical research to Cambridge, working since 1892 with the 25-in. telescope presented to the University by his father.[64] He had made spectroscopic observations of

[62a] Eddington was knighted in 1930 and received the Order of Merit in 1938. National and international scientific societies honoured him with memberships and prizes: he was elected a Fellow of the Royal Society of London in 1914, and received the Society's Royal Medal in 1928; he was President of the Royal Astronomical Society from 1921 to 1923 (Fellow since 1906, Gold Medal in 1924). The International Astronomical Union had elected him President in 1938. Eddington died in Cambridge on 22 November 1944.

[63] For a biographical sketch of J. H. Jeans, see Volume 1, footnote 115.

[64] H. F. Newall was born on 21 June 1857 at Gateshead-on-Tyne, England. He attended Rugby School (1872–1876) and Cambridge University (1876–1880), being Maxwell's first undergraduate student at the Cavendish. After his graduation in mathematics he went to Wellington College as an assistant master, but returned to Cambridge already in October 1881 to become assistant and later Demonstrator of Experimental Physics (1886–1890) at the Cavendish. During that period he published papers on hydrodynamics and the conduction of electricity through liquids together with J. J. Thomson. He resigned from the demonstratorship in 1890 and assumed the unpaid position of the Newall Observer. He superintended the installation of the 'Newall Telescope' to which a prism slit spectroscope was soon added. In 1905 the munificent bequest of £5000, offered by Frank McClean, allowed Newall to add more special equipment to the telescope for the observation of the sun. A few years later the telescopes and spectroscopes of the Royal Society, which had been used earlier by Sir William Huggins, were transported to Cambridge. The University constructed the Astrophysical Building and created a new chair without stipend, electing Newall to it on 15 June 1909. Newall retired from his Cambridge position in 1928. He died at Cambridge on 22 February 1944.

the sun and stars and had participated in several solar eclipse expeditions (22 January 1898 to Pulgaon, India; 28 May 1900 to Bouzareah, Algeria; 17–18 May 1901 to Sawah Lorento, Sumatra; 30 August 1905 to Guelma, Algeria; 20–21 August 1914 to Feodosia, Crimea), gathering a large amount of data. Since April 1913, when the Solar Physics Observatory, used previously by Norman Lockyer, had been transferred from South Kensington to the enlarged Astrophysics Building of Cambridge University, Newall and his assistants had available to them unique equipment for the observation of stellar atmospheres. However, the man who would exploit the rich data gathered with it was still to come.

In 1919 Newall invited Edward Arthur Milne to be Assistant Director of the Solar Physics Observatory, thereby opening the latter's distinguished career as an astrophysicist. Milne, a former Cambridge student, had just returned from war service and was recommended for his new position by Archibald Vivian Hill, his wartime supervisor.[65] He assumed the duties of his new position only after taking a year off to attend courses on mathematics and physics because he wanted to fill the gaps in his interrupted education. Then Newall put him to work on a subject of primary importance for astrophysics: the theoretical study of stellar atmospheres. By combining and extending the results obtained by Arthur Schuster and Karl Schwarzschild on radiative transfer and equilibrium, Milne investigated two major types of questions: first, how is the radiation transferred through the atmosphere of a star; second, what is the state of ionization of the atmosphere? For this purpose he formulated and solved the equation of radiative transfer; he devised a particular model of the solar atmosphere, which described the observations satisfactorily. He also treated the inverse problem of using the crucial data about stellar atmospheres in order to determine the parameters of the theoretical models. Thus, for example, he derived the temperature distribution in the solar atmosphere from the observed darkening towards the limb of the emitted radiation. In collaboration with Ralph Fowler he exploited Megh Nad Saha's theory of thermal ionization in order to fix the temperature scale of the stellar spectral sequence. In his Bakerian Lecture of 1929 Milne summarized these pioneering investigations performed during the previous decade; he also

[65] E. A. Milne, who was born in Hull on 14 February 1896, the son of a headmaster of St. Mary's Church of England School, attended Lambert Street Council School in Hull (1901–1905), Hessle National School in Hessle (1905–1908), and Hymers College in Hull (1908–1913). At Hymers College he was attracted to the study of mathematics, especially through the influence of C. H. Gore, the headmaster, and C. Chaffer, the mathematics master and afterwards Director of the Admiralty Research Laboratory, Teddington. After winning an open scholarship in mathematics and natural science at Trinity College, he went up to Cambridge in December 1913. His defective eyesight excluded him from active military duty during the war, but early in 1916 he joined the Munitions Inventions Department, where he met and collaborated with R. H. Fowler. In 1919 Milne returned to Cambridge and was elected a Fellow of Trinity College the same year, having written three theses: one on maximum values of integrals, the second on sound waves in the atmosphere, and third, a theoretical study of the composition, ionization and viscosity of the earth's atmosphere at great heights.

discussed the connection between the structure of the atmosphere and opacity, the main observed property (Milne, 1929). Milne had arrived at the first reliable quantitative theory of the intensity of the spectral lines emitted from a star. It was for this work that he received several honours, notably the Smith's Prize in 1922 and the Gold Medal of the Royal Astronomical Society in 1935. Milne continued his work in Manchester, where he moved in 1925 as Beyer Professor of Applied Mathematics, succeeding his close friend Sydney Chapman.[66]

While physics and various branches of applied mathematics were represented at Cambridge in the 1920s by many brilliant personalities, this was less so in pure mathematics. The tradition founded by Cayley was not continued quite at the same level, especially after Godfrey Harold Hardy left Cambridge in 1919 to become Savilian Professor of Geometry at Oxford. The first two decades of the twentieth century had been very exciting for mathematics at Cambridge, though the successors of Cayley in the Sadleirian Chair, Andrew Russell Forsyth and Ernest William Hobson, were not directly responsible for this fact.[67] The great deeds were performed by men belonging to the next generation, with one exception, and these outstanding Cambridge mathematicians contributed to the foundations of mathematics in a unique way. The oldest among them was Alfred North Whitehead, who was educated at Trinity College and became a lecturer there until he left Cambridge for London in 1911.[68] He made his reputation in mathematics through the three monumental volumes of *Principia Mathematica* (Whitehead and Russell, 1910, 1912, 1913). His collaborator in this extensive and

[66] Milne held the Observatory post from 1920 to 1924 and was University Lecturer from 1922 to 1925. After his resignation from the Observatory he became Lecturer in Mathematics at Trinity College. When he was appointed to the professorship in Manchester (1924), he took leave of absence until he was free of his Cambridge duties. In 1928 Milne was elected Rouse Ball Professor of Mathematics at Oxford. He took up this position in January 1929; he was also elected to a Fellowship at Wadham College. He remained there for the rest of his life. In later years he worked on the structure of stars and on his theory of 'kinematic relativity.' Milne died on 21 September 1950 in Dublin.

[67] A. R. Forsyth (1858–1942), who was educated at Cambridge University and became Fellow of Trinity College in 1881, succeeded his teacher Arthur Cayley in 1895. After retiring from Cambridge in 1910, he joined Imperial College at South Kensington, London, in 1913 and stayed there until 1923. He contributed papers and wrote books on differential geometry, theory of functions and differential equations.

Forsyth's successor in the Sadleirian chair, E. W. Hobson (1856–1933), worked on geometry and function theory. He became a Fellow of the Royal Society in 1893 and won a Royal Medal for mathematical investigations in 1907. He was President of the Mathematics and Physics Section of the British Association for the Advancement of Science in 1910, and recipient of the De Morgan Medal in 1920.

[68] A. N. Whitehead was born on 15 February 1861 at Ramsgate, Isle of Thanet, England. He received his B.A. in 1884, M.A. in 1887 and D.Sc. in 1905, all from Cambridge University. After leaving Cambridge he became a lecturer in applied mathematics and mechanics and later Reader in Geometry at University College, University of London, from 1911 to 1914. Then he served as Professor of Applied Mathematics, and finally as Chief Professor of Mathematics, at Imperial College of Science and Technology until 1924, when he assumed the position of a Professor of Philosophy at Harvard University. He retired in 1936 and died in Cambridge, Massachusetts, on 30 December 1947. He was the recipient of numerous honours, including the British Order of Merit in 1945.

fundamental study, which attempted to derive mathematics from a few logical axioms, was Bertrand Arthur William Russell, more than ten years his junior, whom he had discovered as a Trinity examiner.[69] From 1910 onwards Russell held a lectureship at Trinity College until, as a result of his pacifist views and activities, he was dismissed from the College in July 1916. Among his students were the American Norbert Wiener and the Austrian Ludwig Josef Johann Wittgenstein (1889–1951). Wittgenstein was Russell's most brilliant disciple. He continued the logical researches; he was later brought back to Cambridge by Russell, obtained his doctorate and a Fellowship of Trinity in 1929, became a lecturer (1930–1936, 1937–1939) and finally Professor of Philosophy (1939–1947) at Cambridge University.

While Russell, like Whitehead, abandoned mathematics and turned to Philosophy, his younger friend Godfrey Harold Hardy, born in Cranleigh, Surrey, on 7 February 1877, became the brightest star in pure mathematics at Cambridge. He had entered Trinity College in 1896, had been Fourth Wrangler in Part I of the Mathematical Tripos in 1898 and First in Part II a year later. Since 1900 a Fellow of Trinity, he was appointed lecturer in 1906. In 1911 he began his collaboration with John Edensor Littlewood, who had been a student at Trinity from 1903 to 1907. Together they wrote nearly a hundred papers on many different topics, such as the theory of numbers, special functions, integrals and trigonometric series. This, perhaps the most enduring collaboration in the history of mathematics, was not interrupted when the two were separated and only Littlewood stayed on in Cambridge.[70] In 1914 another mathematician, who also worked principally on number theory, arrived from India: this was Srinivasa Ramanujan. Ramanujan had received no proper mathematical training, but Hardy and Littlewood had recognized his genius and Hardy had brought him to Trinity College, where he first received a scholarship and was later elected a

[69] Bertrand Russell, son of John Lord Amberley, a descendent of the sixth Duke of Bedford, was born at Trellek on 18 May 1872. He was brought up by governesses and tutors until he went into residence at Trinity College, Cambridge, in fall 1890. After obtaining high degrees in mathematics (he was Seventh Wrangler) and philosophy, he was elected Fellow of Trinity College in 1895. He spent some time in Paris and Berlin, and then lived in Haslemere, devoting his time to the study of philosophy and mathematics. In 1903 he published his first book on the principles of mathematics (Russell, 1903), and then proceeded with A. N. Whitehead to develop and extend the mathematical logic of Gottlob Frege and Giuseppe Peano. He was elected a Fellow of the Royal Society in 1908.

In 1910 Russell obtained a special lectureship in logic and the philosophy of mathematics at Trinity College. In 1914 he was invited to Harvard University as a visiting professor. Later, he lectured only occasionally: National University of Peking, 1920–1921; University of Chicago, 1938; University of California at Los Angeles, 1939. He wrote many books on philosophical subjects. After World War II he led campaigns for nuclear disarmament. He received the Sylvester Medal of the Royal Society (1934), the Order of Merit (1949) and the Nobel Prize in Literature (1950). Bertrand Russell died on 2 February 1970 at Plas Penrhyn, near Penrhyndendreath, Wales.

[70] J. E. Littlewood, born on 9 June 1885 in Rochester, England, became Fellow of Trinity College in 1908. He was Richardson Lecturer of the University of Manchester (1907–1910) before he returned to Cambridge in 1910 as Lecturer (1920–1928, Cayley Lecturer). He was elected Rouse Ball Professor of Mathematics in Cambridge University in 1928 and occupied this position until he retired in 1950. He died on 6 September 1977 in Cambridge.

Fellow.[71] He worked on number theory, combinatorial analysis and modular functions. Hardy found his 'insight into algebraic formulae, transformations of infinite series . . . most amazing' and he compared him in that respect 'only with Euler or Jacobi' (Hardy, 1921, p. xxviii). However, in 1919 the romance of Hardy's collaboration with Ramanujan came to an end; so did to some extent the great decade of pure mathematics in Cambridge.[72]

In the 1920s in Cambridge the dominant mathematical field was analysis and its leading figure was J. E. Littlewood. Henry Frederick Baker, since 1914 Lowndean Professor of Astronomy and Geometry, represented geometry.[73] Baker was born in Cambridge on 3 July 1866, had entered St. John's College in the Michaelmas term of 1884, and had been Senior Wrangler in 1887. He had received the Fellowship of his College in 1888, had won a Smith's Prize in 1889, and had become College Lecturer in 1890. Baker had still attended the lectures of Arthur Cayley, but had later come under the influence of Continental

[71] Srinivasa Ramanujan was born on 22 December 1887 at Erode in South India, the son of an accountant to a cloth merchant. He first went to school at five, then held a free scholarship at the Town High School in Kumbakonam, where his mathematical abilities were soon recognized. Owing to his weakness in English he did not succeed in obtaining further education; however, he continued to work independently on mathematical problems. He got married in 1909 and in 1911 he took up an appointment as a clerk in the Madras Port Trust. There he was brought to the attention of G. T. Walker, the visiting head of the Meteorological Department in India, who had formerly been a Fellow and Mathematical Lecturer at Trinity. Upon Hardy's recognition of his genius, and Walker's recommendation, Ramanujan received a Research Studentship for two years and finally a scholarship of £250 from Madras to enable him to go to England. This was augmented by an Exhibition of £60 from Trinity College, Cambridge. Thus, free of any duties, he was able to devote himself entirely to studies. He was elected to the Fellowship of Trinity, and in 1918 he was elected a Fellow of the Royal Society. His work was hindered considerably by a long illness starting in spring 1917, from which he seemed to recover two years later. He returned home to India and died at Kumbakonam on 26 April 1920.

[72] In 1919 G. H. Hardy became Savilian Professor of Geometry at Oxford. He returned to Cambridge and Trinity only in 1931, succeeding E. W. Hobson as Sadleirian Professor of Pure Mathematics. The Cambridge mathematicians 'were delighted to have him back: he was a *real* mathematician, they said, not like those Diracs and Bohrs the physicists were always talking about, he was the purest of the pure' (C. P. Snow in Hardy, 1967, p. 9). He retired in 1942 and died in Cambridge on 1 December 1947.

Hardy received many honours. He was elected a Fellow of the Royal Society in 1910 and received a Royal Medal in 1920, the Sylvester Medal in 1940, and the Copley Medal was due to be presented to him on the day of his death. The London Mathematical Society, of which he was twice Secretary and President, awarded him the De Morgan Medal in 1929, and the Paris Academy of Sciences elected him *Associé Etranger* in 1947.

About his collaboration with Littlewood and Ramanujun, Hardy wrote:

> The real crises of my career came . . . in 1911, when I began my long collaboration with Littlewood, and in 1913, when I discovered Ramanujan. All my best work since then has been bound up with theirs, and it is obvious that my association with them was the decisive event of my life. I still say to myself when I am depressed, and find myself forced to listen to pompous and tiresome people, "Well, I have done one thing *you* could never have done, and that is to have collaborated with both Littlewood and Ramanujan on something like equal terms" (G. H. Hardy, *A Mathematician's Apology*, Cambridge University Press, 1967 Reprint, p. 148).

[73] The six previous occupants of the Lowndean Chair, which was founded in 1750, had been astronomers, the last two being John Couch Adams (1859–1892), known for his codiscovery of the planet Neptune—independently of Leverrier—and Robert Stawell Ball (1892–1913).

mathematicians, especially the Italian geometers and Felix Klein, having twice visited the latter in Göttingen. Though his first publications were in algebra, invariant theory and algebraic functions, he had turned to geometry in 1911, and this occupation made him an obvious candidate for the Lowndean Professorship.[74] After his appointment to the Lowndean Chair, Baker worked mainly on projective geometry. Many results of his investigations were incorporated into the six volumes of *Principles of Geometry*, which Baker published between 1922 and 1933. The first volume dealt with the foundations of projective geometry; the second and third volumes were devoted to projective geometry in two and three dimensions; in the fourth volume the principal configurations in spaces of four and five dimensions were treated from an advanced point of view, while the fifth and sixth volumes were concerned with special problems. In the *Principles* Baker's method of research came through particularly well. They showed that he was interested in geometry as a general scheme which enabled one to deal in a simple way with mathematical objects 'whose properties can otherwise be developed only at great length, with complicated analysis' (Baker, 1922, p. viii).

Through his books and his teaching Henry Baker created a school of geometry, whose recruits were eager to devote themselves completely to geometrical reasoning. His influence was deepened through the so-called 'tea-parties' which he held at his home each Saturday afternoon during term time. His successor in the Lowndean Chair, William Vallance Douglas Hodge, who went up to Cambridge as a student in 1923, described these tea parties as follows:

> The party, sometimes numbering fifteen to twenty, and having in common only their devotion to geometry, met at 4:15 p.m. for tea, and all were expected to attend —a 'blue' who combined athletic prowess with an interest in geometry was allowed to come late, but he was expected to put in an appearance as soon as he could. After tea one member of the party gave a talk on a geometrical topic, and then the matter was open to discussion. The audience had to keep on its toes, for if there was any pause in the discussion Baker was liable to call on anyone for an opinion. At the height of its popularity there was no difficulty in finding speakers, as there were always two or three young men ready with new results that they wished to describe, but in early days, when numbers were smaller, any member of the party might be detailed to get up a paper and read it the following week. For some newcomers it was quite an ordeal to address the tea-party for the first time, for in the discussion that followed no quarter was given, but, though Baker was not prepared to tolerate any nonsense, he was always ready to shelter the innocent speaker when his contemporaries became too fierce. Through these tea-parties one soon learned to look on Baker not merely with admiration, but with genuine affection, and to this much of his success was due. (Hodge, 1956, p. 53)

[74] At this time H. F. Baker was already well recognized. He had been elected a Fellow of the Royal Society in 1898 and had become the first Cayley Lecturer in the University of Cambridge in 1903. He had published a book on *Abel's Theorem and the Allied Theory, Including the Theta Functions* in 1897 and another on *Theory of Multiply Periodic Functions* in 1907. He had edited the four volumes of the mathematical papers of J. Sylvester between 1904 and 1912 and had been President of the London Mathematical Society. He had received the De Morgan Medal in 1905 and the Sylvester Medal of the Royal Society in 1910. He retired in 1936, but continued to write original contributions and books on mathematics. Baker died in Cambridge on 17 March 1956.

Henry Baker's tea parties were part of the Cambridge environment in which talent was nourished and genius allowed to grow unhindered. At the same time as William Hodge, Paul Dirac arrived in Cambridge; he was also drawn into the tea parties and they would have a lasting influence upon him.

II.2 Quantum Theory in Cambridge

Because of the pioneering preparatory work of J. J. Thomson and J. Larmor and the active interest of E. Cunningham and A. S. Eddington, relativity theory had come to enjoy a favourable, even privileged position at Cambridge University. The conditions for the reception of the other major fundamental theoretical conception of the twentieth century, quantum theory, were much less favourable at least in the beginning. For more than one decade there existed in Cambridge an opposition to Planck's treatment of blackbody radiation, the leaders of this opposition being Lord Rayleigh and James Jeans. They had themselves worked out a distribution law for blackbody on completely classical grounds. Jeans defended this law as late as fall 1911, until he was forced by the accumulated empirical and theoretical evidence to accept Planck's law and the quantum theory.

This story began in summer 1900, a few months before Max Planck arrived at his radiation law. At that time Lord Rayleigh proposed a formula to describe the intensity distribution of what he called 'complete radiation,' which was the same as blackbody radiation by another name (Rayleigh, 1900b). The subject was taken up again in 1905 by Jeans who, on the basis of his earlier work in statistical mechanics, treated the partition of energy between matter and ether (i.e., radiation) and obtained what was later called the Rayleigh–Jeans law (Jeans, 1905a). It predicted that in thermodynamic equilibrium a finite amount of energy can be contained only by radiation of infinitely high frequency.[75] This result naturally contradicted the available experimental data; however, Jeans explained the discrepancy by assuming that in the experiments on blackbody radiation the state of equilibrium between matter and radiation was never quite attained. He claimed that in all observed distributions the energy was still largely associated with the degree of freedom of matter and that the transfer of energy to the

[75] In Rayleigh's formula of 1900 not only did a constant factor remain undetermined but also an exponential factor, depending on temperature and frequency, which avoided the above-mentioned 'ultraviolet catastrophe,' was put in without further justification. The first explicit calculation of all factors contained in the radiation law for long wavelengths was performed by H. A. Lorentz (1903). In 1905 Lord Rayleigh returned to the law of complete radiation; he expounded the universal validity of the formula for long wavelengths, i.e., his previous law *without* the exponential factor, and calculated the constant factor (Rayleigh, 1905b). His result was, however, wrong by a factor of 8. This error was not contained in Jeans' independent calculation (Jeans, 1905a), which Rayleigh accepted as being correct (Rayleigh, 1905c). The Rayleigh–Jeans law was also found by Einstein as the law that followed from classical electromagnetic theory and kinetic theory. (This derivation was contained in Einstein's paper on the light-quantum and the photoelectric effect, which he submitted to *Annalen der Physik* in March 1905: Einstein, 1905b.)

vibrations of the ether had occurred only for small frequencies. The reason for this seemed to him to be evident by examining the energy distribution among the degrees of freedom of complex gas molecules. He argued that in this case the translational and all rotational degrees of freedom (except the ones around an axis of symmetry) were 'active,' that is, they obtained their equilibrium share of energy in a short time through the collisions between the molecules; the oscillational degrees of freedom, on the other hand, were 'passive,' as the corresponding coordinates could only be influenced slowly by the collisions, and it should thus take a very long time to transfer energy to them. In the ether the long wavelengths belonged to the active, and the short to the passive degrees of freedom, with a continuous transition between the two extreme situations. Hence the time necessary to reach complete equilibrium as described by the Rayleigh–Jeans formula 'must be measured in millions or billions of years unless the gas is very hot' (Jeans, 1905b, p. 97).

Jeans' way out of the discrepancy between experiment and theory was not accepted by Max Planck, who simply declared that 'the difficulty in question arises only on account of an unjustified application of the law of equipartition to all independent state variables' (Planck, 1906, p. 176). On the other hand, Hendrik Antoon Lorentz repeated Jeans' argument concerning the nonequilibrium between matter and radiation in a blackbody at the International Congress of Mathematicians in Rome (Lorentz, 1908a), and finally Jeans advocated it again on the occasion of the first Solvay Conference in fall 1911. In support of his thesis Jeans devised a hydrodynamical analogue of the situation encountered in the experiments on heat radiation (see Jeans' report, Section 11, in Langevin and de Broglie, 1912). In a system of vessels containing water, which are connected by thin pipes, the level of the fluid in all vessels will be the same after a sufficiently long time has elapsed for equilibrium to be established, provided the entire system is conservative. If, however, there are leaks in the pipes in between, through which a certain amount of water is lost in course of time, not all the vessels will eventually show the same level. The vessels which are connected to the rest of the system by extremely thin and highly dissipative pipes —so that not enough fluid can be supplied from other vessels—will in fact be empty. By suitably selecting the vessels and the dissipative pipes in between, one may obtain a continuously decreasing water level in subsequent vessels. Thus, for example, the heights of the water columns may follow the same curve as is found for the ratio of the observed and the equipartition energies of selected frequencies in blackbody radiation. In this respect the hydrodynamical system represents an analogue of the blackbody, while the water supply mechanism would correspond to the mechanism by which radiation is created from the energy removed from matter. Jeans concluded that the latter mechanism would perfectly be able to account for Planck's energy distribution and that the quantum of action need not be introduced at all.

However, Jeans' ingenious proposal did not appeal to his audience at the Solvay Conference. The sharpest criticism was expressed by the mathematician Henri Poincaré who, at the end of the discussion of Jeans' talk, remarked: 'It is evident that by a suitable choice of the dimensions of those connecting pipes between the vessels, on the one hand, and the magnitude of losses, on the other, Mr. Jeans can account for any experimental result. However, that is not the role of physical theories. One must not introduce as many arbitrary constants as there are phenomena to be explained. The goal of physical theory is to establish a connection between diverse experimental facts, and above all to predict' (Poincaré in Eucken, 1914, p. 64; see Mehra, 1975c, p. 24). Jeans' obstinate arguments against the quantum hypothesis led Poincaré to investigate the theoretical foundation of Planck's law for blackbody radiation. In his last but one published paper he concluded that a consistent derivation of this law was not possible unless one assumed the existence of an essentially discontinuous motion in nature (Poincaré, 1912a).

In contrast to Jeans' assumption of nonequilibrium, Planck's distribution was found to hold good over long periods. As a result of the discrepancy between the predictions of classical theory and the experimental observations, on the one hand, and of Poincaré's analysis, on the other, Jeans accepted quantum theory.[76] He became such a staunch convert that he began to believe that quantum theory would explain many other phenomena, besides the radiation law, which the classical theory had not been able to explain. Thus, he became—already in 1913—one of the first to support Niels Bohr's theory of spectral lines.[77] A year later he published his *Report on Radiation and the Quantum Theory* for the Physical Society of London, in which he drew special attention to the successful application of quantum theory to the problem of atomic structure (Jeans, 1914b). But this report meant much more: it was the first complete account of all known aspects of quantum theory. Jeans discussed not only the law of radiation and the explanation of the origin of spectral lines, but also the photoelectric effect and the behaviour of specific heats of solids at low temperatures. Finally he discussed what he considered to be the physical basis of quantum theory: the atomicity of matter, radiation and electricity. Jeans' report rendered an invaluable service, as it introduced the British scientific public to the new quantum concepts which had been developed since Planck derived his radiation law in 1900. 'It did more than any other work published in English language,' remarked a biographer of James Jeans, 'to secure acceptance for the quantum theory, owing to its brilliant expository qualities and even more to its being rooted in the prejudices and habitual modes of thought of the English school of mathematical physics descending from Newton' (Crowther, 1952, pp. 107–108). For Jeans, however, the report ended an involvement of one and a half decades in the problems of

[76] Jeans had arrived at the conclusion that Planck's law demanded some discontinuity in the description of natural phenomena already in 1910 (Jeans, 1910). However, in 1910 Jeans was not convinced about the validity of Planck's law. Poincaré's derivation was more complete.

[77] Immediately after Bohr had given a short account of his theory at the Birmingham Meeting of the British Association for the Advancement of Science in September 1913, Jeans remarked in his report summarizing the problems of radiation theory that 'Dr. Bohr has arrived at a most ingenious and suggestive, and I think we must add convincing, explanation of the laws of spectral lines' (Jeans, 1914a, p. 379).

radiation theory and kinetic theory of matter. From then on he turned his attention to cosmogony.[78]

In the years during which Jeans changed from being an opponent to becoming an advocate of quantum theory, other scientists in Cambridge became interested in it too. In this context one does not think of J. J. Thomson, who only expounded certain arguments in favour of a 'patchy' or molecular structure of radiation (Thomson, 1904c).[79] Joseph Larmor, on the other hand, addressed himself directly to the problems of quantum theory in his Bakerian Lecture of 1909 on 'The Statistical and Thermodynamical Relations of Radiant Energy' (Larmor, 1909). There he proposed to derive Planck's law of blackbody radiation on the assumption that in phase space (of statistical mechanics) there existed finite cells among which the energy was distributed with equal probability. One of the consequences of this assumption was that subdivisions of Planck's energy unit—that is, $h\nu$, the energy quantum, with ν being the standard frequency of the oscillator—might occur in the theoretical description of phenomena. Such subdivisions showed up in two investigations that were published in 1911: first Max Planck, in connection with his 'second quantum hypothesis,' introduced half quantum numbers for the stationary state of the oscillator (Planck, 1911a); second, in a formula, by which Walther Nernst and his British student Frederick Alexander Lindemann in Berlin sought to account for the deviations of the observed specific heats of solids at low temperatures from the values predicted by Albert Einstein's theory of 1906, they used both energy-quanta and half energy-quanta (Nernst and Lindemann, 1911a, b).[80]

The ideas of Thomson and Larmor about atoms and energy-quanta influenced John William Nicholson, another Cambridge man, in his pioneering work on the quantum theory of the atom.[81] He viewed the atom, in agreement with Thomson's model, as consisting of electron rings which, however, are arranged

[78] It should be mentioned, however, that Jeans prepared a second edition of his report on quantum theory in 1924, in which he tried to cope with the rapid growth of the theory (Jeans, 1924). For example, he included Einstein's statistical derivation of Planck's law of 1916 as well as certain confirmations and extensions of Bohr's theory of atomic structure.

[79] Apart from the fact that Thomson arrived at this conclusion without referring to the quantum theory, he did not really adhere to the view that light is composed of constant units. Later, he proposed to explain the quantum effects of radiation on the assumption that 'the mechanism by which the transformation of radiant energy into thermal energy or thermal energy into radiant is effected is a process which has an element of discontinuity in it,' and he tried to construct such mechanisms based on classical mechanics (Thomson, 1912b, p. 643).

[80] F. A. Lindemann and his brother studied in Berlin to obtain their doctorates under Nernst's direction. In 1910 Nernst had found the first results on the specific heat of solids that seemed to confirm Einstein's formula (Einstein, 1906g). F. A. Lindemann then derived a theoretical formula relating the characteristic frequency (which occurred in Einstein's equation) to the melting point of the solid under investigation (F. A. Lindemann, 1910). Since the subsequent, more detailed measurements of the specific heats could not be described completely by Einstein's formula, Nernst and Lindemann proposed a modification. By trial and error they arrived at a new formula in which Einstein's frequency term was replaced by two terms of the same structure, one containing the characteristic frequency ν, the other $\nu/2$.

[81] For a biographical sketch of J. W. Nicholson, see Volume 1, footnote 261.

around a small concentrated nucleus, for he believed—having worked earlier in electron theory—in the electromagnetic origin of the heavy mass of positive charge. The stability of the atom should be due to the fact that the total acceleration of the circulating electrons adds up to zero; hence, according to Larmor's condition for the accelerated electron, no radiation is emitted unless the electron rings are excited to perform transverse vibrations. In his third communication on these questions, which appeared in June 1912 in the *Monthly Notices of the Royal Astronomical Society*, Nicholson introduced h, the quantum of action, as follows: first, he assumed that the angular momentum of the electron rings takes on values $nh/2\pi$, where n is an integer; second, the emission and absorption of light by an atom is connected with 'discontinuous changes' in its angular momentum (Nicholson, 1912b). Thus, he made use of two essential features of quantum theory: the existence of a finite quantum of action h and the discontinuity of motion. 'It is readily seen,' he suggested, 'that this view presents less difficulty to the mind than the more usual interpretation, which is believed to involve an atomic constitution of energy itself' (Nicholson, 1912b, p. 679).[82] With his new theory Nicholson was not only able to explain many of the discrete lines observed in astronomical spectra, but also to predict new ones. His theory was immediately welcomed, for, as the astrophysicist H. F. Newall, Nicholson's Cambridge colleague, remarked: it encouraged 'the hope that another large step has been taken towards a working atom' (Newall, 1912, p. 327).

Although Nicholson's theory was soon superseded by Niels Bohr's atomic theory, his work exerted an important influence. Bohr had gone to Cambridge and the Cavendish Laboratory in fall 1911, when Nicholson was still around, and the two even had some discussion on the problems of electrical conduction by electrons. After Bohr left Cambridge for Manchester a few months later he became interested in atomic structure. He had read Nicholson's papers on the subject, but he changed one of his assumptions radically; while Nicholson had been anxious to avoid the atomicity of energy, Bohr reinforced it by means of his frequency condition. On the other hand, Bohr accepted the idea of quantizing the angular momentum of electron rings or orbits. He thus advanced concepts which did not seem to harmonize, at least for someone who had grown up in the Cambridge tradition of Maxwell, Larmor and J. J. Thomson. Nicholson therefore continued to prefer his own atomic theory over Bohr's; he calculated new results and eagerly pointed out the difficulties which Bohr's model of many-electron systems encountered. To a certain extent he was responsible, for some time, for Bohr's theory receiving less attention in England than on the Continent. On the other hand, Nicholson also contributed to the progress of Bohr's theory in an indirect way: his colleague William Wilson at King's College, London, was motivated by Nicholson's work in his discovery of the quantum conditions for multiply periodic systems (Wilson, 1915). Indeed, the relation of the quantization

[82] In this context see, for example, the discussion of the difficulty of employing the conception of radiation quanta by J. J. Thomson in various lectures and papers around 1911 (e.g., Thomson, 1912b) with which Nicholson was familiar.

of the integrals, $\int p\,dq$, to the quantization of the angular momentum possessed by an electron ring was immediately obvious. Wilson also extended the quantum condition to the case in which a magnetic field acts on the orbiting electron (Wilson, 1923).

Although these considerations and results obtained in Cambridge and London threw some light on quantum phenomena, they did not have the consequences and impact of the atomic theory which Niels Bohr began to develop during his stay at Manchester from March to July 1912. It was probably decisive that at this critical time in his career Bohr did not stay on in Cambridge; in Manchester he received encouragement from Ernest Rutherford who was open to new ideas if they promised to explain the observed phenomena adequately and opened new horizons in physics. In fall 1911 Rutherford had attended the first Solvay Conference in Brussels and upon returning to Manchester he had given an account of his impressions of the reports on quantum theory at an occasion where Bohr had been present. Bohr had been inspired to combine Rutherford's nuclear model of the atom with the concept of the quantum of action to develop a theory of atomic structure in which the transition of electrons from one stable (or stationary) orbit to another accounted for the characteristic spectral lines.[83] This theory found immediate support in the brilliant experimental investigations of the X-ray spectra of atoms undertaken by Henry Gwyn Jeffreys Moseley at Manchester and Oxford, before it was taken up and extended by Arnold Sommerfeld at Munich. When in 1919 Rutherford, who had remained friendly to Bohr, went to Cambridge as Cavendish Professor it was evident that modern atomic theory would have his support, although he himself was working on nuclear problems to which quantum theory had not yet been applied. Rutherford generally kept away from detailed theoretical reasoning, but at the Cavendish the theoretical part of the activities was in any case organized by Charles Galton Darwin and Ralph Howard Fowler. Soon these two secured a good position for quantum theory in teaching and research.

The older of the two, Charles Darwin, a grandson of the naturalist Charles Robert Darwin and son of the astronomer George Howard Darwin, had already been associated with Rutherford and atomic physics at Manchester University, where he held the Schuster Readership of Mathematical Physics from 1910 to 1914.[84] There he had worked on the theory of scattering of α-particles by matter,

[83] Bohr completed his first papers on the atomic constitution in Copenhagen, from where he sent his manuscripts to Rutherford. In fall 1914 he returned to Manchester University to take up the appointment as Schuster Reader in Mathematical Physics. He stayed there until summer 1916 when he was appointed to the newly created Professorship of Theoretical Physics in the University of Copenhagen.

[84] C. G. Darwin, born on 19 December 1887 in Cambridge, was a graduate of Cambridge University: he had entered Trinity College in 1906 and been Fourth Wrangler in the Mathematical Tripos of 1909. Immediately after taking Part II of the Tripos in 1910, he was appointed to his position at Manchester. He stayed there till the First World War broke out. During the war he was attached to the Royal Engineers as an officer in a field survey unit in France, engaged primarily on sound ranging and flash spotting.

Darwin returned to Cambridge in 1919 and became Lecturer of Mathematics at Christ's College, Cambridge, then Professor of Natural Philosophy at the University of Edinburgh (1923–1936),

thus providing a sound background for the theoretical interpretation of the experiments. Moreover, he and Moseley had collaborated on the diffraction of X-rays by crystals, from where Moseley had gone on to investigate systematically the discrete X-ray lines emitted by atoms. After his return from war service in 1919 Darwin was elected Fellow and Lecturer in Christ's College, Cambridge, where his illustrious grandfather had also been a member. While his work in Manchester was founded completely on classical theory, he now began to take into account the ideas of quantum theory. This seemed to be necessary especially for the study of the partition of energy in assemblies of atoms, which he undertook with Fowler. This problem already had a distinguished history in Cambridge because of the pioneering work of James Jeans in classical statistical mechanics.[85] But Darwin, who had witnessed the great success of Bohr and Moseley in atomic physics, was convinced that progress could be achieved beyond what Jeans had done by incorporating the most recent advances of quantum theory. This opinion was shared by Fowler, his collaborator, who became an enthusiastic quantum physicist and continued to work on the theory even after Darwin left Cambridge in 1923 to become Tait Professor of Natural Philosophy in the University of Edinburgh. Darwin and Fowler broke with the tradition of the Cambridge school of applied mathematics, whose representatives —like Larmor and Jeans—had worked along the lines of perfecting the methods by which detailed consequences from the accepted ideas of Newton and Maxwell could be derived. Fowler, in particular, without giving up the rigour of his mathematical training, put the still controversial concepts of the emerging quantum theory in the centre of his investigations. He thus initiated a new era of theoretical physics in Cambridge.

In order to understand the change brought about by Fowler in the Cambridge attitude to mathematical physics it is necessary to deal with his background and personality. Ralph Howard Fowler was born on 17 January 1889 at Fedsden, Roydon, Essex, the eldest son of Howard and Ena Fowler, *née* Dewhurst.[86] After being educated by a governess until he was ten and then at a preparatory school, he was elected in 1902 to a scholarship at Winchester where he remained until 1908. Fowler received a balanced education: he excelled in such different

---

Master of Christ's College (1937–1939) and Director of the National Physical Laboratory (1938–1949). Darwin became a Fellow of the Royal Society in 1922, and received many other honours, including: Royal Medal, 1935; Hughes Medal of the Royal Society, 1936; and the U.S. Medal of Freedom, 1948. He died in Cambridge on 31 December 1962.

[85] Jeans had summarized his investigations in this field in a treatise on *The Dynamical Theory of Gases* (1904). The second edition of this work appeared in 1916 and served Darwin and Fowler as the starting point of their work (see Milne, 1945, p. 68).

[86] Fowler's father had been educated at Clifton and New College, Oxford. He was a good sportsman who played cricket and rugby; his athletic abilities were passed on to Ralph and his sister Dorothy, who became a distinguished golfer in England. Howard Fowler was called to the bar but did not practice as a barrister, for he went into business after getting engaged to his future wife. There was also another son, Christopher (six years younger than Ralph), who was prevented from entering New College, Oxford, by the outbreak of the war and was killed on the Somme in April 1917. The family life of the Fowlers was very happy, and even as a student Ralph used to spend his vacations regularly in Essex, or Norfolk, or, later, in Somerset, where his parents eventually moved.

subjects as mathematics and classics; he played cricket, football and golf. Thus, he developed his intellectual and physical abilities simultaneously; he became a big and healthy man with a strong character and engaging charm. As his friend E. A. Milne noted: 'Fowler always made up for what must be described as his sometimes deliberate downright rudeness, when he was speaking plainly, by the most abject and contrite apologies afterwards, and none of his many friends ever bore him the least malice for occasions when he had wounded their feelings' (Milne, 1945, p. 62).

In December 1906 Ralph Fowler was elected to a Major Scholarship at Trinity College, and he went up to Trinity in the Michaelmas term of 1908. He studied mathematics and passed the mathematical Tripos, Part I, in 1909 taking a First Class, and Part II in 1911, with a mark of distinction. He became B.A. in 1911 (M.A. in 1915) and began to do research in pure mathematics, being attracted to problems of the behaviour of solutions of second-order differential equations. In 1913 he was awarded a Rayleigh Prize for Mathematics and in October 1914 he was elected to a Fellowship at Trinity. Fowler kept on contributing to pure mathematics even after he had turned to theoretical physics. He wrote papers on the approximate solutions of differential equations, in particular on Emden's equation, which determines the equilibrium conditions in gaseous stars, and he published a tract on *The Elementary Differential Geometry of Plane Curves* (Fowler, 1920).

The outbreak of World War I interrupted Fowler's scientific career, and he immediately obtained a commission in the Royal Marine Artillery. He took part in the unfortunate Gallipoli campaign of 1915 as a gunnery officer and was badly wounded in the shoulder. During his convalescence he was invited by the physiologist Archibald Vivian Hill, a former Fellow of Trinity College, Cambridge, and then a Captain in the Cambridgeshires, to help him test the Darwin–Hill Mirror Position Finder, an instrument invented by Hill and Horace Darwin.[87] Hill had been asked by Darwin, fifth son of Charles Robert Darwin, to form what became the 'Anti-Aircraft Experimental Section' of the Munitions Inventions Department. Fowler joined in early 1916, together with Edward Arthur Milne, then a student at Trinity. The war work changed his outlook considerably. He learned to know and collaborate with scientists from various places and of different specializations, many of whom later became his friends

[87]A. V. Hill, born on 26 September 1886 in Bristol, studied mathematics at Trinity College, Cambridge, and was Third Wrangler in 1907. Under the influence of the Cambridge physiologist John Newport Langley he turned to physiology. He studied in Germany from 1910 to 1911 and returned to Trinity College, where he had been elected Fellow in 1910. After his military service he became Brackenbury Professor of Physiology at Manchester University in 1920, then moved to University College, London (1923–1925), and finally became Foulerton Research Professor of the Royal Society (1926–1951). In 1922 Hill won the Nobel Prize for Medicine and Physiology (together with Otto Meyerhof) 'for his discovery relating to the production of heat in muscles.' He received numerous other scientific and public honours: for instance, the Copley Medal of the Royal Society, whose member he was since 1918 (1935–1945, Secretary; 1945–1946, Foreign Secretary); and the Order of the British Empire in 1918. He represented Cambridge University in Parliament from 1940 to 1945 and became a Companion of Honour in 1946.

and associates, such as Douglas Rayner Hartree.[88] The extensive instrument trials that were carried out on Whale Island, beginning in October 1916, taught him to see the connection between mathematical research and experiments. Fowler became the Assistant Director to Hill, being responsible for the experiments. As Milne described their collaboration,

> The two fitted together like hand and glove. Hill was the inspirer of most of the researches that were conducted, whilst Fowler executed them in the field. It was a most friendly community—sometimes bullied by Fowler (though he always assuaged its feelings afterwards), and then released from its tension by a visit from Hill. But what is significant to record . . . is the important scientific influence which Hill had on him [Fowler], and he in turn on the members of the section. At directing research, both were superb; to behold the way they set about a new problem, often in a new terrain with new material, to see the way they made inferences and came to conclusions and sound judgments and to take part in it all, was far better training than most universities can offer to aspirants in research. (Milne, 1945, p. 66)

The association with Hill had a great influence on Fowler's later career. It brought him closer to problems of applied physics and provided him the possibility of directing talented people towards planned research.[89] Moreover, he had a great share in writing professional reports, performing laborious numerical calculations and carrying out detailed observations to confirm theoretical results. When he left military service in April 1919, honoured by the award of the Order of the British Empire, and returned to Cambridge and Trinity College, Fowler was a mature and fully developed personality. He had gained invaluable experience, which he could not have obtained in leading the undisturbed life of a Cambridge don. He had also made up his mind to work on problems of theoretical and mathematical physics rather than pure mathematics.

Fowler was now thirty years old, and the appointment as College Lecturer in Mathematics at Trinity in 1920 provided him financial security. But instead of settling down to the pursuit of his earlier involvements he was eager to learn new fields. Towards the end of his military service at Portsmouth he had begun to study certain recent topics of theoretical physics: he had read Jeans' *Dynamical Theory of Gases* and Eddington's *Report on the Relativity Theory of Gravitation* (Eddington, 1918). Back in Cambridge he turned himself into a student again. He attended lectures, e.g., E. Cunningham's 9 A.M. lectures on Special Relativity, together with Milne, his friend and former collaborator. However, the greatest

[88] To Hill's team belonged W. Hartree and C. W. Hartree, father and brother of D. R. Hartree, A. C. Hawkins, and H. W. Richmond; the mathematicians G. T. Bennett, T. L. Wren, R. A. Herman, C. E. Haselfoot, J. G. Crowther, W. E. H. Berwick, W. R. Dean and S. Pollard, the physicist C. Mayes, and many others.

[89] Fowler published the scientific results of his military work in several papers on problems of ballistics and siren harmonics, which he wrote in collaboration with C. N. H. Lock, E. G. Gallop, H. W. Richmond and E. A. Milne.

source of new physical ideas was the experimental work at the Cavendish, and it was Rutherford who provided inspiration to Fowler. The two started a close collaboration which lasted until Rutherford's death. As Milne remarked:

> A wonderful friendship sprang up between the two and it extended to all of Rutherford's pupils and co-workers. I say *wonderful*, because *a priori* it was rather unlikely that there should exist a strong sympathy between Rutherford, with his depth and search for simplicity and his impatience with abstract mathematical theory, and Fowler, with his breadth and search for generality and his insistence on mathematical rigour. But Rutherford's robust and rugged genius appealed to Fowler, and Fowler's amazing versatility and power of becoming absorbed in problems put to him by others appealed to Rutherford. (Milne, 1945, p. 68)

The two men also established close family ties when in 1921 Fowler married Eileen, only daughter of Sir Ernest and Lady Rutherford.[90]

Unlike most other mathematicians in Cambridge, Fowler very much became a part of the Cavendish Laboratory. Not only did he play golf with Rutherford, F. W. Aston and G. I. Taylor, but he became interested and involved in the problems of the experimentalists. For example, when Aston made improvements in his mass spectrograph Fowler helped him in analyzing mathematical problems connected with the focusing of the ions in electric and magnetic fields (Aston and Fowler, 1922). Fowler performed many calculations to explain the observations on the passage of α-particles and electrons through matter that had been made at the Cavendish, and published his results in several papers (Fowler, 1923b, 1924b, 1925b). On the whole he became the principal teacher and advisor in theoretical modern physics at the Cavendish. He attracted able students. The serious and competent discussion of experimental and theoretical problems, and the hearty hospitality of the Rutherfords and the Fowlers, made Cambridge a mecca of physics in those days, which outstanding scientists like Niels Bohr, Paul Ehrenfest and James Franck visited repeatedly.

Fowler began his work at Cambridge by publishing several papers on the results of his wartime research as well as his mathematical investigations on plane curves. But soon he began to contribute to quantum theory and statistical mechanics. In June 1921 he submitted a note in which he discussed the extension of Fourier's integral theorem and its application to quantum theory (Fowler, 1921). This problem occurred to Fowler from reading Poincaré's paper in which he had shown that the quantum hypothesis was the only one that led to Planck's radiation law (Poincaré, 1912a). Poincaré had made use of Fourier's integral theorem in order to invert the infinite integral,

$$\Phi(kT) = \int_0^\infty \exp\left\{ -\frac{\eta}{kT} \right\} w(\eta)\, d\eta, \tag{2}$$

[90] Fowler's marriage was very happy. They had four children, two sons and two daughters. Eileen Fowler died in December 1930, shortly after the birth of their fourth child.

describing the partition function of a resonating electron, and to obtain $w(\eta)$, the *a priori* probability,

$$w(\eta) = \frac{1}{2\pi i} \int_C \Phi(\alpha) \exp(\alpha\eta)\, d\alpha, \tag{3}$$

when the energy lies between η and $\eta + d\eta$; C denotes a contour in the complex α $[= (kT)^{-1}]$-plane, with a positive real part and an imaginary part, going from $-i\infty$ to $+i\infty$. Fowler noticed the difficulty that in the case of the functions employed by Poincaré, Fourier's integral theorem did not apply at all. This was especially so for $w(\eta)$, which should be zero everywhere except when η assumed values that were integral multiples of an energy quantum $p \cdot \epsilon$, such that the integral, $\int_{p\cdot\epsilon-\delta}^{p\cdot\epsilon+\delta} w(\eta)\, d\eta$, was finite and equal to unity. Fowler introduced the necessary rigorous mathematical steps to ensure that Poincaré's conclusion was correct.[91] Though this note represented just a skillful mathematical exercise, it was important for a special reason: James Jeans, in his 1914 report on quantum theory, had assigned a crucial role to Poincaré's mathematical proof for the assertion that the description of certain natural phenomena required the element of discontinuity.

Fowler's first paper on kinetic gas theory and statistical mechanics developed from his study of Jeans's *Dynamical Theory of Gases*. He noticed that there arose a discrepancy in analyzing the empirical data concerning the viscosity and compressibility of gases according to the standard formulae, and he suggested that the accepted formula for viscosity was not a good approximation. He also discovered a new procedure for calculating the constant a in Van der Waals' equation of state for a real gas; this a is related to the mutual attraction of the molecules, and Fowler noticed that it is not really constant but varies slowly with temperature because the tendency of the molecules to form clusters in gases depends on the temperature. Fowler also showed that the basic integrals occurring in kinetic gas theory converged only if the neighbouring molecules interacted with one another with a force proportional to a higher inverse power of the distance than -4 (Fowler, 1922). Fowler's work was continued in this direction about two years later by John Edward Jones, or Lennard-Jones as he called himself later, then 1851 Exhibition Senior Research Student at Trinity College.[92] In a series of papers starting with a note on the equation of state

[91] Fowler also generalized his extension of Fourier's integral theorem—which essentially involved the replacement of Riemannian by Stieltjes' integrals—to the case of Mellin transforms. He then gave the physical application of the new mathematical methods, e.g., to the absorption of nonhomogeneous beams of X-rays by matter.

[92] J. E. (Lennard-)Jones, born on 27 October 1894, was educated at the University of Manchester and at Trinity College, Cambridge. The First World War interrupted his studies and he became a flying officer in the Royal Flying Corps and later experimental officer at the Armament Experimental Station, Orfordness. In 1919 he returned to the University of Manchester as Lecturer in Mathematics. Until 1925, when he became Reader in Theoretical Physics at the University of Bristol, he also held the Exhibition Senior Research Studentship at Cambridge. Upon his marriage in 1925 to Kathleen Mary Lennard he changed his name by deed poll to Lennard-Jones. He was inspired to work on kinetic theory by his teacher Sydney Chapman—who had been Lecturer at Trinity College from 1914 to 1919 and then became professor at Manchester until 1924 when he went to Imperial College of Science and Technology, London.

(Lennard-Jones, 1924), he arrived at the conclusion that the available data on viscosity, the equation of state of gases and the properties of crystals, led to the conclusion that forces between the atoms and molecules consisted of an attractive term, which was proportional to the inverse fifth power of the distance, and of a repulsive term, which was proportional to a higher inverse power.[93]

Fowler's first paper on kinetic gas theory was communicated to the *Philosophical Magazine* by his senior colleague C. G. Darwin, the other 'theoretical physicist' in Cambridge and at the Cavendish. Darwin and Fowler now joined forces to collaborate on a major project, a theory of the partition of energy in atomic systems. The problem of how the energy is distributed among the degrees of freedom of systems (that are treated by the methods of statistical mechanics) had occupied a prominent place in the theoretical work of Maxwell, Rayleigh, Larmor and Jeans at Cambridge. But this work was done before the breakdown of classical concepts in dealing with quantum phenomena had been demonstrated. Once Jeans had become convinced of the validity of quantum theory, he had advocated its incorporation into the foundations of statistical mechanics. Jeans gave an account of 'the difficulties arising out of the classical treatment' in the second edition of his *Dynamical Theory of Gases* and indicated 'how these difficulties disappear in the light of the new conceptions of the Quantum Theory' (see the preface to Jeans, 1916). In the third edition of the book, however, which he completed in October 1920, he also did not undertake to develop statistical mechanics on the basis of quantum theory, though in a chapter entitled 'Quantum Dynamics' he outlined the recent results on atomic theory. Thus, the two schemes, classical statistical mechanics and quantum dynamics, remained two separate topics and Jeans still claimed years later that 'it cannot be said that any great progress has yet been made in bringing quantum-dynamics into its proper relation to general statistical mechanics' (Jeans, 1925b, Dover reprint of the *Dynamical Theory of Gases*, 1954, p. 436).

This was more or less the situation in 1921 when Darwin and Fowler entered the field. The Bohr–Sommerfeld theory had been developing so rapidly that definite hopes existed that an adequate dynamical description replacing classical

[93] In 1929 Lennard-Jones initiated the wave mechanical study of molecular structure in the form of molecular orbitals (Lennard-Jones, 1929). Since 1927 he had been Professor of Theoretical Physics at Bristol and worked at the Henry Herbert Wills Physical Laboratory. In 1932 he was invited to Cambridge as the first theorist to fill the newly established John Humphrey Plummer Chair of Theoretical Chemistry. He continued to work on the nature of the chemical bond and on various other difficult problems of theoretical chemistry. He created a vigorous school of theoretical chemistry in Great Britain. During the Second World War he served as Superintendent of Armament Research and Director-General of Research, and after the war he served on numerous committees and councils. He retired from the Plummer Chair in 1953 and became second Principal of the new University College of North Staffordshire at Keele. However, he died a year later on 1 November 1954.

Lennard-Jones received many honours. In 1933 he was elected Fellow of the Royal Society, and in 1953 he was awarded the Davy Medal. In 1953 he also received the Hopkins Prize of the Cavendish Philosophical Society for the best original work in connection with mathematico-physical or mathematico-experimental science during the period 1945–1948, and a little later the Longstaff Medal of the Chemical Society.

mechanics for atomic physics would soon be found. On the other hand, although statistical ideas had contributed importantly to the development of quantum theory and a proper statistical mechanical treatment of atomic systems seemed desirable to account for many observed phenomena, little had happened. There were some papers on gas theory, concerned especially with the problem of calculating the so-called chemical constant which determines the entropy of a volume of gas under given conditions; Paul Ehrenfest and his student Viktor Trkal, in a critical discussion of the earlier work on this subject, had tried to anticipate certain fundamental features of a future quantum statistical mechanics (Ehrenfest and Trkal, 1920). The essence of Ehrenfest and Trkal's procedure was to divide the phase space of the system under investigation into cells of volume h^f, where f is the number of degrees of freedom, to each of which some weight is attached. Their exposition of the problems provided the starting point for the work of Darwin and Fowler.

In order to deal with atomic systems, Darwin and Fowler began with the usual procedure of classical statistical mechanics, consisting of two steps: first, one calculates the probability of any microscopic state of a given system and selects the maximum of this probability; this maximum probability is then connected with the thermodynamic, that is, macroscopic entropy of the system in equilibrium. However, both steps involve approximations which are difficult to justify. For example, in the evaluation of the maximum probability Stirling's approximation for factorials is used, which is bad if the numbers under consideration are small, as might often be the case in quantum systems. Darwin and Fowler planned to avoid any of the mathematically questionable methods, for as they stated in the introduction of their first paper:

> The object . . . is to show that these calculations [in statistical mechanics] can all be much simplified by examining the average state of the assembly instead of its most probable state. The two are actually the same, but whereas the most probable state is only found by the use of Stirling's formula, the average state can be found rigorously by the help of the multinomial theorem, together with a certain not very difficult theorem in the theory of the complex variable. By this process it is possible to evaluate the average energy of any group in the assembly, and hence to deduce the relation of the partition to temperature, without the intermediary of entropy. The temperature here is measured on a special scale, which can be most simply related to the absolute scale [of thermodynamics] by the use of the theorem of equipartition, and we shall also establish the same relationship directly by connecting it with the scale of a gas thermometer. Throughout the paper the analysis is presented with some attempt at rigour, but it will be found that apart from this rigour it is exceedingly easy to apply the method of calculation. (Darwin and Fowler, 1922a, p. 450)

The new procedure took into account the essential requirements of statistical mechanics and quantum theory. In order to treat a particular system of M atoms (or molecules) of type A and N atoms (or molecules) of type B, one divides the phase space of each atom (or molecule) into cells of equal size; with each cell is associated a particular weight factor, p_r or q_s, respectively, $(r, s = 0, 1, 2, \ldots)$, and an energy ϵ_r^A or ϵ_s^B that depends on certain external parameters of the system. Then C, the total number of complexions (or possibilities for distributing atoms into cells), is given by

$$C = \sum_{a,b} M!\,N! \prod_r \frac{p_r^{a_r}}{a_r!} \prod_s \frac{q_s^{b_s}}{b_s!}, \tag{4}$$

where a_r and b_s are the occupation numbers of the corresponding states, and the sums over a and b extend over all possible occupation numbers compatible with the conditions, $\sum_r a_r = M$, $\sum_s b_s = N$, and $\sum_r \epsilon_r^A a_r + \sum_s \epsilon_s^B a_s = E$, where E is the total available energy of the system. $\overline{E}_A$, the average energy of the atom of type A, is then obtained from the equation

$$C\overline{E}_A = \sum_{a,b} M!\,N! \sum_r \epsilon_r a_r \prod \frac{p_r^{a_r}}{a_r!} \prod_s \frac{q_s^{b_s}}{b_s!}. \tag{5}$$

In order to evaluate the products containing factorials on the right-hand side of Eqs. (4) and (5), Darwin and Fowler proceeded as follows. They first introduced $f_A(z)$ and $f_B(z)$, the partition functions of the molecules of types A and B, by the equations

$$f_A(z) = \sum_r p_r z^{\epsilon_r^A} \qquad \text{and} \qquad f_B(z) = \sum_s q_s z^{\epsilon_s^B}, \tag{6}$$

where z is a quantity less than unity that is related to the total energy E as

$$E = Mz \frac{d}{dz} \ln f_A(z) + Nz \frac{d}{dz} \ln f_B(z). \tag{7}$$

They noticed that C and $C\overline{E}_A$, in Eqs. (4) and (5), are identical with the coefficients multiplying the factor z^E in the expansions of the functions $[f_A(z)]^M [f_B(z)]^N$ and $\{z(d/dz)[f_A(z)]^M\}[f_B(z)]^N$, respectively, in powers of the variable z. However, these coefficients can also be obtained by taking contour integrals around a circle with the centre at the origin and radius less than unity, that is,

$$C = \frac{1}{2\pi i} \oint \frac{dz}{z^{E+1}} [f_A(z)]^M [f_B(z)]^N \tag{4a}$$

and

$$C\overline{E}_A = \frac{1}{2\pi i} \oint \frac{dz}{z^{E+1}} \left\{ z \frac{d}{dz} [f_A(z)]^M \right\} [f_B(z)]^N. \tag{5a}$$

The points in the neighbourhood of $z = \theta$, where the integrand on the right-hand side of Eq. (5a) assumes the maximum, contribute most to that integral. This is especially so when the maximum is strong, such as the one that occurs in the case of sufficiently large particle numbers and energies; then it becomes legitimate to substitute θ for z in all factors of the integrand that do not themselves have strong maxima at θ or at other points. In particular, one may remove the factor $\{z(d/dz)[f_A(z)]^M\}$ from under the integral sign and substitute in it the value θ for z. Thus, for the average energy $\bar{E}_A$ one obtains the result

$$\bar{E}_A = M\theta \left\{ \frac{d}{dz} \ln f_A(z) \right\}_{z=\theta}. \tag{8}$$

Darwin and Fowler showed that the quantity θ appearing in Eq. (8) is related to the absolute temperature of the system via the equation

$$\theta = \exp\left\{ -\frac{1}{kT} \right\}. \tag{9}$$

They were also able to define S_{stat}, the 'statistical entropy' of a given system of M particles with the distribution function $f(z)$, by the equation

$$S_{\text{stat}} = k\left\{ M \ln f(\theta) + E \ln\left(-\frac{1}{\theta} \right) \right\}, \tag{10}$$

which has the properties of the thermodynamic entropy and may be identified with it (Darwin and Fowler, 1922b, p. 833). Finally, they also derived an expression for the statistical fluctuations of the properties of atomic systems from their theory (Darwin and Fowler, 1922c).

Though Darwin and Fowler's new method in statistical mechanics appeared to be more consistent and less plagued by doubtful mathematical approximations than the ones which had been used earlier in atomic systems, the authors applied it only to a few simple cases, notably to an assembly of free (i.e., noninteracting) Planck oscillators and to systems of free atoms. To deal with a system of free atoms Darwin and Fowler considered the classical limit; thus, they assumed the size of the cells in phase space to be infinitely small, which allowed them to replace the sum in the partition function, Eq. (6), by an integral. The new method unified the formerly different approaches to classical and quantum systems and, in simple cases at least, the well-known results of quantum theory and statistical mechanics were reproduced. However, the main purpose was to be able to calculate new results for more complex atomic systems, and in his next paper Fowler turned to the discussion of equilibrium in a system containing dissociating molecules (Fowler, 1923a). That problem had been treated earlier by Ehrenfest and Trkal under the special assumption that the degrees of freedom of the molecules considered are either completely excited or completely unexcited (Ehrenfest and Trkal, 1920). In reality this assumption is never satisfied, and Fowler encountered no difficulty in going beyond the work of Ehrenfest and Trkal. Thus, in the case of a system which is made up of hydrogen molecules and atoms in dissociation equilibrium, he obtained the ratio

$$\frac{\bar{N}}{(\bar{M})^2} = \frac{h_2(\theta)b(\theta)}{\sigma\theta^{\chi}\{h_1(\theta)\}^2}, \tag{11}$$

where $\overline{N}$ is the average number of hydrogen molecules and $\overline{M}$ is the average number of hydrogen atoms. The factors in the numerator and denominator on the right-hand side are as follows: $h_1(\theta)$, the partition function of the hydrogen atoms; $h_2(\theta)$, the partition function of the hydrogen molecules without the rotational and oscillational degrees of freedom, the latter being described by the distribution function $b(\theta)$; σ, the symmetry number of the molecules, which is equal to 2 for hydrogen; and χ, the work required to separate the hydrogen molecule into two hydrogen atoms. As a result, the degree of dissociation in a given volume under constant given pressure is easily found, provided one knows the function $b(\theta)$. Though Fowler did not have an exact form for $b(\theta)$, he proposed several reasonable physical approximations for it.

Another important physical situation which Fowler was able to treat satisfactorily was the ionization of hydrogen atoms at high temperatures (Fowler, 1923a, §7). The equilibrium ratio of $\overline{N}$, the number of hydrogen atoms, over $\overline{M}^2$, the square of the number of hydrogen nuclei (or of the liberated electrons), is again given by Eq. (11); only now σ is to be taken as unity, and the distribution functions for the translational degrees of freedom of the hydrogen atom and the nucleus are practically identical. Hence the complete expression on the right-hand side reduces to the quotient of the partition function of the internal degrees of freedom of the hydrogen atom and the product of θ^{χ} times the partition function of the free electron, i.e., $b(\theta)\{\theta^{\chi}h(\theta)\}^{-1}$. Assuming that the hydrogen atoms and the free electrons satisfy the (classical) equation of state of perfect gases, Fowler derived the equation

$$\ln \frac{x^2 \cdot P}{1-x^2} = -\frac{\chi}{kT} + 2.5\ln(kT) + 3\ln\left\{\frac{\sqrt{2\pi m_e}}{h}\right\} - \ln B(T), \tag{12}$$

which yields x, the fraction of ionized hydrogen atoms under a total pressure P and at the absolute temperature T. On the right-hand side of Eq. (12), χ is the ionization energy of the hydrogen atom, m_e the mass of the electron, h Planck's constant, and $B(T)$ is identical with $b(\theta)$, with θ being replaced by T according to Eq. (9). If $B(T)$ can be identified with 1, Eq. (12) goes over into the ionization equation which Megh Nad Saha had derived earlier on the basis of thermodynamical arguments (Saha, 1920, p. 486). For Bohr's hydrogen atom Fowler found that the function $B(T)$ could now be expressed as a series starting with a constant term of value 2, which would be the only contribution whether the hydrogen atom is supposed to be in the ground state or in the ionized state. He suggested, however, that the discrepancy between his statistical result and the thermodynamical one of Saha might be due to the fact that he had applied the theory of perfect gases to atoms and electrons.[94]

[94] Later development would indeed confirm Fowler's suspicion, as the electrons were not only found to obey nonclassical statistics but also to be in a degenerate state.

Fowler noticed another difficulty with the quantum-theoretical expression for $B(T)$; it diverges unless one introduces a cutoff. However, the cutoff can be motivated physically, as the conditions of the system do not allow the electrons of the hydrogen atom to stay in the outer orbits.

The application of statistical mechanics to the problems of ionization was of particular interest in Cambridge, for Edward Arthur Milne of Trinity College had at once recognized the far-reaching consequences of Saha's theory of thermal ionization for the problem of determining the temperatures in stellar atmospheres. Now that Fowler had established the ionization formula, Eq. (12), for the hydrogen atom, and an extension to more complex atoms was straightforward, he and Milne decided to collaborate on the question of finding the rigorous connection between the intensities of the absorption lines observed in the spectra of stars and the temperatures of the absorbing atmospheres (Fowler and Milne, 1923). Many years later Milne recalled how this work was motivated and how it proceeded:

> Saha had concentrated on the marginal appearances and disappearances of absorption lines in the stellar sequence, assuming an order of magnitude for the pressure in a stellar atmosphere and calculating the temperature where increasing ionization, for example, inhibited further absorption of the line in question owing to the loss of the series electron. As Fowler and I were one day stamping round my rooms in Trinity and discussing this, it suddenly occurred to me that the *maximum* in the intensity of the Balmer lines of hydrogen, for example, was readily explained by the consideration that at the lower temperatures there were too few excited atoms to give appreciable absorption, whilst at the higher temperatures there were too few neutral atoms left to give any absorption; the actual intensity must depend on the product of two factors, the excitation factor and the ionization factor, of which the first factor, the fraction excited, must increase from almost zero, whilst the second factor, the fraction un-ionized, must decrease to almost zero. Thus the product, being zero at the two limits of temperature, low and high, must have at least one maximum in between, and probably only one. I rapidly expounded this to Fowler, and we said we would work it out for our joint paper. That evening I did a hasty order of magnitude calculation of the effect and found that to agree with a temperature of 10,000° for the stars of type AO, where the Balmer lines have their maximum, a pressure of the order of 10^{-4} atmosphere was required. This was very exciting, because standard determinations of pressures in stellar atmospheres from line shifts and line widths had been supposed to indicate a pressure of the order of one atmosphere, or more, and I had begun on other grounds to disbelieve this. I walked round to Fowler's house in the Chesterton Road, and finding him out dropped a short note in his letterbox telling him of this result. The next two or three days I was busy over other matters, and when I next saw Fowler he told me he had read my note and had worked out an exact and elegant formula for the pressure at which the lines of a subordinate series of a neutral atom should have their maximum. I then assembled the astrophysical data at the Solar Physics Observatory, and produced a draft of the paper we eventually read to the Royal Astronomical Society. (Milne, 1945, pp. 70–71)[95]

[95]The work on the theory of absorption lines in stellar spectra was continued by both authors in the following years. In early 1924 Fowler and Milne wrote a second paper in which they used Alfred Fowler's new data on the lines of triply-ionized silicon and ionized carbon in order to calculate the temperatures prevailing in the atmospheres of O and A stars (Fowler and Milne, 1924). A year later Fowler corrected certain details of the theory (Fowler, 1925c). At the same time he extended it together with his student Edward Armand Guggenheim so as to apply also to the interior of stars (Fowler and Guggenheim, 1925).

The discussions of Fowler and Milne concerning the mechanism that causes ionization in atomic systems brought the two of them a little later to pay special attention to an idea that Oskar Klein and Svein Rosseland had previously employed in atomic physics. Klein and Rosseland had concluded that there must exist a process which is reverse of the excitation of atoms by electron collisions; in such processes, called 'collisions of the second kind,' an electron collides with an excited atom and leaves it in its normal state, carrying away the surplus energy (Klein and Rosseland, 1921). Fowler and Milne now stated that this is only one of many examples of reciprocal processes in atomic physics because the principles of statistical mechanics require that for every elementary process involving atoms and radiation the reverse process had also to exist.[96] They called such pairs of processes 'unit mechanisms' and considered them in detail. For example, Milne studied the pair of processes, of which one constituent is the photoelectric effect and the reverse process is the capture of an electron by an ion, resulting in the emission of radiation (Milne, 1924a). Fowler, on the other hand, examined a more complex process 'of second kind,' one which occurs in systems where the ionization of atoms or molecules by impacts contributes to the equilibrium. He postulated that in such systems three-body encounters—which result in the binding of two of the constituents to form a single atom or molecule, the superfluous energy being carried out by a third body—must take place as frequently as the ionization processes. 'We then recognized,' Milne recalled, 'the complete generality of the "principle of detailed balancing"; the phrase was, I think, due to Fowler, though the importance of the principle was being simultaneously recognized elsewhere' (Milne, 1945, p. 72).[97] In 1924 Fowler was indeed

[96] Fowler stated the argument as follows:

> We believe in fact that these laws [of statistical mechanics] are true whatever the mechanism of such [atomic] processes may be. It seems quite certain that, for example, the law of distribution of velocity of free electrons must be the same in any enclosure in thermodynamic equilibrium. Accepting it, we can consider various assemblies at the same temperature, in which different processes of exchange are going on, or the same processes of rates which can be altered relatively to one another by altering the concentrations. Since the resulting distribution laws must be the same in form, it follows that any process of exchange acting in a particular direction must be invariably accompanied by an analogous reverse process. The pair must be capable in combination of maintaining by themselves the regular laws. This combined action can never require the assistance of any other process, unless such process can be and is its invariable accompaniment, and then in fact forms part of one group of processes which may conveniently be referred to as a *unit mechanism*. (Fowler, 1924a, pp. 257–258)

[97] The hypothesis of the detailed balancing of individual processes, or the 'principle of detailed balancing' (Fowler, 1924c, p. 253) had already had a distinguished history in physics. Einstein had applied it in 1916 to derive Planck's law of blackbody radiation. Then came the work of Klein and Rosseland leading to the recognition of collisions of second kind. Apart from Fowler and Milne, Hendrik Kramers had used the idea of detailed balancing in his paper on continuous X-rays (Kramers, 1923d), as did Eddington in his treatment of the problem of the stellar absorption coefficient (Eddington, 1922a). Finally, Wolfgang Pauli (1923b) and Einstein and Ehrenfest (1923) discussed the Compton effect along the same lines. In 1925 Gilbert N. Lewis enunciated a new principle of equilibrium, or the 'law of entire equilibrium' in chemical reactions as follows: '*Corresponding to any individual process there is a reverse process, and in a state of equilibrium the average rate of every process is equal to the average rate of its reverse process*' (Lewis, 1925, p. 181). However, Christiansen and Kramers (1923) had already applied a similar idea in the theory of monomolecular gas reactions. Richard C. Tolman enunciated the 'principle of microscopic reversibility' (Tolman, 1925a; see also Fowler and Milne, 1925).

very enthusiastic about this principle, for it proved to be very useful in many theoretical problems, as for example in describing G. H. Henderson's experimental results on the capture and loss of electrons by fast α-particles, which were discussed intensively at the Cavendish Laboratory (Fowler, 1924b).

The papers on quantum statistical mechanics and its wide application in physics and astronomy constituted the main part of Fowler's work until 1925, including a comprehensive essay on 'Statistical Mechanics' for which he was awarded the 1923–1924 Adams Prize of the University of Cambridge.[98] This work established his reputation as a theoretical physicist. But in the meantime he had also become deeply involved in the problems of quantum theory. In March 1922 Niels Bohr had outlined the current status of atomic theory in lectures at Cambridge.[99] Ten months later Rutherford reported that 'Fowler is giving a course of lectures this term on "The Application of Quantum Theory to Spectra." He seems to know much more about that question than anyone in the country' (Rutherford to Bohr, 8 January 1923). Soon Fowler entered into the discussion of specific problems of atomic theory. For example, in the 'General Discussion' of the Faraday Society on 'The Electronic Theory of Valency' on 13 July 1923, he talked about the relation of Bohr's atomic theory to the question of covalent binding of molecules (Fowler, 1923c). And then, in the beginning of 1925, after his visit to Bohr's Institute, he got into the most recent problems of atomic theory that were being discussed at the time in Copenhagen, Göttingen and Munich. Fowler's visit to Bohr's Institute had been envisaged for a long time, but it did not occur until 1925, when he could take a leave of absence from Cambridge. During the time he spent in Copenhagen, from the end of January to the beginning of March 1925, Fowler witnessed the last stage in the development of the approach based on the correspondence principle and its most refined application in calculating the properties of spectral lines. He submitted two papers dealing with the intensities of spectral lines: in the first he worked out the consequences of the correspondence principle for band spectra (Fowler, 1925a); in the second he gave a correct theoretical interpretation of the summation rules for the intensities of multiplets of spectral lines (Fowler, 1925d).

With the above-mentioned work Ralph Fowler joined the front-rank of those who represented the latest trends of atomic theory in early 1925. He showed that he was in command of all modern methods for obtaining new results. He was excited about the whole field of atomic theory, and his enthusiasm infected his students and many other physicists at the Cavendish Laboratory. Soon a large group of students and research scholars, who were interested in quantum theory, gathered there. Among the older ones were Douglas Rayner Hartree, Fowler's

[98] Fowler published the material of this essay as a book on *Statistical Mechanics* (Fowler, 1929).

[99] Bohr gave a series of lectures on atomic theory in Cambridge that covered the same topics which he discussed later in June of the same year in Göttingen at the famous 'Bohr-Festival.' References to Bohr's lectures in Cambridge in March 1922 occur in the Bohr–Rutherford correspondence (Bohr to Rutherford, 4 February 1922; Rutherford to Bohr, 11 February 1922; Rutherford to Bohr, 22 February 1922; Bohr to Rutherford, 22 May 1922).

former associate during the war, then a research fellow at St. John's, and Edmund Clifton Stoner, who had received his B.A. already in 1921.[100] Stoner made an important contribution to atomic theory: in a paper published in the *Philosophical Magazine* in October 1924 he proposed a new distribution of electrons among the atomic levels, which replaced Bohr's earlier assignment and paved the way for the exclusion principle and the concept of electron spin (Stoner, 1924). In establishing the latter, another student of Fowler's, Llewellyn Hilleth Thomas, would supply a crucial relativistic argument (L. H. Thomas, 1926).[101] Fowler not only supervised and helped the members of the theoretical group, all the experimental collaborators of Rutherford could also count on his advice if they became involved in questions of atomic theory.[102] Fowler was a most enthusiastic helper, collaboration with whom was 'tremendous fun.' As Milne recorded: 'There was a thrill about the investigation—it was as though one were physically in love with the particular problem when Fowler was present . . . His presence always helped one to see what was the really important aspect of the matter under consideration, and one came away refreshed and with new heart' (Milne, 1945, p. 67). Fowler was indeed the pioneer of quantum theory in Cambridge; he propagated it with all the vigour of his strong and healthy personality.[102a]

II.3 Activities of a Student: Lectures, Seminars and Private Study

As in Bristol, Dirac was attached to the mathematics department, for in Cambridge theoretical physics, which Fowler, Cunningham and others taught, belonged to applied mathematics. Dirac got the impression that the mathematical

[100] D. R. Hartree, born in Cambridge on 27 March 1897, was seventeen years old when the First World War broke out. At the end of the war he returned to Cambridge to take the Natural Sciences Tripos. In 1924 he became a Research Fellow at St. John's College, but changed to Christ's College in 1928. He obtained his doctorate in 1926. Three years later he was appointed Beyer Professor of Applied Mathematics at the University of Manchester and, in 1937, he moved to the chair of theoretical physics at Manchester; he stayed there until 1946, then returned to Cambridge as Plummer Professor of Mathematics.

Hartree worked on the propogation of radio waves in magneto-ionic media and, especially, on atomic structure. He became particularly well known through his later work on the computation of atomic wave functions (in wave mechanics). He was elected a Fellow of the Royal Society of London in 1932. He died in Cambridge on 12 February 1958.

For a biographical sketch of E. C. Stoner, see Volume 1, footnote 1003.

[101] For an account of the scientific background of L. H. Thomas, see Volume 1, footnote 1072.

[102] For example, Fowler assisted P. M. S. Blackett in his study of energy-momentum conservation in the ionization or excitation of atoms by electron oscillation (Blackett, 1924, p. 66).

[102a] Ralph Fowler became Plummer Professor of Theoretical Physics at Cambridge in 1932. Six years later he was appointed to the directorship of the National Physical Laboratory, succeeding Sir Lawrence Bragg (who, in turn, succeeded Rutherford in the Cavendish Chair at Cambridge); due to illness, however, he resigned this position after a few months, but retained the Plummer Chair. During World War II he served in several positions; in particular, he established relations between Great Britain and Canada with respect to research projects on war problems. Fowler had been elected a Fellow of the Royal Society in 1925 and had received its Royal Medal in 1936. He died in Cambridge on 28 July 1944.

examination in Cambridge, the Tripos, was considerably more advanced than the one he had passed in Bristol to obtain his B.A.; this referred not only to the level but also to the fact that in Cambridge certain subjects, such as electrodynamics, thermodynamics and statistical mechanics, were included that were new to him. In Bristol he had learned a little about the Lagrangian and Hamiltonian methods and quite a lot about potential theory, but not much about electromagnetic waves.[103] Dirac had therefore to make up for the gaps in his knowledge as quickly as possible, both by studying and attending lectures.

To begin with, Dirac learned classical mechanics systematically by studying Whittaker's *Treatise on the Analytical Dynamics of Particles and Rigid Bodies* (Whittaker, 1904).[104] He found it a useful book, for it gave many examples, in the text and at the end of chapters, the solution of which provided a deeper insight

[103] In Bristol, H. R. Hassé's course on applied mathematics was largely about potential theory and the solution of equations such as

$$\left(\frac{d^2}{dx^2} + \frac{d^2}{dy^2} + \frac{d^2}{dz^2}\right)V = 0,$$

where the potential V is a function of the space coordinates x, y and z. Not even Maxwell's electrodynamics, and certainly no relativity theory, was discussed in the lectures. The Cambridge student of physics and applied mathematics of those days was, of course, familiar with these topics.

[104] Edmund Taylor Whittaker, the author of *Analytical Dynamics*, born on 24 October 1873 in Southport, Lancashire, was himself a Cambridge graduate. Having received his secondary school education at the Manchester Grammar School, he gained an entrance scholarship to Trinity College in December 1891. He was Second Wrangler in the Mathematical Tripos of 1895, was elected Fellow of Trinity College in October 1896, and was awarded the First Smith's Prize in 1897. Having served on the staff of Trinity College until 1906, he became Professor of Astronomy at the University of Dublin, where he mainly gave courses of advanced lectures in mathematical physics; one of his students was Eamon de Valera, later on President of Eire. From 1912 until his retirement in 1946, Whittaker taught as Professor of Mathematics in the University of Edinburgh. He died on 24 March 1956 at Edinburgh.

Whittaker contributed to many fields of pure and applied mathematics, notably function theory, algebra, interpolation, potential theory and special functions (Lamé and Mathieu), analytical dynamics, electromagnetic theory and relativity, and quantum theory. His original work is also reflected in the monographs he published, the first one being *A Course on Modern Analysis* (Whittaker, 1902; later editions with G. N. Watson). After his retirement he wrote the two-volume *History of the Theories of Aether and Electricity* (Whittaker, 1951 and 1953). In the course of his life Whittaker received many honours. In 1905 he was elected Fellow of the Royal Society, and he was foreign member of several academies, e.g., *Accademia dei Lincei* (1922), Pontifical Academy (1936), and *Académie Royale de Belgique* (1946). He was awarded the Sylvester Medal (1931), the De Morgan Medal of the London Mathematical Society (1935), and the Copley Medal of the Royal Society (1954).

Whittaker's *Analytical Dynamics*, first published in 1904, went through several editions (second edition, 1917; third edition, 1927; fourth edition, 1937) and remained a standard textbook on this subject because of its thorough organization and lucid presentation of the material. Its organization did not change in the course of time, but Whittaker kept on incorporating the new results of current research in the successive editions. The first chapter dealt with the kinematics of rigid bodies; the equations of motion were then discussed and cast very soon into their Lagrangian form. In the following three chapters integration methods were presented, and exactly soluble systems of particles and rigid bodies were discussed. Chapter VII dealt with the general theory of vibrations, and Chapter VIII with nonholonomic and dissipative structures. Whittaker then devoted four chapters to the principle of least action and Hamiltonian mechanics including transformation theory. Finally he turned to the theory of many-body systems, discussing in some detail the three-body problem.

Whittaker's book was written for the advanced mathematician and it contained many examples which had been taken out of research papers published in journals. Moreover, the author included exercises that had been given as problems in Cambridge college examinations, about which his

into the methods of mechanics. It also discussed Hamiltonian methods and transformation theory in great detail, satisfying Dirac's need for a beautiful and general theory. He carefully worked through *Analytical Dynamics*, soon mastering the Hamiltonian methods to such an extent that he could apply them to atomic systems. He also discovered certain points in Whittaker's book that did not seem to him to be correct.

One such point, which he wanted to improve upon at that time, was the question of whether the Hamiltonian variational principle could also be applied to the so-called nonholonomic systems, that is, systems which are described by nonintegrable, velocity-dependent equations for the constraints. Whittaker had stated that 'it is only for holonomic systems [whose constraint equations do not contain the velocities of the particles] that the varied motion is a possible motion [that is, it satisfies the equations of motion and the constraint equations]; so that if we compare the actual motion with the adjacent motions which obey the kinematical equations of constraint, Hamilton's principle is true only for holonomic systems' (Whittaker, 1904, p. 246). Dirac was not convinced by Whittaker's reasoning; he even had an argument about it with Fowler, who thought that what was in the book was all right. Although Fowler became a bit impatient with him, Dirac did not give in. This was the only disagreement he remembered to have had with Ralph Fowler. Twenty-five years later he decided the argument in his own favour: he proved in 1949 that nonholonomic systems can also be treated by Hamiltonian methods, provided one allows that the equations of constraint are valid only as, what he called, 'weak' equations; that is, when the motion of a system satisfying the equations of constraint is changed to an adjacent motion, the constraint equations are violated by a term of infinitesimally small order of magnitude.[105] 'I think that was really straightening out that early difficulty,' he recalled (Dirac, Conversations, p. 43). Though he

---

Cambridge colleague, George Hartley Bryan, the first reviewer of the book wrote:

> Some of the questions . . . may give foreign mathematicians a little insight into the unpalatable nuts which Cambridge students are expected to waste time in trying to crack for examination purposes. The antics of insects crawling on epicycloids, or the vagaries of particles moving along the intersections of ellipsoids with hyperboloids of one sheet, are of no scientific interest, and the time spent in "getting out" problems of this character might better be employed in learning something useful. Moreover, Cambridge college examiners have a habit of endowing bodies with the most inconsistent properties in the matter of perfect roughness and perfect smoothness . . . Those who have the ability to do more difficult work should pass on to the advanced parts of a book like Mr. Whittaker's and learn what foreign mathematicians have been doing; this is much more useful. (Bryan, 1905, p. 603)

While Bryan's criticism of some of the exercises presented in Whittaker's book was justified, a student like Dirac did not need such warning. From his engineering training he was well acquainted with the fact that ideal situations, such as infinitely smooth surfaces, etc., do not exist in nature and would only occasionally provide a useful approximation to reality. Thus, he certainly did not overestimate the value of college examination questions, which Bryan opposed so vehemently, although they might have provided the opportunity to check whether he had developed the skill required of the Wranglers in the examinations for the Mathematical Tripos.

[105] Dirac discussed his solution first in a lecture given at the Canadian Mathematical Seminar in Vancouver in August 1949 (Dirac, 1950). Several years later he came back to the generalized Hamiltonian theory of nonholonomic systems (Dirac, 1958a) and used it to cast Einstein's theory of gravitation into Hamiltonian form as a preliminary step to quantize the gravitational field (Dirac, 1958b).

was not able to present this solution in the discussion with Fowler, Dirac had come to believe firmly in the power and applicability of Hamiltonian methods in all dynamical problems in his early days in Cambridge.

To learn topics of theoretical physics, other than classical mechanics, Dirac went to lectures. Most of the lectures and seminars in which Dirac was interested at Cambridge took place at the Cavendish Laboratory and he spent most of his time there.[106] He attended Cunningham's course on electrodynamics; Maxwell Newman, another Fellow of St. John's, lectured on thermodynamics, which Dirac followed.[107] In statistical mechanics he attended the lectures of Fowler and of John Edward Lennard-Jones. Lennard-Jones introduced him to the Boltzmann equation as the starting point of statistical mechanics. However, he did not like statistical mechanics as much as the other physical theories mainly because of the kind of approximations connected with the Boltzmann equation. 'You have this collision term coming in,' he complained, 'which means that all really important things are lumped together in one term which is not explained very well. I dislike that very much' (Dirac, Conversations, p. 11). In one of his early papers, the one on detailed balancing (Dirac, 1924d), which grew out of the lectures of Fowler and Lennard-Jones, Dirac made an effort to improve this unsatisfactory definition of the collision term. As for the general theory of statistical mechanics, he preferred the approach of Gibbs.

Ralph Fowler introduced Dirac to quantum theory, and it was in Fowler's lectures that he first learned about the Bohr atom. In Bristol he had not heard about Bohr's theory at all, perhaps because he was in the mathematics department; the physicists were housed in a different building and he did not have any contact with them. Thus, it came to Dirac as a surprise that atomic theory was already so far advanced. For learning quantum theory Dirac was lucky to have Fowler as his supervisor. In Cambridge, Fowler was the man most interested in quantum theory and, through his father-in-law, Ernest Rutherford, he had close connections with Niels Bohr in Copenhagen.[108] Lennard-Jones also took some

[106]The classes of all University and College lecturers in physics had been moved to the Cavendish, following its enlargement in 1908. Dirac also preferred to use the small Cavendish library, where most of the important periodicals and a number of monographs and reference books were available. As Oliphant recalled: 'Books and periodicals could be borrowed overnight, but could not be taken from the library before 4 P.M., and had to be returned before 10 A.M. next day' (Oliphant, 1972, p. 52).

[107]Maxwell Herman Alexander Newman, born on 7 February 1897 in London, studied like Dirac in St. John's College, Cambridge, and completed his M.A. in 1924. Three years later he became Lecturer in Cambridge University, staying on until 1945 when he was appointed Professor of Mathematics at the University of Manchester, from where he retired in 1964. Newman devoted his main research to geometrical topology, on which he also wrote a book. He became Fellow of the Royal Society in 1939; he was a member, and became President, of both the London Mathematical Society and the Mathematical Association. He was awarded the Sylvester Medal in 1959, and the De Morgan Medal in 1962.

[108]With another supervisor Dirac might have studied quantum theory to a much lesser extent or perhaps not at all. For example, one of his fellow students at St. John's College, R. Schlapp, was supervised by J. Larmor. Larmor was still interested in some old, classical ideas and he set Schlapp to work on classical problems. Schlapp got a chance to work on quantum mechanics only when he went to work with C. G. Darwin at Edinburgh after completing his doctorate in Cambridge; then he treated Stark effect with Dirac's electron equation (Schlapp, 1928).

interest in quantum theory, especially in connection with his work on intermolecular forces, but it was under Fowler's stimulating and persuasive influence that Dirac made the transition from his initial emphasis on relativity to a new interest in quantum theory. One reason, though not the only one, that quantum theory became attractive for Dirac was the central role played in it by the Hamiltonian methods. He had ample opportunity of seeing in Fowler's lectures how Hamiltonian methods worked in quantum theory; no wonder that he enjoyed it more than the 'ugly' statistical mechanics with its Boltzmann equation.

In spite of the marked shift of his interest towards quantum theory, Dirac did not give up the study of relativity theory altogether. He went to Maxwell Newman's lectures on the geometrical aspects of general relativity. Having already read Eddington's *Space, Time and Gravitation* (Eddington, 1920) in Bristol, he now carefully worked through his second book on this subject, *The Mathematical Theory of Relativity*, which had just appeared when Dirac went to Cambridge (Eddington, 1923b).[109] He was very glad that Eddington was around: 'Sometimes I met him and had some discussion with him on the question of kinematic and dynamical velocity, which led to a little note of mine published in the *Philosophical Magazine*. It was really a wonderful thing for me to meet the man who was the fountainhead of relativity so far as England was concerned. Einstein was just too remote to count' (Dirac, 1977, p. 115). From Eddington's book Dirac learned relativity thoroughly, and he made use of this knowledge in his early papers whenever possible. However, in spite of his great interest in relativistic methods he did not publish at that time, apart from the short note mentioned above, any contribution to an open question in special or general relativity theory. Yet he liked to ponder about such questions. For instance, he was quite interested in Hermann Weyl's formulation of electrodynamics, as he thought that it might be a correct description of nature. In particular, he made a detailed study of the field around an electron on the basis of Weyl's theory and found that the equations required the mass of the electron not to be a constant, but to vary in time with a coefficient depending on the charge. Hence Dirac concluded that Weyl's theory was untenable; everyone seemed to agree with him at that time, 'though for different reasons' (Dirac, Conversations, p. 69).

While the other mathematical celebrities in Cambridge, like Joseph Larmor, the Lucasian Professor of Mathematics, and James Jeans, had no active interest in relativity theory, this was not so with Edward Arthur Milne, the University Lecturer in Astrophysics. He was a close friend and collaborator of Ralph Fowler and became Dirac's supervisor during a term when Fowler was in Copenhagen in spring 1925. Milne worked on the application of statistical mechanics to astrophysical problems; he also made use of the modern theoretical ideas of relativity and the quantum theory, even contributing some original work to the latter. Dirac attended Milne's lectures on the interiors of stars. During Fowler's absence he even worked on a problem suggested by Milne—the calcula-

[109] Dirac found this book 'rather tough at first but eventually mastered it' (Dirac, 1977, p. 115).

tion of the shift of spectral lines in the stellar atmosphere—and published a paper on it (Dirac, 1925b).

Besides becoming acquainted with these various fields of theoretical physics—or applied mathematics, as it was called in Cambridge—Dirac kept up his interest in geometry, not only in connection with relativity but also as a purely mathematical subject. The man who attracted Dirac's attention was Henry Baker, Lowndean Professor of Astronomy and Geometry, the teacher and friend of Peter Fraser, Dirac's mathematics teacher at Bristol. Dirac did not attend Baker's lectures, but he was invited to the tea parties at Baker's home on Saturday afternoons and he went there quite regularly with other students who were keen on geometry. After tea someone would give a talk on a geometrical subject. 'It was always the geometry of flat space,' Dirac recalled. 'It was always handled by the methods of projective geometry. Everyone considered then that projective geometry was the only kind of geometry worth studying. It was so much more powerful than the geometry in which one was confined to Euclid's axioms' (Dirac, 1977, p. 115). As a result of the impact of relativity it had become very popular among geometers to consider four-dimensional space rather than the three-dimensional one, and a good deal of work was devoted to the geometry of four or more dimensions. This fact was also reflected in the talks given by the participants in Henry Baker's tea parties. As Dirac recalled: 'They worked very often in a number of dimensions more than three, four, five or six, and studied the various figures which one could construct in these spaces of a higher number of dimensions, and I was much impressed by the power of their methods. By studying these figures in spaces of higher dimensions one was often able to get quick proofs of results in the ordinary three-dimensional Euclidean space, results which would be very tedious to get by other methods' (Dirac, 1977, pp. 115–116).

Dirac had already been fascinated by the methods of projective geometry in Bristol, and he was very happy to learn more about them in Cambridge. 'These tea parties did very much to stimulate my interest in the beauty of mathematics,' he said much later. 'The all-important thing there was to strive to express the relationships in a beautiful form, and they were very successful. I did some work on projective geometry myself and I gave one of the talks at one of the tea parties. This was the first lecture I ever gave, and so of course I remember it very well. I dealt with a new method of handling these projective problems' (Dirac, 1977, p. 116). Dirac also applied the methods of projective geometry to physics. For example, he found them useful in obtaining certain results in relativity. Since all directions in the four-dimensional Minkowski space can be pictured as points in a three-dimensional projective space, any relations between four-vectors, null vectors, etc., follow at once as the corresponding relations between points in the projective space. Dirac was thus able to arrive in many cases at a visual geometrical picture, and therefore at a better understanding of certain relationships in relativity. The same methods would again help him enormously in his first steps at establishing his quantum-mechanical scheme in fall 1925. In view of the care he always took in presenting mathematically elegant formulations, it is surprising that Dirac never referred explicitly to the projective method in his publications.[110]

Apart from his special interest in geometry, Dirac kept up with the study of purely mathematical methods only to the extent he needed them for application in the problems of theoretical physics he worked on. On the other hand, acquaintance with experimental facts played an increasingly important role in his further education, and Dirac came into contact with experimental physics for the first time in Cambridge. Theory and experiment in physics were rather separate fields in Cambridge. Administratively there was a close connection between pure and applied mathematics, and theoretical physics came under the latter, whereas experimental physics belonged to physics or experimental philosophy. The separation of the specialties applied essentially to the undergraduates. The students in mathematics or theoretical physics remained together a long time, specializing only very late in the curriculum. Fowler's closeness to Rutherford affected their students, and Dirac, for instance, would go to the experimental colloquia at the Cavendish Laboratory. Rutherford dominated the Cavendish. Sir J. J. Thomson, the former Director, was still around but no longer took an active part in the colloquia. With brilliant students and collaborators, Rutherford had given the Cavendish a new dimension and created a school of experimental physics which, for a long time, was far ahead of theory in atomic and nuclear physics. At the Cavendish, theoreticians and experimentalists met frequently to listen to distinguished visitors who came to see Rutherford. On one of these occasions, in spring 1925, Dirac first met Niels Bohr. The visitors were often invited to give informal talks at the Kapitza Club, which Dirac had joined in fall 1924. There also existed another club, the '$\nabla^2 V$ Club,' which consisted jointly of experimental and theoretical people. The meetings of these clubs provided Dirac with an up-to-date knowledge of what was going on in atomic physics.

The Kapitza Club was founded by the young Russian electrical engineer Peter Leonidovich Kapitza, who had been admitted as a research student at the Cavendish in July 1921.[111] Kapitza was an unusual figure in the Cavendish because of his informal and uncommon habits. For example, 'he would race his sports car at top speed over the country roads, bathe in the nude in the Cam and frighten the swans by his imitation of birds' alarm cries' (Larsen, 1962, p. 61). He would also work in the same style and perform crazy experiments, the kind that

[110] Dirac did not publish the results he had presented on projective geometry at Baker's tea party. In his papers on quantum mechanics of fall 1925 and summer 1926 he sometimes made indirect reference to the geometrical methods. We shall discuss these features of Dirac's work in Sections IV and V.

[111] Peter (Petr) L. Kapitza was born on 8 July 1894 in Kronstadt, Russia, the son of a Tsarist general. He studied electrical engineering at the Polytechnic Institute of St. Petersburg. After graduation in 1918 he continued there as an associate until 1921. In 1918 Kapitza had been sent as an attaché with a Soviet purchasing mission to England; on that occasion he also visited the Cavendish. He became so interested in scientific research that he asked the authorities for permission to study in England and work with Rutherford and Chadwick. He was first admitted as a research student at the Cavendish; in 1923 he became Clerk Maxwell Scholar.

Rutherford called 'fool experiments' because they appeared as if they had been started by foolish intentions but turned out to give rather important results. At the Cavendish, Kapitza was first put to work on measuring the energy loss of α-particles when penetrating through matter, and then he collaborated with T. Shimizu on cloud chamber researches. This work stimulated his interest in producing very high magnetic fields, essentially short-circuiting accumulator batteries, in order to deflect energetic α-particles. One of the mottos of his research was: 'Find out what other scientists are doing, then exceed their bounds, and see what happens' (Larsen, 1962, p. 61). Thus, he became responsible for the investigations at the Cavendish that were connected with the production and application of strong magnetic fields, advancing in 1924 to the position of Assistant Director for Magnetic Research. From then on he studied the properties of matter, such as magnetostriction and susceptibility, in extremely high fields; he also extended his research to low temperatures. Later, when his experimental equipment needed more space, he persuaded Rutherford of the necessity for a new laboratory devoted to magnetic and low-temperature research. Rutherford esteemed Kapitza highly and had him elected to a Fellowship of the Royal Society in 1929 (in spite of his Soviet Russian citizenship Kapitza was elected an ordinary Fellow and not a Foreign Member); he turned to the Royal Society and obtained funds for a new laboratory from the bequest of the chemist Ludwig Mond. The Mond Laboratory was opened in February 1933 by the British Prime Minister Stanley Baldwin, then Chancellor of the University of Cambridge, and Kapitza became its Director and the Royal Society's Messel Professor.[112]

Kapitza's tremendous energetic drive and self-confidence as well as his great desire to understand physical problems, not only appealed to Rutherford but had a great impact on the scientific atmosphere in the Cavendish. The Kapitza Club grew from its founder's desire to know more about the major problems and advances both in theoretical and experimental physics of the day. Kapitza thought that the scientific life in Cambridge, both teaching and research, was too formalized and he regarded his Club as an opportunity to promote informal discussion in the style of the uninhibited Russian tradition of argumentation to which he was accustomed. Since young students are often afraid to express their

[112]The new laboratory was designed by Kapitza, with much help from Cockcroft. The architect was H. C. Hughes. Two pieces of art, which Kapitza ordered from the sculptor Eric Gill, decorated the building. On the outer wall there was carved a crocodile and in the small, round entrance hall a plaque showing a portrait of Rutherford was mounted. When these decorations were unveiled at the opening ceremony, the university people were rather puzzled. But Kapitza explained the presence of the crocodile as follows: 'In Russia the crocodile is the symbol for the father of the family. But this animal is also regarded with awe and admiration because it has a stiff neck and never turns its head. It just goes straight forward—like science, like Rutherford' (Larsen, 1962, p. 62). In any case, it was known before to some in the Cavendish that Kapitza's pet name for Rutherford was 'crocodile.' While the crocodile was approved of in general, the plaque of Rutherford disturbed the conservative observers, Aston in particular. On Rutherford's suggestion Kapitza wrote to Niels Bohr to ask his opinion of the portrait. Bohr replied, 'The carving of Rutherford looks to me most excellent, being at the same time thoughtful and powerful' (Bohr to Kapitza, 15 March 1933). The case was thus settled; Kapitza was very happy about it and presented Bohr a copy of the carving by the sculptor.

opinions in front of their elders, being afraid to make fools of themselves, senior people were excluded from the membership as a rule. The Kapitza Club met on Tuesday evenings after dinner, in the early days, in Kapitza's rooms at Trinity College, and, after Cockcroft became a member in 1924, in his room at the Cavendish. 'Attendance was compulsory,' recalled John Desmond Bernal. 'If you missed two sessions running you were thrown out' (Larsen, 1962, p. 62). Before somebody read a paper on some recent development in physics, Kapitza would open the session with apparently random remarks, and during the sessions 'he sometimes made intentional howlers, which even the youngest and least experienced would not be afraid of correcting' (Crowther, 1974, p. 198). Thus, he added a genuinely new institution to the Cambridge scene, which made the discussion of those fresh topics possible for which no experts were available at the Cavendish Laboratory.

The Kapitza Club was opened on 17 October 1922 with a talk by Kapitza himself. He also took over about half of the first sixteen sessions until 20 April 1923.[113] Other than Kapitza the speakers during this period were: D. R. Hartree; Henry De Wolf Smyth, then U.S. National Research Fellow at the Cavendish, who obtained his Ph.D. under Rutherford in 1923, and returned to Princeton University to pursue a distinguished career; E. S. Bieler, then 1851 Exhibition Scholar from McGill University, Montreal, Canada, who obtained his Ph.D. in 1923 and died soon afterwards; H. W. B. Skinner, E. G. Dymond, J. E. (Lennard-)Jones and, as first guest speaker from abroad, S. Boguslawski from Russia, a pre-World War I student of Max Born's in Göttingen. The list of distinguished visitors who gave talks on their work at the Kapitza Club is long: Paul Ehrenfest visited the Club in spring 1923; for the 28th Meeting on 27 July 1923 Gilbert N. Lewis from Berkeley, who at about that time (13–14 July 1923) also attended a meeting of the Faraday Society in London on 'The Electronic Theory of Valency,' was invited; and later in the fall, at the 36th Meeting on 20 November 1923, Svein Rosseland from Copenhagen spoke. Kapitza was very persuasive in inviting visitors to his Club; he thus provided his friends and collaborators unique opportunities of hearing from experts about what was going on elsewhere. However, the main bulk of the talks was given by the members themselves; they reported on their own work as well as on interesting papers published in British and foreign scientific journals. The following people were listed as members of the Kapitza Club in the year 1924–1925: N. Ahmad, P. M. S. Blackett, J. C. Boyce, J. A. Carroll, J. Cockcroft, R. W. Ditchburn, K. G. Emeléus, P. A. M. Dirac, R. W. Gurney, D. R. Hartree, J. E. (Lennard-)Jones, P. Kapitza, L. H. Martin, M. Rogers, H. W. B. Skinner, W. L. Webster, W. A. Wooster and L. H. Thomas. Of these, all except Hartree and the two newcomers, Paul Dirac and Llewellyn Hilleth Thomas, were experimentalists.

[113]The Kapitza Club was continued at the Cavendish for more than thirty-five years. The sessions were numbered sequentially and, as a rule, the speaker would write down the topic of his talk in the *Minute Book* and sign his name (AHQP, Microfilm No. 38).

As we have mentioned above, the discussions at the Kapitza Club were very free. 'It was a kind of grand inquisition on all important questions of physics,' recalled Bernal. 'Men with great names were "summoned," heckled and interrupted; Kapitza did most of the heckling, but no one minded because of his enthusiasm' (Larsen, 1962, p. 62). The topics discussed were both experimental and theoretical. For example, G. N. Lewis spoke on 'What Is the Origin of Radioactive Substances?'; J. E. Jones on 'The Temperature and Composition of the Upper Atmosphere' (32nd Meeting, 17 October 1923); K. G. Emeléus on 'Abnormal Cooling Function of Dry Ether' and M. H. A. Newman on 'A Property of Quantum Orbits on Weyl's Relativity Theory (Schrödinger [1922c])' (both in the 33rd Meeting, end of October 1923); L. F. Curtiss on 'Davisson and Kunsman's Experiments on Scattering of Electrons' (38th Meeting, 12 December 1923). In 1924 and 1925 special emphasis was placed on discussing the latest developments in atomic and quantum physics. Thus, James Franck delivered three talks in October 1924: the first on 'New Experiments on the Resonance Fluorescence of Mercury Vapour' (62nd Meeting, 14 October); the second on 'Relations Between Quanta and the Disintegrating Heat of Diatomic Molecules' (63rd Meeting, 16 October); and the third on 'Relations Between Electron Impacts and the Dissociation of Molecules; Ionization by Impact of Positive Ions' (64th Meeting, 17 October). The members of the Club reviewed such brand new topics as Millikan and Bowen's extension of the laws of X-ray doublets to optical spectra (H. W. B. Skinner at the 65th Meeting, 21 October 1924), Wentzel's theory of dispersion (J. C. Boyce at the 72nd Meeting, 13 January 1925), Bohr's paper on resonance radiation (L. H. Martin at the 78th Meeting, 10 March 1925), Breit's theory of the photoelectric effect (R. W. Gurney at the 81st Meeting, 31 March 1925), Einstein's theory of monatomic perfect gases (L. H. Thomas at the 93rd Meeting, 21 July 1925), and Bothe and Geiger's experiment on the Compton effect (R. W. Ditchburn at the 96th Meeting, 12 August 1925). During the academic year 1924–1925 there also came two visitors who were well known for their important contributions to quantum theory: on 18 May 1925, at the 86th Meeting of the Kapitza Club, Niels Bohr presented his views about the fundamental problems and difficulties of quantum theory; and on 28 July 1925, at the 94th Meeting, Werner Heisenberg talked about 'Term Zoology and Zeeman Botany,' the new approach for masterminding the details of atomic spectroscopy. A week after Heisenberg's visit, on 4 August at the 95th Meeting, Paul Dirac spoke on 'Bose's and L. de Broglie's Deductions of Planck's Law.'

Beyond any doubt, the Kapitza Club served a most important purpose for the research students at the Cavendish. No theoretical result, which Kapitza or the other members of the Club found interesting, was left out of the discussions. For example, Heisenberg's work on the quantum-theoretical re-interpretation of kinematics, which ushered in the new quantum mechanics, was discussed immediately after the paper came out by T. M. Cherry ('and the gallery,' as the *Kapitza Club Minute Book* reported) at the 105th Meeting on 24 November 1925. Cherry also reviewed Schrödinger's work on wave mechanics at the 124th

Meeting on 8 June 1926. Some weeks later, on 29 July at the 130th Meeting, Max Born presented his collision theory including the statistical interpretation of the wave function. It is no exaggeration that the members of the Kapitza Club were well informed about the important theoretical and experimental developments of the day; this information, in addition to the thorough training they received at the Cavendish under the leadership of Rutherford and Fowler, enabled them to contribute many new and fundamental results to quantum physics.

Due to his personality and his different background Kapitza was the most colourful figure in the Cavendish. In spite of his eccentricities he was well liked by the people in the Laboratory and he made many friends there. One of these friends was Paul Dirac, whom he even persuaded to take up some experimental work later on.

Dirac hit upon the idea that pure rotation of gases at high speeds might be exploited for separation of the isotopes contained in them, using an apparatus without moving mechanical parts. His experimental arrangement consisted of three pipes, such that a mixture of a light gas and some heavy gas was pumped through one pipe, which branched into two others. When he performed the experiment and checked the composition of the gases in the two pipes, he observed only 'a negligible amount of separation,' but he discovered a 'thermal' effect in the process: the gas came out of the two pipes with widely different temperatures. For instance, when air at room temperature was blown into the initial pipe under a pressure of 6 atm, the temperature at one end would be below the freezing point and the other about 100°C higher. This effect arose entirely because of the viscosity of the gas, by which energy from the inner layers of the gas in the first pipe was transferred to the outer layers, and the layers of different temperatures were separated by the geometrical arrangement of the two branching pipes. Dirac stopped the work on isotope separation when Kapitza returned to Russia in 1934.[114] Alone he did not have enough enthusiasm to carry on the experimental work, and nothing more happened to it until World War II. Then great interest existed in separating the isotopes of uranium for nuclear fission and, among many other possibilities, Dirac's method was explored further with some modified apparatus at Oxford. Dirac went there from time to time to discuss with the people working on the experiments. However, the separation of

[114]Kapitza had continued to be a citizen of Russia, and from 1926 onwards he visited there several times. He went to Russia again in 1934 to attend the conference held to celebrate the 100th anniversary of D. I. Mendeleev's birth. A few days before he was due to return to the Cavendish, Kapitza was informed that he must stay and work in Russia, 'as the Soviet authorities thought he was able to give important help to the electric industry' (Rutherford to Baldwin, 29 April 1935, quoted in Oliphant, 1972, p. 102). On Rutherford's appeal that Kapitza be allowed to return to Cambridge and his laboratory to complete his work, the Soviet Government responded that he could do so just as well in Russia. Negotiations began and, among others, Dirac went to Russia to talk to Kapitza and the authorities. Finally the Russian Government bought Kapitza's apparatus for £30,000; Cockcroft had it packed and dispatched to Kapitza. A part of this money went to Mond Laboratory and was used for projects which did not compete with the work of Kapitza. Kapitza settled permanently in Russia. He became Director of the Institute for Physical Problems (of the Academy of Sciences) in Moscow and soon had the necessary laboratories to continue his research work on low-temperature physics.

isotopes by the rotation of gases in devices without moving mechanical parts was never performed on a major scale, for it could not compete in efficiency with the diffusion method.

This was Dirac's only actual involvement in experimental work after his engineering days in Bristol, as well as his only work, theoretical or experimental, which had some connection with nuclear physics. There was, however, another experiment which he and Kapitza proposed to perform: In a joint paper they discussed the theoretical possibility of obtaining electron diffraction from a grating of standing light waves (Kapitza and Dirac, 1933). They estimated that continuous light sources would give too low intensities to observe the effect and suggested using an intense mercury-arc source. No one was happier about this work than Rutherford, who wrote to Enrico Fermi: 'You may be interested to hear that Professor Dirac is also doing some experiments. This seems to be a good augury for the future of theoretical physics' (Rutherford to Fermi, 23 April 1934). The Dirac–Kapitza effect had to wait almost forty years to be observed experimentally by using a laser source of light.

However, in 1924, when he became a member of the Kapitza Club, Dirac was still at the beginning of his life in scientific research. He adapted himself quickly to the high standards of physics in Cambridge. He absorbed all the information he needed from lectures, colloquia and other scientific activities. On the other hand, he sought and found much time and solitude for work and contemplation. Altogether Dirac's nature and habits did not change very much after Bristol and his schedule remained simple. During the week he would attend four or five lectures. Morning and evening were devoted to studying, with short walks in the afternoon. Only rarely would he see other research students, except at dinner in the evenings. Occasionally he was invited to tea. He read very little literature and never went to the theatre. He found the greatest relaxation in taking long walks. Every Sunday he would go for a whole day's walk, taking his lunch with him. He would not intentionally think about his work during those walks, although he might review it now and then. Often new ideas would come to him. It was on one of those long walks, on a Sunday in September 1925, that it occurred to him that the commutators—which he had discovered in Heisenberg's quantum-theoretical re-interpretation of the anharmonic oscillator—might correspond to the Poisson brackets in classical mechanics.

Chapter III
The Making of a Quantum Physicist

In the first two years of his stay in Cambridge, Paul Dirac became a serious quantum physicist. This process was hastened by the inspiring guidance of Ralph Fowler and his own research work, which acquainted him with the problems of atomic theory and gave him many insights into the ways solutions were being sought at that time. Dirac's early work may be organized into three parts. The first contains the beginner's papers, through which the research student learned how to formulate and solve theoretical problems and to write about them. Then, after spending only half a year in Cambridge, Dirac became involved in two important aspects of the atomic theory of the day: he published two large papers, one on the principle of detailed balancing and the other on the adiabatic principle. In these papers Dirac already demonstrated several qualities of his later research: his interest in a general treatment of the problems under investigation; his ability to use new mathematical tools; and his skill in being able to pick out the essential details of the real physical situation. His work until summer 1925 was then devoted to less weighty problems, and he published only one paper on an astrophysical subject. However, during the two years from fall 1923 to fall 1925 Dirac studied the full range of quantum problems and he became aware of the existing fundamental difficulties. His knowledge and experience prepared him to identify immediately the crucial step which Heisenberg introduced into quantum theory in July 1925, and to build upon it an elegant structure of quantum mechanics in the classical mold.

III.1 First Steps in Research

Paul Dirac had not received any training in performing research in theoretical physics when he came to Cambridge University. Unlike almost any graduate in physics from Cambridge at that time, he had had no close interaction with research before. He also did not know much about physics. Hence his programme for the start was clearly given: he had first to learn about the important physical theories and the current research work in them, and then to try to find in what he learned, definite problems on which to work. Now the lectures that Dirac attended in Cambridge provided him with the general ideas of what was going on in various fields of theoretical physics. He 'seldom heard everything, every word that a lecturer said,' but tried to obtain the major drift of ideas and filled the gaps by reading (Dirac, Conversations, p. 2). Frequently he found the notation used by other people needlessly awkward and, in order to understand a piece of work thoroughly, he would first put it in his own notation and try to make the calculations as simple as possible. Every now and then he saw a chance to improve something he had read; such improvements led to some of his early papers. However, his first research problem did not come about in this way, but was given to him by his supervisor, Ralph Fowler.

Dirac met his supervisor regularly. He would go to see Fowler once a week and discuss his work with him. Fowler did not wait until Dirac had found a suitable problem to work on from his own reading, but put him right away on a research project: to treat the dissociation of a gas under the action of a temperature gradient. Dirac performed the calculations without delay and wrote his first paper, which Fowler communicated to the *Proceedings of the Cambridge Philosophical Society*; it was received on 3 March 1924 and duly published (Dirac, 1924a).

The paper on the 'Dissociation under a Temperature Gradient' dealt with a rather complex problem of statistical mechanics in which Fowler had become interested towards the end of 1923. When a gas consisting of large 'double' molecules (say hydrogen) and smaller 'single' molecules (atomic hydrogen)—the double molecules being capable of dissociating into two similar single molecules —is kept in a long tube, whose axis points in the z-direction, a state of equilibrium is achieved. This state comes about because of the physical fact that the rate of dissociation of double molecules into single molecules equals the recombination rate of single molecules to form double molecules at each space point, and its properties depend on the temperature conditions applied. If the temperature is uniform throughout the gas, the equilibrium state is the same at all points in the tube. If, on the other hand, the two ends are kept at different temperatures, a temperature gradient will exist everywhere in the tube and the equilibrium between single and double molecules will change from point to point and diffusion will set in. A steady state will be obtained when the diffusion of double molecules into any region of the tube equals the excess of the rate of their dissociation over their recombination in that region. Fowler asked Dirac to study these processes in detail and, in particular, to calculate the change of concentration at the end of a tube, of length z_0 and unit cross section, when a given temperature difference is applied. Fowler hoped that the result of the calculation would answer the practical question of whether one could make use of an observation of the difference of concentration to determine H, the so-called 'velocity' of the dissociation and recombination process, a quantity of considerable interest in kinetic theory.

To solve the problem, Dirac had to consider the equations of statistical mechanics that describe the behaviour of the gas mixture within the tube. These equations enabled him to find the relations for two quantities which determine the steady state: the constant energy flow, A, and the excess number of single molecules, 2σ. He considered four equations: the equation for the conservation of matter, which requires that one double molecule always dissociates into two single molecules; the equation for the energy flow which is caused by thermal conduction, on the one hand, and the motion of single molecules, on the other (because of the positive heat of dissociation, the two single molecules in the z-direction carry greater kinetic energy than one double molecule); the diffusion equation in which the two terms due to the differences of concentrations and temperatures are considered; and finally an equation expressing the equilibrium balance between the net loss of single molecules through recombination and the excess flow of single molecules into the same region. With the help of these equations Dirac derived two differential equations for the quantities A and σ. He integrated them by using reasonable approximations and obtained the following results: first, the temperature distribution, $T = T(z)$, is given by the relation

$$T = -\frac{A}{\kappa'}z - \frac{\mu_1\mu_2}{\mu_1 + 2\mu_2}\frac{\tilde{k}D}{\kappa'}B \cdot \exp\{\pm\beta(z - z_0)\} + \text{constant}; \qquad (13)$$

second, the excess number of single molecules at the end of the tube is calculated to be 2σ, where

$$\sigma = \pm\frac{\mu_1 + 2\mu_2}{\mu_1 + \mu_2} \cdot \frac{\kappa' - \kappa}{\kappa}\frac{A}{D\tilde{k}\beta}. \qquad (14)$$

In these equations all quantities are known: at the local temperature T, μ_1 and μ_2 are the equilibrium values of the numbers (per unit volume) of single and double molecules at the point z; κ and κ' denote the thermal conductivity and the apparent thermal conductivity, respectively; D is the diffusion coefficient and $\tilde{k}$ the heat of dissociation. Finally, β and B are constants which can be expressed in terms of other quantities; for example, β is proportional to the square root of H, the 'velocity' of the dissociation and recombination process.

With the help of Eqs. (13) and (14) Dirac was in a position to answer Fowler's question whether a measurement of the excess number of single molecules or that of the temperature distribution within the tube would allow one to determine H. He had discovered that any observable effects should be most pronounced at the end of the tube containing the dissociating gas molecules. To obtain an estimate of the order of magnitude of σ, he substituted the typical values for the quantities entering on the right-hand side of Eq. (13), that is, $D = 0.1$, $\kappa = 5 \times 10^{-5}$, $\tilde{k} = 5000$ cal./gram-molecule, and put κ' equal to its value at $z = z_0$. Assuming an average temperature of 300°K and a mean temperature gradient A/κ' of 10°K/cm, Dirac arrived at the conclusion that at the end of the tube the relative deviation of the number of single molecules (from its equilibrium value in the case of uniform temperature) is of the order of several percent, provided the dissociation constant H has the order of magnitude of $1\ \text{s}^{-1}$. Hence this deviation, as well as the deviation of the temperature dependence from the linear curve (described by the first term on the right-hand side of Eq. (13)), may be observed *in principle*. In practice, however, the situation is different and Dirac noted that: 'The difficulty in making the required measurements at the end of the tube is that the quantities to be measured fall off rather rapidly according to an exponential law with space constant β^{-1}, which is about 2 mm with the numerical values previously taken when H is 1 s^{-1}' (Dirac, 1924a, p. 137).

Thus, the outcome of Dirac's calculation was not very promising for the purpose Fowler had in mind. But it did not matter very much, for one goal had been achieved: the young research student had shown that he was perfectly able to solve a given problem in a field that was new to him by making careful use of the well-known equations. Fowler found the results interesting enough to be published in a paper, and that was all that counted for the moment. For Dirac the problem of dissociation meant just a kind of exercise. It was not one he liked very much, for he thought he was not good in carrying through lengthy algebraic calculations. However, even in this beginner's work he had appealed to general principles, such as the conservation laws, to arrive at the result. He liked making use of general principles, and it would become one of the dominant features of his future work.

On completing his first research paper, Dirac turned to problems which were more in his line and which arose from his studies and reading. The following six papers, which he wrote from spring 1924 to fall 1925, had one thing in common: in each of them he treated subjects that were relatively standard in the literature; he would criticize some existing result and put it on a firmer basis than it had before. Two notes, each only two pages long, which Dirac submitted next in April and May 1924, respectively, were typical of his method of finding problems for research by seeking to improve on certain statements existing in the scientific literature. The content of these notes and the circumstances of their writing show how Dirac developed as a research scholar and became ready for more important and more difficult tasks.

In the first of these, 'Note on the Relativity Dynamics of a Particle' (Dirac, 1924b), Dirac elaborated on a statement which Eddington had made in his book *Mathematical Theory of Relativity* concerning the notation of velocity in relativity theory. The velocity of a particle in that theory may be defined in two ways: either one follows the motion of the particle in space and time and derives the 'kinematical velocity' from the observed world-line, or one can measure the four-momentum of the particle and find from it the 'dynamical velocity.'[115] By

[115] An isolated particle of finite extension in space may be considered in relativity theory as a narrow tube in four (space-time) dimensions inside which $T_{\mu\nu}$, the energy-momentum tensor, is nonzero, while surrounding it the energy-momentum tensor is zero. The momentum and mass of the particle are then obtained by integrating the components $T_{\mu 4}$, $\mu = 1, 2, 3$, over a three-dimensional volume. This yields the three-vector (p_x, p_y, p_z) or $(Mu_{\text{dyn}}, Mv_{\text{dyn}}, Mw_{\text{dyn}})$, where p_x, p_y, p_z are the components of the momentum of the particle, M its mass, and $u_{\text{dyn}}, v_{\text{dyn}}, w_{\text{dyn}}$, the components of the 'dynamical velocity.' Hence the latter are obtained by dividing the corresponding momentum components by M, i.e., $u_{\text{dyn}} = p_x/M$, etc. The kinematical velocity of the particle, on the other hand, is given by the equation

$$(u_{\text{kin}}, v_{\text{kin}}, w_{\text{kin}}) = \left(c\frac{dx_1}{dx_4}, c\frac{dx_2}{dx_4}, c\frac{dx_3}{dx_4}\right),$$

where $x_1, \ldots, x_4$ are the space-time coordinates of the centre of mass and c the velocity of light *in vacuo*.

comparing the two notions of velocity, Eddington had arrived at the following conclusion: 'It does not seem to be possible to deduce without special assumptions that the dynamical velocity of a particle is equal to the kinematical velocity. The law of conservation [i.e., the conservation of the Cartesian components of the momentum, Mu, Mv, Mw, and of energy, E, where M is the mass and (u, v, w), the dynamical velocity of the particle; Eddington put c, the velocity of light *in vacuo*, equal to unity] merely shows that (Mu, Mv, Mw, E) is constant along the tube [i.e., the spatially extended world-line of the particle] when no field of force is acting; it does not show that the direction of this vector is in direction of the tube' (Eddington, 1923b, p. 125). Since the idea of two *different* velocities of a particle was difficult to accept, Eddington continued: 'I think that there is no doubt that in nature the dynamical and kinematical velocities are the same, but the reason for this must be sought in the symmetrical properties of the ultimate particles of matter' (Eddington, 1923b, p. 125). That is, since he had been able to prove the desired identity of kinematical and dynamical velocities for symmetrical particles only, Eddington speculated that all fundamental constituents of matter should have three mutually perpendicular planes of symmetry.[116]

Eddington's speculation led to a difficulty with respect to certain accepted views in atomic physics. Many particles occurring in nature do not seem to possess the symmetrical constitution required by Eddington's argument. For example, the molecules treated in statistical mechanics often have preferential axes. For someone in atomic theory, like Fowler, it was not acceptable that the kinematical and dynamical velocities should be different in these cases. His student Dirac was happy to pick up the relativistic problem from Eddington; here was the opportunity to improve upon something stated in a famous expert's book. He went ahead and showed the equivalence of the two velocities 'without any postulates concerning the form of the particle, by considering the conditions that hold at surfaces of discontinuity in a distribution of matter' (Dirac, 1924b, p. 1159).

What Dirac had to prove in particular was that the directions of both velocities for an assymetric particle were the same, or that the two four-vectors had only a time-like component when the particle was at rest. He proceeded in two steps. First, he discussed the case when a three-dimensional space-like surface of discontinuity separates for all times a region of continuous distribution of matter on one side from a region containing no matter on the other side. Choosing this surface to be $x_1 = \text{constant}$, Dirac considered the equation of conservation of $T^{\mu\nu}$, the energy-momentum density tensor of general relativity, which he found in Eddington's book (Eddington, 1923b, p. 118). This equation

[116]Eddington proved his statement in the following manner. A symmetrical particle, which is kinematically at rest, cannot have any momentum since there is no preferential direction in which the momentum could point. Hence the tube and, therefore, the energy-momentum four-vector must be along the t-axis. However, to apply this argument he found that it was not necessary to assume complete spherical symmetry, but that three perpendicular planes of symmetry would suffice. He concluded that: 'The ultimate particle may for example have the symmetry of an anchor-ring' (Eddington, 1923b, p. 125).

may be written as

$$\frac{\partial T^{\mu 1}}{\partial x^1} = -\frac{\partial T^{\mu 2}}{\partial x^2} - \frac{\partial T^{\mu 3}}{\partial x^3} - \frac{\partial T^{\mu 4}}{\partial x^4} - \{\alpha\nu, \mu\} T^{\alpha\nu} - \{\beta\nu, \nu\} T^{\mu\beta}, \qquad \mu = 1, 2, 3, 4, \tag{15}$$

where { } denotes the Christoffel symbol with the appropriate summations over double indices. Now Dirac argued as follows: Since all terms on the right-hand side of Eq. (15) are finite at the space boundary $x_1 = \text{const.}$, $T^{\mu 1}/x_1$ on the left-hand side must be finite as well. If, in addition, $T^{\mu 1}$ is zero beyond the boundary, it follows that it is also zero within the region containing matter. Hence u_{dyn}, the dynamical velocity normal to the boundary, like u_{kin}, the kinematical velocity, vanishes. Turning now to a particle at rest, enclosed for all times by the surfaces $x_1 = \text{const.}$, $x_2 = \text{const.}$, $x_3 = \text{const.}$, one shows in the same way that all $T^{\mu\nu}$, except T^{44}, the double time-like energy density, are zero. As a result, the space-like components of the dynamical velocity of a particle of *any shape* vanish like the corresponding kinematical velocities. This result completed the proof.

The work on the problem of the two velocities satisfied Dirac, for it not only allowed him to continue his interest in relativity but gave him the opportunity of meeting Eddington and discussing with him. After he had found the solution of the problem, he wrote it up and showed it to Eddington. Eddington found it all right; however, before communicating Dirac's note to *Philosophical Magazine* he suggested a few changes. These were principally to put in some introductory remarks and to replace the original mathematical argumentation at the end by the simple sentence: 'Hence the dynamical velocity vector has the same direction as the kinematical velocity vector, viz. along x_4' (Dirac, 1924b, p. 1159).[117] Dirac was very glad to follow the advice of the famous man. He took over Eddington's sentence and also added an introduction in which he explained the problem and stated what the previous author had done.

Dirac supplied a similar extensive introduction to his next paper, 'Note on the Doppler Principle and Bohr's Frequency Condition,' which Fowler communicated to *Cambridge Philosophical Society*, where it was received on 19 May 1924, read on 14 July, and duly published (Dirac, 1924c). Dirac had found this problem from his private reading of quantum theory. At that time he was studying Niels Bohr's 1923 review article, 'On the Application of the Quantum

[117] Dirac had completed his original manuscript with the mathematical remarks: 'Putting [see Eddington, 1923b, p. 103] $T^{44} = A(dx_4/ds) \cdot (dx_4/ds)$, then $T^{\mu\nu} = A(dx_\mu/ds) \cdot (dx_\nu/ds)$; multiplying by $g_{\mu\nu}$, $A = g_{\mu\nu}T^{\mu\nu} = T = \rho_0$, where ρ_0 is the proper-density of a particle, and is invariant. Hence $T^{\mu\nu} = \rho_0(dx_\mu/ds)(dx_\nu/ds)$, and since this is a tensor equation it is also true for any other system of coordinates.' Whereupon Eddington remarked in the margin of the manuscript: 'I think the reader will scarcely understand what you are getting at without further explanation. If you put in the introductory remarks that I suggest, it will be sufficient to say: Hence the dynamical velocity vector has the same direction as the kinematical velocity vector, viz. along x_4' (AHQP, Microfilm No. 36, Section 9).

Theory to Atomic Structure,' in which the author had expounded his most recent views on the fundamental postulates of atomic theory. That article, first published in *Zeitschrift für Physik* (Bohr, 1923a), had been translated into English by Leon Francis Curtiss—then National Research Fellow (U.S.A.) at the Cavendish—and had become available in early 1924 as a separate supplement to *Proceedings of the Cambridge Philosophical Society*. This paper served for Dirac as an authoritative introduction to atomic and quantum theory. In reading it he came across a problem which he felt he ought to tackle, for it had to do with his beloved subject of relativity theory.

In his review article Bohr had mentioned the problem of deciding as to which frame of reference was most suitable for formulating the condition

$$h\nu = E_1 - E_2, \tag{16}$$

connecting ν, the frequency of light emitted by an atomic system, with E_1 and E_2, its energies before and after the process of emission, respectively. Bohr had come to the conclusion that the natural frame of reference is one in which the atomic system as a whole is at rest; in all other frames of reference the frequency of the radiation emitted in various directions of space exhibits relativistic Doppler effects. Bohr had assumed that if free electrons participate in the radiative process, then the frame of reference is determined by the motion of the electron before and after the collision with the atom, the latter being relatively at rest. But he had also noted that 'an uncertainty on this point may arise if we would assume that the process of radiation is connected with a change of momentum' (Bohr, 1923a; English translation, p. 28, footnote ‡). A way out of this uncertainty had been pointed out in the literature and Bohr referred to it. Erwin Schrödinger had treated the case of an atomic system whose velocity changes from v_1 to v_2 by the emission of radiation and had found that ν, the frequency of this radiation, is given by the relativistic Doppler formula

$$\nu_0 = \nu \left\{ \frac{c - v_1 \cos\theta_1}{\sqrt{c^2 - v_1^2}} \cdot \frac{c - v_2 \cos\theta_2}{\sqrt{c^2 - v_2^2}} \right\}^{1/2}, \tag{17}$$

where ν_0 is the frequency of radiation emitted by atoms at rest, c is the velocity of light *in vacuo* and θ_1 and θ_2 are the angles between the direction of the velocities and the propagating radiation (Schrödinger, 1922a, p. 303, Eq. (12′)). However, in deriving Eq. (17) Schrödinger had invoked Einstein's concept of directed light-quanta, which, in Bohr's opinion at that time, seemed to contradict fundamental correspondence arguments. Hence Bohr, though he agreed with Schrödinger's result, was not happy about his method of deriving it.

Dirac immediately noticed that here was a point to improve upon and he set out to establish Eq. (17) without using the light-quantum argument. He found the solution by a simple and elegant relativistic argument. For that purpose he had only to consider the four-momentum of the atom before and after emission, and he remarked: 'The change of this vector during the transition process is another [four-vector] δE^i whose space and time components are equal, so that its direction in space-time is the same as that of the radiation' (Dirac, 1924c, p. 433). He then introduced what he called the contravariant 'frequency vector' for plane waves defined as

$$\nu^i = \frac{c}{2\pi} g^{ik} \frac{\partial \phi}{\partial x_k}, \tag{18}$$

which lies in the direction of propagation of radiation. Here ϕ is the phase angle of the wave at the space-time point x_k and g^{ik} is the metric tensor. From the validity of the relativistically covariant equation,

$$\delta E^i = h\nu^i, \tag{19}$$

he deduced that ν^i, given by Eq. (18), is a four-vector whose time component, ν^4, indeed changes in accordance with Eq. (17) under a Lorentz transformation.[118] Thus, by a straightforward application of relativistic principles Dirac was able to derive all of Schrödinger's results *without departing from the wave theory of radiation.* In particular he established the covariant formulation of the frequency condition (see Schrödinger, 1922a, p. 302, Eq. (5a)), a step which he considered as being essential in quantum theory.

With these papers, which he wrote during the first six months of his stay in Cambridge, Dirac proved himself to be a productive research student. And yet, in the beginning at least, he did not enjoy writing a paper. When he mentioned his dislike to Fowler, the latter answered: 'Well, if you are not going to write your work up, you might as well shut up shop' (Dirac, Conversations, p. 43). With this down-to-earth reply of Fowler's, Dirac decided that he would just have to force himself to write. Words did not come to him easily, and he started by making two or three successive drafts from rough notes until he was satisfied. It was from this lack of desire for writing that Dirac developed his precise and concise style of communication into a model of linguistic and technical accuracy. He would always work out his ideas pretty well in his mind before writing, making only the final changes in the course of writing. As a rule he presented the ideas in the same order in which they occurred to him. He would deviate from this rule only if the logical order and justification of the steps in a paper required

[118]To make this fact more explicit, Dirac eliminated v_2 and θ_2 from Eq. (17), obtaining the relation,

$$\nu' = \nu \frac{1 - (v_1/c)\cos\theta_1}{\sqrt{1 - (v_1^2/c^2)}},$$

where ν' is a relativistic invariant. This equation shows that ν has the properties of the fourth component of a relativistic vector.

a different arrangement. He would not discuss his results or show them to anybody before they were fully fixed in his mind and written up. And then he did not like changes![119] Of course, in the beginning the senior people helped and advised him in his research and they also occasionally improved upon his writing and the presentation of results. And, especially, they assisted him in getting his papers published, for most scientific journals accepted only papers communicated by reputed people. For example, the contributions to *Proceedings of the Royal Society* had to be communicated by a Fellow of the Society, and the same was true of *Proceedings of the Cambridge Philosophical Society*. As a rule, Fowler would suggest to Dirac where to send his papers, the more important ones going to *Proceedings of the Royal Society*. The *Proceedings of the Cambridge Philosophical Society* too had a high standing and were widely read, and Dirac published there also.

Dirac's two short notes of spring 1924, which we have discussed, contained certain features that would occur again and again in his later work. One, which would become a habit with him, was the writing of rather long introductions. In his early papers, in which he sought to improve upon certain points in the literature, such introductions were necessary in order to inform the reader about what was going on. Later, when Dirac developed his own ideas in sequences of several papers, each connected with the previous one and continuing where the other stopped, he would still write long introductions. In these he would describe the methods, which he had laid down in a previous work and which he intended to use again in the paper under discussion, especially if he had improved upon the earlier methods. Sometimes the improvement consisted not in correcting or elegantly reformulating the mathematics but in giving a clearer exposition of the earlier ideas. Thus, the introductions served two purposes: first, to satisfy the author himself about the scope and logical presentation of what he had undertaken to discuss; second, to help the reader understand the author's premises and goals. Dirac was always concerned about getting things clearly in his own mind, to the extent that the explanation of his ideas to other people, though important, was secondary.

Another aspect of Dirac's early papers was the central role played in them by simple relativistic arguments. The latter arose, of course, from his early interest in relativity, an interest which he had developed by systematic study upon arrival in Cambridge. Thus, even when obviously relativistic problems did not have to be

[119]Bohr once remarked to Heisenberg, 'Whenever Dirac sends me a manuscript, the writing is so neat and free of corrections that merely looking at it is an aesthetic pleasure. If I suggest even minor changes, Paul becomes terribly unhappy and generally changes nothing at all' (Heisenberg, 1971, p. 87).

Bohr himself used to 'construct' his papers by dictating them to his scientific collaborators or a member of the family, making corrections, writing and rewriting until at last he was satisfied. Once, in the late 1920s, when Dirac was in Copenhagen, Bohr had to write a paper in English and asked Dirac to help him. Bohr paced as he dictated, went back and corrected, and so it went on for a long time. Then Dirac, during an impressive pause, said: 'Professor Bohr, when I was at school, my teacher taught me not to begin a sentence until I knew how to finish it' (Dirac, Conversations, p. 43).

treated, as for instance in statistical mechanics or quantum theory, Dirac would still try to make use of relativistic covariance and invariance relations.[120]

Finally, in both of his short notes of spring 1924 Dirac had alighted upon rather fundamental questions to treat. In the note on relativistic dynamics he dealt with the question of whether an actual particle should be 'the nucleus of a symmetrical field [electrical and gravitational],' as Eddington had demanded (Eddington, 1923b, p. 126). In the note on the Doppler principle in quantum theory Dirac worked out the correct formulation of the frequency condition, a fundamental postulate of atomic theory. This interest in fundamental questions would distinguish all of Dirac's future work, beginning with his first major paper on the statistical equilibrium between atoms, electrons and radiation.

III.2 The Principle of Detailed Balancing

At the time when Dirac went up to Cambridge, Ralph Fowler and his friend Edward Arthur Milne were 'impressed by Klein and Rosseland's isolation of the phenomenon known as "collisions of the second kind," in which the reverse of excitation by collision occurred' (Milne, 1945, p. 71). Back in fall 1920 Oskar Klein and Svein Rosseland had discussed the nonradiative encounters between atoms and electrons and had shown that, in order to maintain thermodynamic equilibrium in an enclosure containing atoms and free electrons, a certain reversibility of atomic processes had to be assumed (Klein and Rosseland, 1921). In particular, they had postulated that the reverse of the process of excitation of an atom by electron impact, that is, a process in which the atom transfers energy to electrons and passes without radiating to a stationary state of lower energy, must also exist. In fall 1923 Fowler and Milne became interested in the mechanism of ionization of atoms. They applied in their work the key idea of Klein and Rosseland, namely, the assumption that atomic processes are balanced by reverse processes. Fowler and Milne regarded this assumption as a new, general principle of statistical mechanics.

Dirac heard about the principle of detailed balancing in Fowler's lectures on statistical mechanics. Although he was not especially fond of statistical mechanics as a physical theory, the general principle (of detailed balancing) attracted him and he began to study the available literature on it. His goal was to understand the origin of detailed balancing and its implications for all situations in atomic physics. He was happy to discover that in this work he could apply

[120]It is of interest to note that Erwin Schrödinger and Louis de Broglie discussed quantum-theoretical problems before Dirac on the basis of relativistic arguments. De Broglie had even employed the phase function of electromagnetic waves in a paper submitted to *Philosophical Magazine* in October 1923 (de Broglie, 1924a). Although this paper appeared in print in February 1924, about three months before Fowler communicated Dirac's note on the frequency condition (Dirac, 1924c), it did not influence Dirac. However, a remark in the next paper (Dirac, 1924d) shows that Dirac had seen de Broglie's work a little later. (See Section III.2.)

certain methods that were familiar to him, such as the one he had learned from Wolfgang Pauli's paper on the equilibrium between free electrons and radiation at a given temperature (Pauli, 1923b). Pauli had used essentially the same kind of arguments that Dirac had himself already employed, namely, the consequences of relativistic invariance of the quantities involved. Because of his acquaintance with these methods, Dirac was able to make rapid progress on the new problem. Within less than two months after submitting his last note on the frequency condition (Dirac, 1924c), he completed the paper on 'The Conditions for Statistical Equilibrium between Atoms, Electrons and Radiation' (Dirac, 1924d).[121] It was communicated by Rutherford early in July 1924 to *Proceedings of the Royal Society* and was published in their issue of 1 November 1924. The paper on detailed balancing was not only the longest article which Dirac wrote during his first two years as a research student in Cambridge, but also the most important. In it he collected and organized the previous knowledge about the principle and its applications; he succeeded in deriving the equations to describe the most general situations; and he formulated an approach to the problems of quantum statistical mechanics that would become very fruitful in his future work. In spite of the fact that he had used ideas that were already known, the paper on detailed balancing was Dirac's first original contribution to quantum theory.

Dirac began by stating clearly the fundamental basis of the principle of detailed balancing. In his introduction he summarized the arguments and results of Pauli's paper (1923b); he did the same for the subsequent work of Einstein and Ehrenfest (1923), which extended Pauli's treatment of systems of free electrons and radiation to arbitrary systems consisting of atoms or molecules, free electrons and radiation. Pauli had arrived at the conclusion that relativistic invariance required that the frequency of collisions between light-quanta and electrons should be identical to that of the 'inverse elementary process which arises from the first one by the reversal of time' (Pauli, 1923b, p. 280). However, Einstein and Ehrenfest had noticed that an 'inverse elementary process' might not exist in certain systems as, for instance, in the case of a hydrogen atom in a constant magnetic field (Einstein and Ehrenfest, 1923, p. 304). Therefore, in such cases they had proposed to obtain thermal equilibrium by postulating the balancing of the process of absorption of one light-quantum by an atom and the process of emission by an atom of a light-quantum having the same frequency and the *same direction*—not the opposite direction as in Pauli's 'inverse' elementary process.

Dirac took both considerations into account and defined two processes inverse to a given reaction in atomic physics, the 'converse' reaction and the 'reverse'

[121] Dirac's first four papers were submitted in a rapid sequence: the paper on dissociation under a temperature gradient (Dirac, 1924a) was received on 3 March 1924; the note on relativity (Dirac, 1924b) in April, and the note on Doppler's principle and the frequency condition (Dirac, 1924c) on 19 May 1924. After the paper on detailed balancing (Dirac, 1924d), received on 8 July 1924, there was an interruption until 19 December 1924 when his paper on the adiabatic principle was received by the *Proceedings of the Royal Society* (Dirac, 1925a).

reaction. In agreement with Pauli, he used as the fundamental basis of detailed balancing the assumption that 'all atomic processes are reversible, or, more exactly, that if after any encounter [collision] all the velocities are reversed, then the whole process would just repeat itself backwards, the systems finally leaving the scene of action being the same as the original systems in the first process and having reverse velocities' (Dirac, 1924d, p. 581). This assumption, which expressed the perfect symmetry between past and future in atomic physics, implied that each kind of encounter was just as likely to occur as its 'converse,' the 'inverse elementary process' in the sense of Pauli. But it did not follow automatically that the 'reverse' encounters, corresponding to the processes introduced by Einstein and Ehrenfest, would balance the original ones, though one would expect it. However, Dirac closed the gap. He noticed that if the assemblies were everywhere isotropic, the reverse encounter could be obtained from the converse 'by successive reflections in three mutually perpendicular mirrors at rest relative to the assembly as a whole, and so their frequency of occurrence must be equal . . . . This proves,' he argued, 'that every kind of encounter occurs' with 'the same frequency as the reverse encounter, the two together leaving unaltered both the numbers of the various kinds of systems, and also their distribution in velocity or momentum. This condition is sufficient for statistical equilibrium, and on the above-made assumption of the reversibility of the encounters, it is necessary' (Dirac, 1924d, pp. 581–582).

The clarification of what an inverse process meant in terms of mathematical operations was absolutely necessary for Dirac before he could turn to what Fowler called 'a very general formulation of the laws of detailed balancing of collisions' (Fowler, 1929, p. 453). These laws were the equations governing the thermal equilibrium between matter (consisting of different kinds of atoms, molecules, ions and electrons) and radiation; Dirac specifically treated the collisions between several kinds of atoms, taking into account radiative and nonradiative processes as well as ionization.

As we have already mentioned, the main feature of Dirac's procedure was to employ relativistic arguments. In fact, the very first result in his paper on detailed balancing was obtained in this way. He wanted to determine the behaviour of $\rho(p_x, p_y, p_z)$, the density of mass particles whose momenta lie in the interval p_x, p_y, p_z and $p_x + dp_x$, $p_y + dp_y$, $p_z + dp_z$, in an arbitrary frame of reference, that is, to determine the behaviour of $\rho(p_x, p_y, p_z)$ under a Lorentz transformation. To do so, he argued as follows: $\rho(p_x, p_y, p_z) \cdot dp_x dp_y dp_z$, the number of particles per unit volume having momenta in the infinitesimal interval of magnitude $dp_x dp_y dp_z$, transforms according to the same law as the reciprocal of a volume moving with the velocity of the particles. However, the velocity of the particles, apart from a factor c^{-1}, transforms like E, the energy associated with the particles. On the other hand, Dirac found by a straightforward calculation that $(dp_x dp_y dp_z)/(d\tilde{p}_x d\tilde{p}_y d\tilde{p}_z)$, the ratio of the differentials in two different frames of reference (the second one being denoted by the tilde superscripts), equals the corresponding energy ratio, $E/\tilde{E}$. Thus, he obtained the relation

$$\rho(p_x, p_y, p_z) = \rho(\tilde{p}_x, \tilde{p}_y, \tilde{p}_z), \tag{20}$$

which expresses the invariance of the 'density-in-momentum' of mass particles under an arbitrary Lorentz transformation.

The importance of the result contained in Eq. (20) lies in the fact that it helps to derive in a very efficient manner the equations of detailed balancing in the frame of reference that is adequate for describing the thermal equilibrium of the assemblies under consideration, while the individual processes balancing one another are most simply formulated in what Dirac called, in agreement with Pauli, the 'normal frame of reference' (Dirac, 1924d, p. 584; Pauli, 1923b, p. 277). For example, if one studies assemblies in which the processes are encounters of n atoms or atomic systems giving rise to n' atomic systems and their reverse encounters, the normal frame of reference is different for each encounter.[122] In each encounter the normal frame of reference is identical with that frame in which the center of gravity of the initial n atoms (or the final n' atoms) is at rest. In the normal frame of reference, denoted by the subscript zero, the frequency of the encounter of n particles from which n' particles emerge is given by

$$dN_0 = \rho_{10} dp_{10} \rho_{20} dp_{20} \ldots \rho_{n0} dp_{n0} \, \phi \, dp'_{10} \ldots dp'_{n'0} . \tag{21a}$$

The rth of the n particles possesses momenta between p_{rx}, p_{ry}, p_{rz} and $p_{rx} + dp_{rx}$, $p_{ry} + dp_{ry}, p_{rz} + dp_{rz}$, and the r'th of the n' particles has momenta between $p'_{r'x}$, $p'_{r'y}, p'_{r'z}$ and $p'_{r'x} + dp'_{r'x}, p'_{r'y}, + dp'_{r'y}, p'_{r'z} + dp'_{r'z}$. In writing Eq. (21a) it is assumed that the density of atomic systems in the assembly is sufficiently small, so that the different systems are outside each other's sphere of influence except during an encounter. Hence ρ_{r0} $(r = 1, \ldots, n)$, the densities of the participating particles, just multiply. Every differential factor, dp_{r0} (or $dp'_{r'0}$), is understood to represent $dp_{rx}\, dp_{ry}\, dp_{rz}$ (or $dp'_{r'x}\, dp'_{r'y}\, dp'_{r'z}$), the three-dimensional momentum differential of the corresponding system in the normal frame of reference.[123] Finally, ϕ is a probability coefficient which is determined by the momenta p_{r0} and $p'_{r'0}$ alone (and does not depend on the temperature T). Similarly, the frequency of the reverse encounter is given by the expression

$$dN'_0 = \rho'_{10} dp'_{10} \rho'_{20} dp'_{20} \ldots \rho'_{n'0} dp'_{n'0} \phi' \, dp_{10} \ldots dp_{n0} , \tag{21b}$$

with the corresponding probability coefficient ϕ'.

[122] By an 'atomic system' Dirac denoted 'any molecule, atom, or ion in any stationary state, or a free electron, or anything else that can take part in an atomic process except a quantum of radiation' (Dirac, 1924d, pp. 583–584). Two chemically identical atoms in different stationary states were counted as two different systems; hence the initial and final states could be entirely different, although they were composed of the same constituents. Thus, an electron and an ion might recombine in the encounter to yield a neutral atom, or two molecules might form one molecule, etc.

[123] Although p_r and $p'_{r'}$, the momenta of the particles before and after the encounter, are not independent of each other but restricted by the condition of energy and momentum conservation (and possibly even more by other symmetry requirements), this fact does not play a role in the resulting equation of detailed balancing because the differentials for the encounters and reverse encounters are restricted in the same way. Dirac therefore decided just as well to leave out the restrictions.

The principle of detailed balancing in isotropic assemblies demands the equality of the frequencies dN_0 and dN_0'; hence one obtains the equation

$$\rho_{10}\rho_{20}\ldots\rho_{n0}\phi = \rho_{10}'\rho_{20}'\ldots\rho_{n'0}'\phi', \tag{22}$$

in the normal frame of reference of the encounters. However, due to Eq. (20) and the fact that the probability coefficients ϕ and ϕ' do not depend upon the velocity of the centre of gravity of the colliding particles, Eq. (22) is valid in all frames of reference and the zero suffixes can be dropped. In particular, Eq. (22) holds for that frame in which the assembly as a whole is at rest and in statistical equilibrium; in this case the distribution of momenta is given by Maxwell's law, that is,

$$\rho_r = A_r \exp\left\{-\frac{(E_r+\epsilon_r)}{kT}\right\} \quad \text{and} \quad \rho_{r'}' = A_{r'}'\left\{-\frac{(E_{r'}'+\epsilon_{r'}')}{kT}\right\}, \tag{23}$$

where E_r and $E_{r'}'$ are the internal, and ϵ_r and $\epsilon_{r'}'$ the kinetic energies of the systems denoted by r and r', respectively, at temperature T, and k is the Boltzmann constant. Thus, the detailed balancing equation (22) reduces to the equation

$$A_1A_2\ldots A_n\phi = A_1'A_2'\ldots A_{n'}'\phi', \tag{24}$$

for the temperature-dependent coefficients A_r and $A_{r'}'$ of the Maxwellian distribution.[124]

Equation (24) represents the most general equation for detailed balancing in assemblies in which only mass-particles participate in reactions establishing the equilibrium. Dirac used it at once to derive an important result of statistical mechanics. For this purpose he introduced the concentrations c_r (or $c_{r'}'$) of the atomic systems in an assembly, defined as

$$c_r = A_r \exp\left\{-\frac{E_r}{kT}\right\}\int\int\int \exp\left\{-\frac{\epsilon_r}{kT}\right\} dp_r \tag{25}$$

and found that the following equation was satisfied:

$$\frac{d}{dT}\ln\left\{\frac{c_1'c_2'\ldots c_{n'}'}{c_1c_2\ldots c_n}\right\} = \frac{1}{kT^2}\left\{\sum_{r'=1}^{n'}(E_{r'}'+\bar{\epsilon}_{r'}') - \sum_{r=1}^{n}(E_r+\bar{\epsilon}_r)\right\}; \tag{26}$$

here $\bar{\epsilon}_r$ and $\bar{\epsilon}_{r'}'$ are the average kinetic energies ($=\frac{3}{2}kT$) of the atomic systems r and r', respectively. Since the expression within the curly bracket on the left-hand

[124]Note that Eq. (24) is true only if the systems obey Eq. (23), that is, the Maxwell–Boltzmann distribution for the momentum. For then, the products of the densities, which occur in Eq. (22), contain the factors $\exp\{-\sum_{r=1}^{n}(E_r+\epsilon_r)/kT\}$ and $\exp\{-\sum_{r'=1}^{n'}(E'_{r'}+\epsilon_{r'}')/kT\}$ on opposite sides, respectively; because of energy conservation they are identical and may be dropped.

side is usually called K, the 'equilibrium constant' of the reaction, and the energy difference within the bracket on the right-hand side is called the 'heat of reaction per molecule,' Eq. (26) is just the well-known Van't Hoff 'isochore' equation for chemical reactions.[125] Dirac had made use of this equation in his first research paper on dissociation (Dirac, 1924a, p. 134), and he was happy that it now followed from general considerations of detailed balancing and had a much wider range of application than he had assumed. He found it 'interesting to notice that this equation is still exactly true with relativity mechanics' (Dirac, 1924d, p. 587). This result again convinced Dirac of the fact that the fundamental equations of physics were either also valid under relativistic conditions or they could easily be generalized for that purpose.

On completing the treatment of assemblies with material particles, Dirac went on to discuss those assemblies in the equilibrium of which the emission and absorption of radiation plays a non-negligible part. Simple cases of such assemblies, namely, those in which only the encounters between free electrons and radiation of a given frequency take place, had already been discussed in Pauli's paper (Pauli, 1923b) which, to a large extent, had inspired Dirac's procedure. In that paper Pauli had also formulated the equation of detailed balancing of the elementary processes and their 'inverse' processes in what he called the 'normal frame of reference,' that is, in a frame in which the momenta of the radiation and the free electron are of the same magnitude but opposite directions.[126] In the more general case now considered by Dirac, n atomic systems—by absorbing light of appropriate frequencies—give rise to n' atomic systems. In this case, the 'normal' frame of reference can be defined as that frame in which either the three-momenta of the initial particles and radiation, or the momenta of the final particles leaving the point of collision, add up to zero. In this frame of reference it seemed that 'the most natural assumptions to make are that the frequency of the absorption process is directly proportional to the intensity of the radiation of the right frequency and in the right direction to be absorbed, and that the frequency of the reverse emission process is independent of the intensity of the

[125]An 'isochore' is a curve on the pT-diagram of a thermodynamical system, the pressure and temperature being variable, the volume remaining constant.

In simple cases the equilibrium constant K can be calculated very easily. For example, K is unity if all the collisions taking place in the assembly leave the atomic systems in the same state. In a process in which radiation-free collisions between electrons and atoms (i.e., collisions of the 'first' and 'second' kind) occur, which cause a transition of the atoms from a state E_1 to E_2 and back, the equilibrium constant becomes

$$K = \frac{q_2}{q_1} \exp\left\{ -\frac{E_2 - E_1}{kT} \right\},$$

where q_1 and q_2 denote the statistical weights of the corresponding states.

Equation (26) can be very useful in practical calculations, because if K and one of the probability coefficients (say ϕ, of the reaction under investigation) are known, one can obtain the probability coefficient of the inverse reaction.

[126]This 'normal' frame had also been employed earlier by Schrödinger in his note on the Doppler principle and the frequency condition (Schrödinger, 1922a), which Dirac had studied earlier.

incident radiation $[I_0(\nu_0)]$' (Dirac, 1924d, p. 589). Equation (22) of detailed balancing for pure material systems had then simply to be generalized to the equation

$$I_0(\nu_0)\rho_{10}\rho_{20}\ldots\rho_{n0}\phi=\rho'_{10}\rho'_{20}\ldots\rho'_{n'0}\phi', \tag{27}$$

where ϕ and ϕ' are the probability coefficients of the absorption and emission processes, respectively. However, as against Eq. (22), Eq. (27) is not valid in an arbitrary Lorentz frame of reference, for only the product $I_0(\nu_0)\nu_0^{-3}$, and not just $I_0(\nu_0)$, is a relativistic invariant, as Dirac already knew from Pauli's paper (Pauli, 1923b, p. 282). Therefore, in the case that radiation of frequency ν and intensity $I(\nu)$ is absorbed, Eq. (24) for material systems, which is satisfied in an arbitrary frame of reference, is generalized to the equation

$$I(\nu)\nu^{-3}\nu_0^3\exp\left\{\frac{h\nu}{kT}\right\}A_1A_2\ldots A_n\phi=A'_1A'_2\ldots A'_{n'}\phi', \tag{28}$$

where the A's and A''s are again the temperature-dependent coefficients of the Maxwellian distributions of the atomic systems, given by Eq. (23).

From Eq. (28) Dirac concluded at once that the factor $I(\nu)\nu^{-3}\exp\{h\nu/kT\}$ must be a relativistic invariant; due to Wien's displacement law it must also be independent of the temperature. Hence the equilibrium distribution of radiation, given by the equation

$$I(\nu)=(\text{const.})\nu^3\exp\left\{-\frac{h\nu}{kT}\right\}, \tag{29}$$

seemed to obey Wien's law. Since this result contradicted the actual situation, where Planck's radiation law holds rather than Eq. (29), Dirac had to assume with Pauli (1923b) that the right-hand side of Eq. (27) does not balance the left-hand side and that the reverse process is also stimulated by the incident radiation. In order to ensure Planck's law for the radiation distribution he multiplied the right-hand side of Eq. (28) by the factor $[1+\alpha I_0(\nu_0)]$, where α $(=c/h\nu_0^3)$ is Einstein's 'stimulation' coefficient. The equation for detailed balancing involving the absorption of radiation then became[127]

$$\frac{2h\nu_0^3}{c^2}A_1A_2\ldots A_n\phi=A'_1A'_2\ldots A'_{n'}\phi'. \tag{30}$$

This equation does not depend on the frame of reference, and the only reminder of the normal frame of reference is contained in the occurrence of the frequency ν_0 on the left-hand side.

[127] Equation (30) follows from the equation of detailed balancing by replacing the momentum densities of the mass systems with the help of Eq. (23) and assuming that the radiation obeys Planck's law.

Further generalization to atomic processes, in which l light-quanta of frequencies $\nu_1, \ldots, \nu_l$ are absorbed and l' light-quanta of frequencies $\nu'_1, \ldots, \nu'_{l'}$ are emitted, is straightforward. Dirac wrote the corresponding equation of detailed balancing as

$$\prod_{i=1}^{l}\left(\frac{2h\nu_{i0}^3}{c^2}\right)A_1A_2\ldots A_n\phi = \prod_{i'=1}^{l'}\left(\frac{2\pi\nu_{i'0}'^3}{c^2}\right)A'_1A'_2\ldots A'_{n'}\phi'. \tag{31}$$

That is, for each light-quantum of frequency ν_0 absorbed in the encounter of the n atomic systems, one has to multiply the left-hand side of Eq. (24), the equation for the radiation-free process, with a factor $2h\nu_0^3/c^2$; and for each light-quantum of frequency ν'_0 emitted together with the n' atomic systems, one has to multiply the right-hand side of Eq. (24) with a factor $2h\nu_0'^3/c^2$.

Since Dirac had essentially followed Pauli's argumentation, it is evident that Eq. (31) reduces to Pauli's result in the case of collisions between free electrons and radiation (Pauli, 1923b, p. 281, Eq. (26)). If only one atom participates in a reaction absorbing and emitting radiation, the result of Einstein and Ehrenfest (1923, p. 305, Eq. (14)) is reproduced. While these conclusions merely demonstrated the fact that Dirac had consistently generalized the results of his predecessors, he emphasized another point which became clear in his derivation: If, for an assembly absorbing one frequency of radiation—denoted by ν_0 in the normal frame of the reaction—one investigates all possible changes of Eq. (28) (the original equation of detailed balancing, which does not contain stimulated emission), changes that lead to Planck's law instead of Wien's law, then 'by the present method it can easily be shown that there are no reasonable alternative assumptions [to the ones he had made]' (Dirac, 1924d, p. 592). This may be seen already in the simplest case in which one light-quantum of frequency ν_0 is absorbed. If the right-hand side of Eq. (28) is multiplied by any factor other than $[1 + \alpha I_0(\nu_0)]$, the probability of the reaction will be a logarithmic function of the intensity of the incident radiation, and Dirac argued that 'it is very unlikely that a probability would be of that form' (Dirac, 1924d, p. 593). Therefore he felt confident in stating the 'general law' that: '*Any atomic process which results in the emission of one or more quanta of radiation is stimulated by external incident radiation of the same frequency as that of any of the emitted quanta, the ratio of the amount of stimulated emission to the amount of spontaneous emission being proportional to the intensity of the incident radiation divided by the cube of its frequency and independent of the nature of the process concerned, and the direction of the stimulated radiation being the same as that of the incident radiation*' (Dirac, 1924d, p. 593).[128]

[128] Dirac disagreed with Otto Halpern's treatment of the equilibrium conditions for simple emission and absorption. Halpern (1924) had assumed the probability coefficients as depending on the temperature, which contradicted the principles. Dirac argued: 'The conclusions to which he comes, that the stimulated radiation need not be emitted in the same direction as the stimulating radiation, can hardly be correct, since unless they are in the same direction, their frequencies will not be equal in all systems of coordinates' (Dirac, 1924d, p. 592, footnote †).

In his paper on detailed balancing Dirac derived numerous consequences from his equations. For example, for assemblies consisting of neutral atoms, ionized atoms and free electrons, he obtained by integrating Eq. (26) Saha's equation for the fraction of ionized atoms (Saha, 1920). By including several stationary states of the atoms, Dirac obtained the generalized ionization equation, given earlier by Fowler (1923a, p. 21, Eq. (7.4); see our Eq. (12)). All these practical results were already known; however, Dirac gave them a deeper foundation in a general scheme.

For Dirac another aspect was even more important: this was the fact that he had to search for the first time for equations in which matter and radiation appeared simultaneously. He recognized, especially by using relativistic arguments, that he could treat them consistently. However, at the same time he discovered that radiation exhibits properties which distinguish it from material particles. For example, he had had to consider assemblies in which the reactions included the emission and absorption of radiation, as well as the stimulated transition of atomic systems—a process which did not seem to have any analogue in the encounters of mass-particles alone.[129] He also called attention to the fact that the concentration (i.e., the intensity) of radiation in equilibrium was completely fixed—for a given temperature and frequency—by Planck's law, while the concentration of mass-particles was largely arbitrary.[130] These results seemed to indicate that radiation and matter *did not* obey the *same* laws, contrary to a suggestion that had been made by Louis de Broglie. Although Dirac did not explicitly mention de Broglie's paper, which had been published in February 1924 in *Philosophical Magazine* (de Broglie, 1924a), he had it in mind when he wrote: 'For the discussion of equilibrium problems, quanta of radiation cannot be regarded as very small particles of matter moving with nearly the speed of light' (Dirac, 1924d, p. 594).

Thus, in summer 1924 Dirac came to the conclusion that a unified treatment of radiation and matter, as seemingly encouraged by relativity theory, was not possible. Still, his paper on detailed balancing was a landmark in seeking to achieve this unification, and he would contribute substantially towards it in the following years. A year later Dirac would adopt a more favourable attitude

[129]Dirac was convinced about the importance of stimulated emission. For example, he argued that by it the disintegration of any radioactive substance emitting γ-rays should be increased, provided external radiation of the same frequency is present. At the same time, however, he was aware that 'the effect will be very small owing to the fact that the stimulation coefficient varies inversely as the cube of the frequency' (Dirac, 1924d, p. 593).

[130]In assemblies, in which only radiation-free collisions between nonrelativistic particles of masses $m_1, m_2, \ldots, m_n$ and $m'_1, m'_2, \ldots, m'_n$ occur, the concentrations have to satisfy the equation

$$K = \frac{\phi}{\phi'} \left(\frac{m'_1 m'_2 \ldots m'_{n'}}{m_1 m_2 \ldots m_n} \right)^{3/2} (2\pi kT)^{3(n'-n)/2} \exp\left\{ \sum \frac{E_r - E_{r'}}{kT} \right\},$$

which is obtained by integrating Eq. (26) and taking the nonrelativistic approximation. The absorption and emission of radiation of frequencies ν_0 and ν'_0, respectively, in the normal frame of the reaction, contribute the multiplicative factors $2h\nu_0^3/c^2$ and $((2h\nu'_0)^3/c^2)^{-1}$ to the right-hand side of this equation; in addition, an energy-dependent factor, $\exp\{1/kT)(\sum_r E_r - \sum_{r'} E'_{r'})\}$, occurs that takes into account the fact that the total energy of the mass-particles may be altered by the reaction.

towards de Broglie's hypothesis and the idea of dealing with material particles and light-quanta on the same footing. And after the advent of quantum mechanics he would continue to think about this unification and develop what would become quantum field theory. In the meantime—in summer 1924—he eagerly studied the literature of quantum theory in the hope of finding some suitable research problems on which he could work.

III.3 An Extension of the Adiabatic Principle

On completing his paper on detailed balancing Dirac became deeply involved in quantum theory. He carefully studied Niels Bohr's review article 'On the Application of the Quantum Theory to Atomic Structure' in the supplement to *Proceedings of the Cambridge Philosophical Society* (English translation of Bohr, 1923a). Upon the advice of Ralph Fowler he found a fundamental problem to work on for the next few months. On 19 December 1924, *Proceedings of the Royal Society* received his paper entitled, 'The Adiabatic Invariance of the Quantum Integrals' (Dirac, 1925a). In this work he succeeded in removing the most serious restriction that had existed hitherto in applying the adiabatic principle to quantum systems.

In his review article Bohr had given a detailed discussion of the adiabatic principle, assigning to it a role almost equal to that of the correspondence principle in discovering the true laws of quantum theory. 'The significance of the Adiabatic Principle in the quantum theory is extraordinarily great,' Bohr had written, 'since it leads to the elucidation and development of formal methods for fixing the stationary states' (Bohr, 1923a; English translation, p. 13). That is, the stationary states of an atomic system have to be defined in such a way that certain properties of the motion, the so-called 'adiabatic constants,' do not change when the system undergoes an adiabatic transformation, which might be brought about by an infinitely slow variation of the external forces acting on the atomic system. In practice, the adiabatic principle can be used in the following way: Given a simple system with independent degrees of freedom, say a multidimensional harmonic oscillator which is quantized according to Planck's prescription; then it can be turned by an adiabatic transformation into a general multiply periodic system with appropriate quantum conditions that may be used to describe a realistic atom.

The adiabatic principle had another important application, namely, that it 'can be used to determine the statistical weights of the stationary states of a given atomic system, if the weights of the states of a system are known, which can be transformed continuously into the given one' (Bohr, 1923a; English translation, pp. 16–17). The actual calculation of statistical weights interested Fowler immensely, and it was he who encouraged Dirac to work on the adiabatic principle.[131]

[131] Dirac acknowledged Fowler's role in his work on the adiabatic principle at the end of his paper. He stated: 'The writer is much obliged to Mr. R. H. Fowler for suggesting this investigation, and for his help during its progress' (Dirac, 1925a, p. 734).

The difficulty to which Dirac addressed himself, and which served as his introduction to the problems of atomic theory, was also mentioned in Bohr's review. Bohr had noted that 'the application of the [adiabatic] principle is naturally limited by the demand, that the motion of the system, if it is to be described by the use of classical laws [of mechanics and electrodynamics], shall exhibit at each moment during the transformation the periodic properties which are necessary for the fixation of the stationary states, and that the degree of periodicity remains unaltered during the transformation' (Bohr, 1923a; English translation, p. 12). The last demand could not really be satisfied, as Johannes Martinus Burgers, who had earlier discussed the applicability of the adiabatic principle in detail, had shown. Burgers had found that if one starts from a periodic system with frequencies $\omega_1, \ldots, \omega_f$, which are not commensurable, and subjects the system to an adiabatic change, it will pass infinitely often through states for which the equation

$$\sum_{k=1}^{n} m_k \omega_k = 0, \qquad m_k = \text{integers}, \tag{32}$$

holds. The reason is that the frequencies ω_k are continuous functions of those parameters, a, which are varied infinitely slowly in the adiabatic change. Burgers had concluded that: 'It still has to be investigated whether this may give rise to difficulties' (Burgers, 1917; English translation, p. 169).

Yet no serious investigation had been undertaken to determine as to what properties atomic systems should possess in order that adiabatic principle should apply, until Ralph Fowler proposed to Dirac to study the question. The fact that a fundamental principle was at stake attracted Dirac, and he set out to obtain 'conditions which are rigorously sufficient to ensure the invariance of quantum integrals' and to write them 'in such a form that it is possible to see whether they are satisfied or not without having to integrate the equations of adiabatic motion' (Dirac, 1925a, p. 725). That is, he wanted to give a general criterion for the validity of the adiabatic principle.

As was his custom in those days, Dirac first studied the available literature on the problem. In this case it was the papers of Burgers, to which Bohr had referred, especially his third communication to *Proceedings of the Amsterdam Academy* (Burgers, 1917). Dirac then reformulated the results in his own language and proceeded to improve upon them. This method led him first to derive what he called the 'exact equations of adiabatic motion' (§2 of his paper). With these, by using careful mathematical estimates, he found the conditions for adiabatic invariance (§3), and in the last section he showed how these conditions could be applied to problems in atomic physics.

Dirac considered a multiply periodic system of n degrees of freedom, described by $2n$ Hamiltonian coordinates—i.e., the dynamical variables q_r and p_r—which change periodically with time. In the Hamiltonian of this system, besides the q's and p's, there occur parameters denoting the masses, charges, or the intensity of a field of force imposed upon the system. Burgers, in his first paper (Burgers, 1916a; English translation, p. 151), had defined a reversible adiabatic change of the system in terms of a variation of one of the parameters, say a, having the following properties: (i) da, the change of a over many periods of the motion, is small; (ii) $\dot{a}$, the time derivative, is approximately constant; (iii) the change occurs in such a way that the Hamiltonian equations of motion remain valid. Dirac sharpened requirements (i) and (ii) mathematically by defining an adiabatic change as the limiting case of a slow variation of a, with[132]

$$\dot{a} \to 0 \qquad \text{such that } \frac{\ddot{a}}{\dot{a}} \to 0. \tag{33}$$

Burger's requirement (iii) is evidently satisfied if at any instant during the adiabatic change—i.e., for any value which the parameter a assumes during its variation—there exists a contact transformation of the periodic variables, q_r and p_r, to action-angle variables, generated by the transformation S which depends on the q's, p's and on a. If the original Hamiltonian function can be split into a part, H_0, which is constant under the change of the parameter a, and a part, H_1, which varies proportionally to $\dot{a}$, plus terms proportional to higher powers of $\dot{a}$, then the transformed Hamiltonian, $\overline{H}$, also contains a term linear in $\dot{a}$. $\overline{H}$ may be written, up to terms quadratic in $\dot{a}$ or depending on higher powers, as

$$\overline{H} = H_0 - \dot{a}F, \tag{34a}$$

where

$$F = \sum_{r=1}^{n} p_r \frac{\partial q_r}{\partial a} - \frac{\partial S}{\partial a} - H_1. \tag{34b}$$

Thus, the Hamiltonian of the system remains unaltered in the limit $\dot{a} \to 0$, where both $\overline{H}$ and H become equal to H_0.

However, for discussing adiabatic changes, equations have to be used in which $\dot{a}$ is finite. These 'exact' equations of adiabatic motion for the action variables J_k and the angle variables w_k ($k = 1, \ldots, n$) are (see Dirac, 1925a, p. 727)

$$\frac{dJ_k}{da} = \frac{\partial F}{\partial w_k} \tag{35a}$$

and

$$\frac{dw_k}{da} = \frac{\omega_k}{\dot{a}} - \frac{\partial F}{\partial J_k}, \tag{35b}$$

where ω_k is the time derivative of the angle variable w_k, giving the frequency of

[132] Dirac was aware of the fact that the limit $d\dot{a}/da \to 0$ could not be satisfied in the beginning and at the end of the adiabatic change. However, he suggested instead that Eqs. (33) could be satisfied during most of the process and that the interval during which the second limit is not zero could be made arbitrarily small.

the kth degree of freedom. Equations (35) were not new when Dirac wrote them down. Burgers had been able to derive them, but only on the assumption that H does not contain $\dot{a}$ explicitly. Burgers had further suggested that: 'Probably it will be found sufficient that in passing from $a = \text{const.}$ to $a =$ a given function of time, the Hamiltonian equations remain unchanged only if we neglect terms of the second and higher orders in $\dot{a}$. This has yet to be investigated' (Burgers, 1917; English translation, p. 169). Now Dirac had removed that restriction, for he had made a special point of dealing with Hamiltonians H which contain a term linear in $\dot{a}$. For practical purposes this generalization of the application of the equations of adiabatic motion, Eqs. (35), was a very useful, even necessary step.

The principal aim of Dirac's work on the adiabatic principle was, however, to answer another question raised by Burgers: that is, to what extent does *an intermediate occurrence of degeneracy* in systems, in which one parameter a is changed infinitely slowly from a value a_1 to another value a_2, affect the adiabatic constants, the J_k's? To obtain the answer Dirac again followed and improved upon the methods outlined by Burgers (1917). He began by representing the function F, given in Eq. (34b), whose partial derivative with respect to the angle w_k determines the change of J_k, by an n-fold trigonometric sum, that is,

$$2\pi F = \sum_m C_m \sin\{2\pi(W_m + \gamma_m)\}. \tag{36}$$

The index m in the sum on the right-hand side of this equation actually denotes n indices $m_1, \ldots, m_n$, all of which assume integral values such that the quantity W_m $(= m_1 w_1 + \ldots + m_n w_n)$ remains a positive number. The γ_m $(= \gamma_{m_1 \ldots m_n})$ are arbitrary phases; the Fourier coefficients C_m $(= C_{m_1 \ldots m_n})$ are functions of the action variables and of the parameter a and its time derivative $\dot{a}$, *all higher powers of $\dot{a}$ being neglected in the calculations*. This representation of F is possible, as all quantities on the right-hand side of Eq. (34b) are periodic in the angle w_k.[133] Starting from this basic assumption of Burgers', Dirac sought to obtain an expression for the change of the adiabatic constants that occurs during an arbitrary adiabatic transformation, which includes the passage through the values of the parameter a where Eq. (32) is satisfied. In particular he attempted to isolate in this expression each finite term and to investigate the conditions leading to a finite change of the action variables.

For this purpose Dirac divided the entire range, say the interval from a_1 to a_2, within which the parameter a increases adiabatically, into very small sections.[134] Considering the section in which a goes from a' to $a' + \delta a$ and using careful

[133] The first and third terms on the right-hand side of Eqs. (34b) are already periodic in the angles w_k, hence one has only to show that the action function S has also the same property. Now S is a sum of terms all of which are periodic in the w_k, except one, which depends on the action and angle variables. The latter term can be written as $-\sum_{k=1}^{n}(J_k + C_k)w_k$, with arbitrary constants C_k; by choosing each C_k to be identical with minus the corresponding action variable, it becomes zero.

[134] It can be assumed without any loss of generality that the adiabatic parameters a always increases during the entire course of the adiabatic change. If not, one may divide the whole adiabatic range into parts during which the condition pertaining to the increasing parameter is satisfied. In

estimates, Dirac obtained for the corresponding change of J_k the result

$$\delta J_k = \sum_{m=1}^{M} m_k C_m \int_{a'}^{a'+\delta a} \cos\{2\pi(W_m + \gamma_m)\}\, da. \tag{37}$$

That is, the sum over the indices m (of which m_k is special) goes only up to a large, but finite positive integer M, because the higher terms give a negligible contribution if the quantities δa and M^{-1} are small enough. He further showed that those terms in the right-hand side of Eq. (37), for which the inequality

$$\left| \frac{d \sum_{k=1}^{n} m_k w_k}{da} \right| = \left| \sum_{k=1}^{n} m_k \omega_k \frac{1}{\dot{a}} \right| > \frac{M^{1/2}}{\eta}, \tag{38}$$

with a given number η, holds, alter J_k by an amount having the order of magnitude

$$\delta J_k = \delta a \left\{ \epsilon\left(\frac{\eta^2}{\dot{a}} \right) + \epsilon\left(\frac{\eta^2 \ddot{a}}{\dot{a}^3} \right) + \epsilon\left(\frac{\eta}{\delta a} \right) \right\}. \tag{39}$$

This change can be made arbitrarily small if η tends to zero in such a way that the quantities $\eta^2/\dot{a}$, $\eta^2\ddot{a}/\dot{a}^3$ and $\eta/\delta a$ also go to zero. If, in addition, M is increased so slowly that in the adiabatic limit ($\dot{a} \rightarrow 0$), $(M^{1/2}/\eta)\,\dot{a}$ also becomes zero, then the only contribution to δJ_k stems from terms satisfying Eq. (32). Thus, by using the rigorous mathematical methods he had learned from Peter Fraser in Bristol, Dirac was able to confirm Burgers' result that all transformations of a periodic system, in the course of which an equation of the type of Eq. (32) never holds, leave the action variables J_k constant.

The next problem was to see whether Burgers' conditions for the validity of the adiabatic principle could not be relaxed. Dirac, therefore, studied cases of adiabatic transformations in which for certain values, say a_0, which the parameter a assumes during its change, Eq. (32) is satisfied for specific sets of integers. Let $(m_1^s, \ldots, m_n^s)$ be such a set, then he showed—again by a careful analysis of the mathematical limits involved—that the right-hand side of Eq. (37) is zero for those systems and those adiabatic changes for which the inequality

$$\left| \sum_{r=1}^{n} m_r^s \frac{\partial \omega_r}{\partial a} \right| > \sum_m \left| C_m \sum_{r,k=1}^{n} m_k m_r^s \frac{\partial \omega_r}{\partial J_k} \right| \tag{40}$$

holds at each point a_0 where the system becomes degenerate. The n-fold sum of the terms C_m on the right-hand side of Eq. (40) extends over *all* sets of integers,

those parts of the transformation, where the parameter decreases, one introduces a new parameter which is negative of the old parameter.

With the condition of a monotonously increasing adiabatic parameter being satisfied, Dirac found immediately from Eqs. (35a) and (36) that the absolute value of the differential quotient dJ_k/da remains bounded (by $\sum_m |m_k C_m|$) as $\dot{a}$ tends to zero. He therefore concluded that J_k can alter its value only within finite limits.

$(m_1, \ldots, m_n)$, for which Eq. (32) is satisfied with the parameter a assuming the value a_0.[135] Inequality (40) is actually a bit too restrictive, for in the exact evaluation each term C_m on the right-hand side is multiplied by a factor $\cos\{2\pi(W_m + \gamma_m)\}$, which is mostly smaller than unity. But this was all the sharpening of the inequality that Dirac could admit and he remarked: 'If the improved conditions are not satisfied, the J_k can vary with suitable initial values of the phases of the motion, and will do so in such a way as to make $\sum m_r^s \omega_r$ remain equal to zero while a changes within a finite amount. As this contradicts the postulate of the existence of stationary states, one must conclude that in these cases the motion cannot be completely described by the use of classical mechanics' (Dirac, 1925a, p. 732).[136]

In applying his results to practical cases, Dirac encountered the difficulty that he would have to check the validity of infinitely many inequalities of the type (40). The reason was that for a given physical system and a given value a_0 of the varying parameter a, there seemed to be infinitely many sets of integers, $(m_1^s, \ldots, m_n^s)$, for which the frequencies satisfied the degeneracy equation; and each such set had to be inserted into the inequality. Yet, for atomic systems of interest the situation was not so desperate. For example, in the case of a periodic system of two degrees of freedom Dirac was able to invent a procedure that worked. In particular he showed that inequality (40) could be violated only by *finitely many* values of the index pair (m_1^s, m_2^s), if $(a', a' + \delta a)$, the considered interval of values of the changing parameter a, contains no point at which the corresponding frequencies, ω_1 and ω_2, and their differential coefficients satisfy the relation[137]

$$\omega_1 : \omega_2 = \frac{\partial \omega_1}{\partial a} : \frac{\partial \omega_2}{\partial a} . \tag{41}$$

Hence it is possible to check the validity of the inequalities in practice in twofold periodic systems. For atomic systems having more than two degrees of freedom similar arguments could not be made. However, Dirac noticed another point

[135] Dirac argued as follows: If condition (40) is satisfied, then the variation of $|\sum_{r=1}^{n} m_r^s \omega_r|$ is bounded from below by the product $\delta a \cdot M_s \cdot b_s$, where M_s is the largest number of the set $(m_1^s, \ldots, m_n^s)$ and b_s a positive number which goes to zero as M_s approaches infinity. Thus, the value of the sum, $|\sum_{r=1}^{n} m_r^s \omega_r|$, remains smaller than a given number, say $\zeta = M_s^{1/2}(\dot{a}/\eta)^{-1}$, only for those values of the parameter a which lie in the interval between $a_0 - \zeta^{-1} b_s^{-1}$ and $a_0 + \zeta^{-1} b_s^{-1}$. This interval is also the region from which a finite contribution to the integral on the right-hand side of Eq. (37) must be expected. However, the size of this interval, $2\zeta^{-1} b_s^{-1}$, can be made arbitrarily small provided one observes appropriate limiting procedures. Hence the adiabatic change of J_k will be zero as long as inequality (40) remains valid.

[136] The situation is, of course, different if the multiply periodic system is always degenerate, i.e., when relations of the type (32) hold identically for all values of the J_k. One can then introduce a smaller set of action-angle variables, J_l and w_l ($l = 1, \ldots, n'$; $n' < n$), which are not degenerate. In this case also one is able to prove the adiabatic invariance of the J_l if inequalities of the type (40) hold for the new variables.

[137] Evidently, Eq. (41) is equivalent to the two equations, $\omega_1 + x\omega_2 = 0$ and $\partial(\omega_1 + x\omega_2)/\partial a = 0$, which have a simultaneous solution for the unknown variable x (rational or irrational) only for finitely many values of a, say $\tilde{a}_1$, $\tilde{a}_2$, etc. Now, if the interval, $(a', a' + \delta a)$, excludes these points, the left-hand side of inequality (40) will always have a finite lower bound, say $M_s \xi > 0$, with M_s the

which helped: in all cases usually encountered in quantum theory, C_m, the Fourier coefficients of the series, Eq. (36), vanish for all but a finite number of suffixes, $m_3, m_4, \ldots, m_n$, because the 'selection principles allow only a finite number of changes for each of the quantum numbers except at most two of them.' He concluded therefore that: 'The quantum integrals of such systems are invariant under any adiabatic changes except at a finite set of points where $\omega_r : \omega_s = \partial\omega_r/\partial a : \partial\omega_s/\partial a$, ω_r and ω_s being the frequencies corresponding to the quantum numbers whose changes are unrestricted, and another finite set (or possibly infinite enumerable set tending to points of the previous set) where relations of the type $\sum m_r\omega_r = 0$ hold' (Dirac, 1925a, p. 734).

In his paper on the adiabatic invariance of quantum integrals Dirac had done more than answer the questions raised by Burgers. First, he had established the precise conditions which could be tested practically in those cases in which one sought to apply the adiabatic principle—as, for example, in order to obtain the statistical weights of states, a problem in which Fowler was interested. Second, he had arrived at the conclusion that the adiabatic principle was valid for most atomic systems that had been considered; a real breakdown occurs only in the adiabatic transformations of a system in which the varied parameter a assumes values $\tilde{a}_0$ such that an equation of the type (41) is satisfied. The physical interpretation of this situation is straightforward: In the course of an adiabatic change, which conserves the values of the action variables J_k, the system may pass through states of degeneracy provided they are left immediately; if, on the other hand, the points $\tilde{a}_0$ are reached, the system will stay too long in the degenerate state and, therefore, the values of J_k will be altered.

Max von Laue arrived at the same result as Dirac's in a paper presented on 30 January 1925 at a meeting of the German Physical Society in Berlin (von Laue, 1925). In particular, von Laue showed that if the sum, $\sum m_k \omega_k$, can be expanded around a_0, the value of the parameter at which degeneracy is achieved, in a series containing the positive powers of the difference $(a - a_0)$, then the adiabatic invariants J_k remain constant. This condition of von Laue's and Dirac's inequality (40) expressed the same property of the system under investigation. Richard Becker took note of this fact when he reviewed the papers of von Laue and Dirac for the *Physikalische Berichte*, but he closed his report by saying: 'A closer analysis of the rather intricate considerations would be superfluous, for these [i.e., Dirac's considerations] can be regarded as being superseded by the much simpler and more general proof of von Laue' (Becker, 1925, p. 1480).

---

largest of the m_r^s. (For the values, $\tilde{a}_1$, $\tilde{a}_2$, etc., this lower bound is zero.) Inequality (40) is then satisfied if the quantity ξ is larger than the right-hand side of Eq. (40) divided by the integer M_s; and if it is not satisfied, the relation

$$\xi \leqslant \sum_m \left| C_m \sum_{r,k=1}^{2} m_k \frac{m_r^s}{M_s} \frac{\partial\omega_k}{\partial J_k} \right|$$

holds. Dirac argued that in case there were infinitely many pairs, (m_1^s, m_2^s), for which the last relation is valid, the quantity ξ and, therefore, also the lower bound, $M_s\xi$, must be zero in contrast to the initial statement.

Dirac, however, felt differently about the merits of his work. 'I thought that was an unfair comment on my paper,' he recalled, 'and the person who wrote that summary had not appreciated the difference between mine and the other work. Mainly in mine you could check whether the conditions are fulfilled or not, without having to integrate the adiabatic equations, while in the other paper [von Laue's] one did have to integrate the adiabatic equations. It was the only time I have been dissatisfied with the comments which someone had made on my work' (Dirac, Conversations, p. 67).[138]

Dirac had indeed no reason to be discontent with his paper on the adiabatic principle. He had performed a good piece of work, in which he had skillfully handled rigorous mathematical methods. He had given a complete and definite solution of the problem that had been around for many years. Moreover, this solution helped in establishing further the Hamiltonian scheme in atomic physics. Dirac was happy to use the Hamiltonian methods which he had learned from Whittaker's *Analytical Dynamics*; he liked the Hamiltonian scheme because it was based on a fundamental principle and led to equations which could be solved under clearly defined conditions. And he regarded the adiabatic principle, whose application allowed one to introduce certain order and regularity in the otherwise chaotic status of the dynamics of atomic systems, as a part of that scheme. Dirac believed that by a systematic application of the Hamiltonian methods he would be able to approach more and more complex problems of atomic theory. He was supported in this view by his supervisor, Ralph Fowler, as well as the literature he frequently referred to, such as Sommerfeld's *Atombau und Spektrallinien*. He continued his interest in the adiabatic principle and in November 1925, nearly a year after his first paper on it, he submitted a second paper dealing with its application to atomic systems in magnetic fields (Dirac, 1925c).[139]

III.4 Growing Involvement in Quantum Physics

If one compares the number and variety of Dirac's publications in 1924 with those during the following year until November, one might think that Dirac had become less productive. There was suddenly a marked contrast with his first year

[138] At that time the papers in physics, particularly those on quantum theory were reviewed quite regularly in the German journal *Physikalische Berichte*. Becker's review of the paper on the adiabatic principle was not the first of Dirac's work. Earlier, Cornelius Lanczos had given a short report about Dirac's note concerning relativity dynamics (Dirac, 1924b) in *Physikalische Berichte* (1924, p. 1712). Gregor Wentzel reviewed Dirac's third and fourth papers on quantum mechanics (Dirac, 1926b, c) in the *Physikalische Berichte* (1926, p. 1469). Both Lanczos and Wentzel wrote fair and positive reviews.

[139] L. H. Thomas, another research student of R. H. Fowler's, also worked on the problem of the validity of the adiabatic principle in cases where degeneracy might occur during the transformation (Thomas, 1925). He found that the principle holds statistically for systems which do not become degenerate for a finite range of the parameter. That is, even if the adiabatic principle is violated for single atoms, it may still be applied to assemblies of the latter; for one can then define quantum conditions in such a way that the corresponding phase integrals become statistically adiabatic invariants.

in Cambridge, when he had plunged into writing papers and had hardly lost any opportunity of making a contribution to whatever subject he found interesting. One might wonder whether Dirac was somewhat exhausted from his intense work during 1924. Or was he searching in vain for a suitable topic of research, especially since his ambition to treat important problems had increased? In any case, the work on his two published papers, one submitted in May (Dirac, 1925b) and the other early in November 1925 (Dirac, 1925c), could not have occupied his thinking for a whole year. Fortunately, there exists an unpublished manuscript dealing with the statistical method of Bose and Einstein and its consequences, which says something about Dirac's scientific involvement. Another piece of work is mentioned in a letter from Ralph Fowler to Niels Bohr. From these sources it is possible to get an idea about the problems which attracted Dirac until fall 1925. They confirm that during the period from fall 1924 to summer 1925 he obtained a systematic and profound knowledge of the wide range of problems and difficulties that atomic and quantum theory presented at that time.

Dirac attended the two courses of lectures on atomic theory that Fowler gave in fall 1924 and spring 1925. Fowler was then easily the foremost expert in this field in England. Though in his own research since 1922 he had concentrated almost entirely on those aspects of quantum theory that played a role in statistical mechanics, he was well informed about all other parts of the theory as well. In his lectures on atomic structure he presented a comprehensive account of the subject. He had lectured on the application of quantum theory to atomic spectra already in the beginning of 1923, but it was only in fall 1924 that Dirac attended Fowler's lectures on that subject and worked out a very neat and complete set of lecture notes. In the first part of this course, entitled 'Quantum Theory,' Fowler discussed Planck's law of radiation, Bohr's theory of the hydrogen atom, and the theory of multiply periodic systems; in the second part, under 'Correspondence Principle,' he assembled the problems which had been approached on the basis of a detailed analogy to classical theory.[140] These lectures enabled Dirac to follow the current literature on atomic theory and to penetrate more deeply into its problems.

[140]Dirac's manuscript of R. H. Fowler's lectures on 'The Quantum Theory' is preserved at Churchill College, Cambridge. A microfilm copy exists in the *Bohr Archives* in Copenhagen (AHQP, Microfilm No. 36, Section 8). According to Dirac's notes, Fowler discussed the following topics:

Part I. *Quantum Theory*. Planck's law; experimental facts of spectral lines; Bohr's theory of the hydrogen atom; Hamilton–Jacobi equation; application to the hydrogen atom; fine structure of hydrogen lines; comparison with experimental values; Stark effect of hydrogen lines; transformation theory of dynamics; degenerate systems.

Part II. *Correspondence Principle*. Expansion into Fourier series in quantum frequencies; simple case of nonharmonic oscillator; general case; case of plane central orbit; three-dimensional case when external forces have an axis of symmetry; rules for fixing stationary states; spectra due to valency electrons; theory of perturbations; relativity orbit; more general proof of Rydberg–Ritz formulae; absorption spectra (empirical facts); electronic impacts; resonance radiation; displacement law for spectra; multiple lines; energy levels of doublet and triplet spectra; effect of a perturbing force with axial symmetry; more complex atomic model (core–valency electron model); X-ray levels; quantum orbits of atoms; Stark effect on the theory of perturbations; band spectra; Zeeman effect; general conclusions (anomalous Zeeman effect).

The leading book on atomic theory at that time was Arnold Sommerfeld's *Atombau und Spektrallinien*, the third edition of which was available in what Dirac thought to be a rather poor English translation (of Sommerfeld, 1922d). The fourth German edition of this book appeared in fall 1924 (Sommerfeld, 1924d), and Dirac studied it with the help of a dictionary. He read a good many papers in journals such as the *Proceedings of the Royal Society*, *The Philosophical Magazine* and the *Proceedings of the Cambridge Philosophical Society*, where he also published his own papers. He read the *Zeitschrift für Physik* quite a lot, for it contained many papers on quantum theory. Dirac also occasionally looked up other German journals like *Physikalische Zeitschrift* and *Annalen der Physik*. Whenever necessary he would refer to scientific papers in British or other journals, such as *The Physical Review*, *Proceedings of the Royal Academy of Sciences of Amsterdam*, *Monthly Notices of the Royal Astronomical Society* and *Proceedings of the Physical Society of London*. Though Dirac knew French well enough since his early youth he did not in general follow the French literature on physics. Thus, apart from studying Sommerfeld's book, he read most of the papers that dealt with quantum theory, and since there were not nearly so many papers on a subject as became the custom later on, it was not very difficult to keep up with them.

From his study of the literature Dirac gathered a complete picture of the existing state of affairs. It confirmed what he had also learned in Fowler's lectures, that the Hamiltonian methods were the basis of the dynamical description of atoms. He discovered that certain problems of atomic physics had been solved by applying, in addition to classical mechanics, the Bohr–Sommerfeld method of quantization: that is, by giving the action variables of multiply periodic systems, which in classical dynamics are constants of motion, discrete values according to the quantum conditions. Hence the task of determining the energy states of a system consisted in transforming its dynamical variables to action-angle variables with the help of the Hamiltonian–Jacobi transformation theory. Of course, Dirac was aware that there remained difficulties in this approach that showed up especially when one wanted to treat atoms having a complicated structure. For example, in order to account for the data of the so-called anomalous Zeeman effect of spectral lines, one had to introduce half-integral quantum numbers which one did not really understand theoretically.

In many calculations, especially those dealing with the intensities of spectral lines, results had been obtained by using correspondence arguments and the classical theory. But not all the rules that were derived in this manner seemed to be satisfied in nature. However, Niels Bohr—and a number of theoreticians who closely followed him—hoped that a 'refined' or 'sharpened' application of the correspondence principle might eventually remove all discrepancies. In fact, Dirac learned about these hopes and some of the most recent results in quantum theory from another course of lectures that Fowler delivered immediately after returning from his visit to Copenhagen in April 1925. In these lectures Fowler discussed the developments ensuing from the radiation theory of Bohr, Kramers and Slater (1924), notably the dispersion theory of Hendrik Kramers and Werner Heisenberg (Kramers, 1924a, b; Kramers and Heisenberg, 1925) and the theory of absorption of radiation developed by John Hasbrouck Van Vleck (1924a, b) and by John Slater (1925a).[141] In both dispersion theory as well as the theory of absorption of radiation, the Hamilton–Jacobi theory and the transformation to action-angle variables again played a crucial role. But the quantization conditions that were employed differed only slightly from the previous $\int pdq$-rules; the prescription followed now was that the classical differential quotients of the physical quantities in question with respect to the action variables were replaced by difference quotients. This procedure, which had been worked out in Göttingen and Copenhagen, was in agreement with the correspondence principle; because for large quantum numbers and not too large differences of the action variables, the difference quotient approached the classical differential quotients. By applying these and other ideas for 'sharpening' the correspondence principle, Heisenberg and Ralph de L. Kronig, then in Copenhagen, succeeded in computing the intensities of the multiplet lines and associated Zeeman effects (Heisenberg, 1925b; Kronig, 1925a, b). These results constituted the most recent development of quantum theory, and the work was still continuing when Fowler lectured about it in Cambridge.[142]

Fowler had returned from Copenhagen with great enthusiasm about the new —he called it 'refined'—use of the correspondence principle. In Copenhagen he had become involved in the calculation of the intensity of lines in the band spectra of diatomic molecules (Fowler, 1925a). His work was based on Heisenberg's suggestion that the total intensity of radiation, to a very good approximation, *does not* depend on the angle between the electron's orbital plane and the direction of the total angular momentum (Heisenberg, 1925a). This conclusion holds good in the classical electron theory of radiation, but it had been difficult to establish it in the Bohr–Sommerfeld theory. Heisenberg had made certain arbitrary and artificial models in order to arrive at the classical result also in quantum theory. While Fowler was willing to go along with such model-building, his student Dirac was not attracted at all to work in the same direction. Dirac did admire people who pursued model-building with a certain virtuosity and discovered interesting results that were confirmed by spectroscopic observations, but

[141] Again Dirac worked out a detailed manuscript of Fowler's lectures on 'Recent Developments.' The manuscript is preserved at Churchill College, Cambridge, and a microfilm copy is kept in the *Bohr Archives* in Copenhagen (AHQP, Microfilm No. 36, Section 8). The first part of Fowler's lectures was devoted to radiation theory, starting with a presentation of the theory of Bohr, Kramers and Slater (1924) and its relation to Einstein's light-quantum theory. Then followed a detailed review of dispersion theory, especially of the work of H. Kramers, W. Heisenberg, J. H. Van Vleck and J. Slater. The third part was devoted to the *refined applications of the correspondence principle*. There Fowler discussed the recently obtained sum-rules in atomic and molecular spectra, the phenomena observed with resonance radiation, and finally the latest theoretical attempts to explain the complex structure of atomic spectral lines.

[142] After his return from Copenhagen, Fowler wrote several letters to Niels Bohr requesting copies of the latest work performed in Copenhagen. For example, in his letter of 4 May 1925 he thanked Bohr for the offprints of Kronig's paper on multiplet structure, and in the letter of 17 June 1925 he looked forward to receiving a copy of Kronig's second paper on the same subject.

the entire correspondence approach left him cold for several reasons. First, it did not allow one to handle a *general* atomic system on the basis of which specific cases could be treated. An even more serious defect was that it lacked a property which he regarded as a necessary prerequisite of any theory, for as he recalled later: 'The correspondence principle always seemed to me a bit vague. It wasn't something you could formulate by an equation. All it said was that there was some similarity between the equations of quantum theory and the equations of classical theory. I don't believe it was more definite than that' (Dirac, Conversations, p. 68).

That it was mainly the vagueness of the correspondence principle that prevented Dirac from working upon it is shown by another fact. He was very interested in an idea by means of which Gregor Wentzel had tried to formulate mathematically certain aspects of the correspondence approach (Wentzel, 1924a, d). To account for the behaviour of quantum systems, Wentzel had associated with the variables describing them an exponential factor, $\exp\{-i\phi\}$, whose phase was defined by the expression

$$\phi = \frac{1}{h} \sum_k p_k \, dq_k , \tag{42}$$

where p_k and q_k are the dynamical coordinates of the system and the summation runs over all the degrees of freedom (Wentzel, 1924a). Wentzel had claimed that the phase was primarily responsible for the deviation of the motion of the quantum system from the classical laws. In a multiply periodic system, the phase became a multiple of 2π and the motion of the system was classical. Transitions from one stationary state to another were connected with a change of the phase; in a multiply periodic system the exponential factor would remain the same. Wentzel had also applied the concept of the phase to aperiodic systems and calculated the intensity of radiation emitted by electrons moving in hyperbolic orbits around the nucleus [i.e., *Bremsstrahlung* (Wentzel, 1924d)]. In all these applications results were obtained, which in the correspondence limit—that is, when the phase was put equal to zero—became identical to the classical result. Wentzel had thus succeeded in representing a 'correspondence principle for the transition probabilities' of atomic systems in a mathematical equation (Wentzel, 1924d, p. 283), and this fact stimulated Dirac to study his papers. Fowler reported about it to Niels Bohr: 'I enclose a note on Wentzel's "phase" which might conceivably interest you or Kramers in view of the discussions at the colloquium one week last session. I found here that Dirac has been working into the question, and I asked him to write down his interpretation. It makes Wentzel clearer to me, but I had hardly got hold of him before at all' (Fowler to Bohr, 29 April 1925). Apparently this work did not lead to any new results that Dirac considered worth publishing.[143]

[143] Dirac's notes on Wentzel's treatment of the phase are not available, and we do not know what conclusions he wanted to draw. However, we do know that Dirac considered the Hamiltonian methods as the fundamental basis to treat atomic systems; Wentzel's phase fitted well into the Hamiltonian scheme.

Wentzel's phase concept was not the only idea in the literature on quantum theory that interested Dirac. Since completing his own work on detailed balancing, Dirac followed with interest the new developments in statistical mechanics, and an especially important one was contained in the papers of Bose and Einstein published between summer 1924 and spring 1925. In June 1924 Albert Einstein had received the manuscript of a paper from Satyendra Nath Bose of Dacca University, in which the author proposed a new statistical method for treating light-quanta, on the basis of which he had derived Planck's radiation law without invoking the extra assumption of stimulated radiation. On recognizing its importance Einstein had not only translated Bose's paper (from English into German) and communicated it to *Zeitschrift für Physik* (Bose, 1924a), but he also extended Bose's statistical method for light-quanta to mass-particles (Einstein, 1924c). In a further publication he had established a connection between the Bose–Einstein statistics and de Broglie's hypothesis of matter waves (Einstein, 1925a). In summer 1924 Dirac had not been sympathetic towards Louis de Broglie's proposal to treat mass-particles and radiation on the same footing and to assume, in particular, that radiation consists of particles with exceedingly small rest mass. However, Einstein's arguments persuaded him that enormous progress in the understanding of the nature of matter and radiation had been achieved and that de Broglie's matter-wave hypothesis was a part of it. He carefully worked through the papers of Bose and Einstein and sought to reformulate and improve some of their derivations. His unpublished notes do not contain essential results going beyond previous discoveries, but they nonetheless make up an instructive document that reveals Dirac's reaction to the state of affairs in statistical mechanics before the advent of quantum mechanics.[144]

Dirac covered four points in his manuscript: the first two were derivations of specific results within what he called 'Einstein–Bose Statistical Mechanics'; the third was a critical discussion of Bose's theory of the interaction of matter and radiation (Bose, 1924b); and the fourth point was an outline of de Broglie's wave theory of matter and its connection with the new statistical methods. Dirac improved only slightly upon a point in Einstein's derivation of the density of momentum of mass-particles (Einstein, 1925b).[145] But with respect to the question concerning the calculation of N_s, the number of particles having a given energy E_s, he supplied an interesting alternative. Bose (1924a) and Einstein (1924c) had obtained the result by counting the number of light-quanta or particles in each cell of volume h^3 in the phase space and then summing over all cells containing particles with energy E_s. However, Bose had claimed later on

[144] Dirac's manuscript consisted of eleven handwritten sheets. It is preserved at Churchill College, Cambridge. A microfilm copy is kept in the *Bohr Archives* in Copenhagen (AHQP, Microfilm No. 36, Section 9).

[145] Einstein had stated without proof that under an adiabatic expansion of the walls of a vessel containing gas, by distances Δl_x, Δl_y, Δl_z along perpendicular directions, the momentum components of the gas molecules, p_x, etc., would change by the amounts $\Delta p_x/|p_x| = -(\Delta l_x/l_x)$, etc. Dirac provided its proof on the basis of kinetic theory.

that he could avoid the first step by simply starting from the equation

$$W = \prod_s \frac{(A_s + N_s)!}{A_s!\,N_s!} \tag{43}$$

which defines the probability of a given distribution of particles such that N_s particles of energy E_s are to be found in A_s different cells (see Bose, 1924b, p. 386). Such a derivation was more to Dirac's liking and, since Bose had not given the details of the calculation, he worked them out in his manuscript. Thus he determined the maximum of the distribution, Eq. (43), under the conditions of constant total particle number, N, and constant total energy, E, arriving at the result

$$N_s = \frac{A_s}{\exp\{\alpha + \beta E_s\} - 1}, \tag{44}$$

which agreed fully with the earlier derivation (Einstein, 1925a, Eq. (30a)). While β is easily found to be identical to $(kT)^{-1}$, where T is the absolute temperature and k the Boltzmann constant, Dirac proposed an original way to determine α. He first noted that α is given by N, the total number of molecules in a given volume of gas; hence it may assume a variety of values. However, the spirit of the new statistics—in which massive particles and massless radiation were treated on the same footing—demanded that N is not arbitrary; rather, similar to the number of light-quanta (in a given volume at a given temperature), it is determined by the condition of maximal entropy. Hence the true value of α must be zero, as there is no subsidiary condition that keeps the total particle number constant. No difficulty arises from the possibility of having an arbitrary number of mass-particles in a given volume, for as Dirac recognized: 'If there are more molecules present than the number required to give $\alpha = 0$, the extra ones cannot take part in the thermal agitation, but must all go into the cell of zero energy' (Dirac, unpublished manuscript, 1925, p. 4).[146]

Dirac would build upon these observations one year later when he invented the statistical methods compatible with quantum mechanics (Dirac, 1926f). And they would remain fruitful even afterwards as, for example, in the investigation of relativistic systems where new mass-particles can be created if a minimum amount of energy is available for that purpose. In summer 1925 such far-reaching conclusions were still far ahead in the future, and Dirac contented himself by making a calculation which showed clearly that a unified treatment of matter and radiation was possible. In particular, he inserted in Eq. (44) the

[146]That is, if the number N is as large as to yield $\alpha = 0$ the state is called 'saturated'; if N is larger, the surplus particles behave like condensed gas molecules; and if N is smaller, the situation is similar to that of unsaturated vapour. In the case of radiation the following situation arises: there are always enough light-quanta to yield $\alpha = 0$; a smaller number cannot exist because the available total energy can always be used to produce more quanta of lower energy; a larger number of quanta will not be present either, for the surplus quanta have zero frequency, hence they do not exist at all.

number of phase cells, A_s, which is found by dividing the phase volume by h^3 (h being Planck's constant), and derived from it the energy density of a perfect Bose–Einstein gas and the energy density of light-quanta, respectively. The first of these results modified the expression obtained from classical statistical mechanics, especially for low temperatures, and the second one led to Planck's radiation law. Dirac saw a special advantage in the new statistical method: the 'induced' emission of radiation did not have to be introduced, but followed from it without extra assumptions. As a consequence, the equations of statistical mechanics describing the equilibrium in arbitrary systems consisting of molecules and light-quanta could be founded on fewer principles and postulates than Dirac had invoked a year earlier. This fact pleased him, for he saw a possibility of improving on his previous work on detailed balancing.[147]

In discussing Bose's theory of the interaction of radiation with matter, Dirac presented a beautiful clarification of a point that had arisen between Bose and Einstein. Bose had argued that the formula

$$W(\nu) = \text{const.}\,\rho(\nu)\left\{\frac{8\pi\nu^3}{c^3} + \rho(\nu)\right\}^{-1} \tag{45}$$

describes the probability that an atomic system absorbs radiation of frequency ν —which is also one of the emitted frequencies—and density $\rho(\nu)$ (Bose, 1924b, p. 390; Bose gave a formula equivalent to Eq. (45)). In an addendum to Bose's paper, Einstein had pointed out that this implied that no stimulated emission could occur; he rather proposed to assume that the absorption probability was proportional to $\rho(\nu)$ and the probability of the inverse process was proportional to the denominator in Eq. (45). Dirac now gave a complete analysis of the problem by studying the detailed balancing of the absorption and emission processes. Let $f_1(n)$ be the probability of an atom to absorb one light-quantum from a cell (in phase space) containing n quanta of a given energy, and $f_2(n)$ the probability of emitting one quantum into a cell (previously containing n quanta of the same energy), then the principle of detailed balancing requires that the equation

$$P_n f_1(n) = P_{n-1} f_2(n-1) \tag{46}$$

be satisfied, where P_n and P_{n-1} are the *a priori* probabilities that a cell contains n and $n-1$ quanta, respectively. Since the fundamental assumption of the new statistics was that all cells were equally probable, $f_1(n)$ had to be equal to $f_2(n-1)$. This fact was not at variance with Eq. (45) for the absorption probability, if one assumed all absorption probabilities as well as all emission probabilities to be identical and equal to a constant. Einstein's alternative assumption,

[147] Note that in the correspondence treatment of interacting matter and radiation (e.g., Van Vleck, 1924a, b; Slater, 1925a), which Dirac learned in Fowler's lectures of spring 1925, the induced emission had to be introduced as an extra assumption. From this point of view, the correspondence principle led to less definite conclusions than the statistical method of Bose and Einstein.

which Dirac also verified, yielded the values

$$f_1(n) = n \times \text{const.} \qquad \text{and} \qquad f_2(n) = (n + 1) \times \text{const.} \tag{47}$$

for the probabilities, independent of the atomic system absorbing and emitting the radiation, provided the statistical weight of its energy levels was the same throughout.

The results which Dirac obtained in his examination of Bose–Einstein statistics did not represent new discoveries, but merely improved the formulations of what had been obtained before. However, the heuristic value of these exercises in a known field for Dirac's future work can hardly be overestimated. The same was true of the last point he covered, entitled 'Waves and Quanta.' He sought especially to understand why the six-dimensional phase space of a structureless particle or light-quantum in statistical mechanics had to be divided into cells of volume h^3. If each particle is associated with a wave and the possibility is admitted that several particles share a wave, then the waves play the role of the cells. In the case of light-quanta this is easy to show, as the number of modes in a volume V, having a frequency between ν and $\nu + d\nu$—i.e., $(8\pi V/c^3)\nu^2\,d\nu$—is equal to A_s, the number of cells calculated from statistical mechanics (after taking into account the factor two for the degrees of freedom of polarization). In the case of material particles associated with a matter wave of frequency ν, the number of modes is given by the same expression as for light-quanta if the velocity c is replaced by the phase velocity u of the matter wave. On substituting de Broglie's expression for the phase velocity,

$$u = \frac{h\nu}{|p|}, \tag{48}$$

where $|p|$ is the absolute value of the momentum of the particle, Dirac found for the number of matter modes the result $(4\pi/h^3)\cdot|p|^2 d|p|$, which was clearly identical with the number of cells in phase space. From the equality of the number of wave modes in a given volume, having a given energy, and the number of cells in phase space of statistical mechanics, he concluded that the two theories, de Broglie's wave theory of matter and the statistical theory of Bose and Einstein, 'are mathematically equivalent' and that the 'fundamental assumption of present statistical mechanics now reads: all values for the numbers of atoms (or light-quanta) associated with a given wave have the same a priori probability' (Dirac, unpublished manuscript, 1925, p. 7).

The new development in statistical mechanics thus changed Dirac's earlier negative response to de Broglie's ideas completely. In summer 1925 he saw clearly that de Broglie's ideas, together with the new statistical methods, provided a consistent basis for the treatment of radiation in equilibrium. He discussed some of these questions in his first talk before the Kapitza Club on 'Bose's and L. de Broglie's Deduction of Planck's Law' at its 95th Meeting on 4 August 1925. Although he did not continue to think along the lines of wave theory, as Schrödinger did at about the same time, Dirac had performed a pioneering piece of work: he had considered the first 'equivalence proof' of a particle and a wave theory.

In comparison to the problems about which Dirac wrote in his unpublished notes, the work on his two published papers during this period could not have occupied his thinking for a long time. The first of these papers, entitled 'The Effect of Compton Scattering by Free Electrons in a Stellar Atmosphere' (Dirac, 1925b), came about as follows. While Fowler was away from Cambridge from the end of January to mid-April 1925, Edward Milne took over as Dirac's supervisor. He interested Dirac in an astrophysical problem: the calculation of the shift in frequency of the radiation (emitted from the interior of the sun) which is caused by the scattering in the solar atmosphere. In 1923 Arthur H. Compton had put forth the suggestion that the [Compton] effect discovered by him might account for the observed displacement of the absorption lines in the solar spectrum near the limb of the sun towards longer wavelengths (Compton, 1923e).[148] A quantitative calculation of the red-shift of solar lines, based on Compton's suggestion, was of considerable importance for the validity of general relativity, for the latter also predicted a red-shift and the observations had been taken as a confirmation of Einstein's prediction.[149] If the calculation yielded a large red-shift because of the Compton effect, general relativity would lose one of its alleged experimental supports, and because of this possibility Milne was interested in the problem. On the other hand, Milne knew that the physical situation in the solar atmosphere containing free electrons was rather complex, and many effects could be expected that would lead to the shifting of the spectral lines. For example, one had to consider the thermal motion of the scattering electrons that gives rise to a symmetrical broadening of the original spectral lines due to the Doppler effect. Second, there was an effect leading to an overall reduction of the average wavelength of scattered radiation, which could possibly balance Compton's red-shift. This effect arises from the physical consideration that the incident radiation is more intensive relative to electrons moving towards it; hence these electrons will scatter more often.[150] From his previous work and

[148]Compton had argued that the continuous emission background (on both sides of an absorption line) will be shifted towards the red because of the scattering by the electrons, and thus be responsible for a similar shift of the observed line. The expected displacement is larger at the limb of the sun because the light coming from that region is, on the whole, scattered more than in the central regions.

[149]In the years following the British Solar Eclipse Expedition of 1919, which confirmed the deflection of light, there was much discussion about the question of whether one had evidence for the gravitational red-shift or not—this being the third crucial test of general relativity after Mercury's perihelion motion and the deflection of light. For example, L. Grebe of Bonn reported about the situation at the *Naturforscherversammlung* in Bad Nauheim in September 1920; he had concluded from the available data that 'the gravitational effect of shift of [spectral] lines required by Einstein's theory really exists' (Grebe, 1920, p. 665). Systematic experimental research was continued, for instance, at Mt. Wilson Observatory. Some results were published in 1925 that strongly supported the existence of the red-shift (Adams, 1925).

[150]Milne had in fact concluded the existence of the 'blue-shift' from the general theoretical observation that if blackbody radiation of a given temperature falls on free electrons in statistical equilibrium at the same temperature, then in accordance with the laws of thermodynamic equilibrium the scattered radiation must have the same distribution as the incident radiation (Milne, 1924b).

studies, Dirac had mastered all the theoretical tools necessary to carry out the calculation of the Compton effect and the other effects Milne had in mind. He was also interested in this astrophysical problem, for its outcome promised to provide support for general relativity.

The formulation of the problem posed no difficulty for Dirac. To study the behaviour of light scattered by matter in thermal equilibrium, Dirac used relativistic arguments similar to the ones in his work on detailed balancing (Dirac, 1924d). When radiation, whose distribution among the frequencies ν is given by the function $I(\nu)$ and which is confined to a solid angle $(\Omega, \Omega + d\Omega)$, falls upon dn electrons, then $R(\nu')d\nu' d\Omega'$, the number of scattered quanta per unit time in the frequency range $(\nu', \nu' + d\nu')$ and the solid angle $(\Omega', \Omega' + d\Omega')$, is given by

$$R(\nu')\,d\nu'\,d\Omega' = \frac{h^2}{m_e^2 c^3} \cdot dn \cdot I(\nu)\,d\Omega \frac{\nu' F(\alpha, \beta)}{\nu(E/c)}\,d\nu'\,d\Omega'. \tag{49}$$

In Eq. (49), h is Planck's constant, m_e the mass of the electron whose four-momentum before scattering is $(p_x, p_y, p_z, E/c)$, and $F(\alpha, \beta)$ is a function of α and β, the two independent relativistic invariants determining the scattering process. In thermal equilibrium at temperatures T of several thousand degrees existing in the solar atmosphere, the electrons have to a good approximation the Maxwellian momentum distribution. With the latter for the factor dn, Dirac calculated the scattering of radiation from *all* electrons in the solar atmosphere and obtained

$$R(\nu') = \frac{h^2}{m_e^2 c^3} N(2\pi m_e kT)^{-3/2} \int_0^\infty I(\nu)\,d\Omega\,\psi(\nu, \nu')\,d\nu, \tag{50}$$

where

$$\psi(\nu, \nu') = \int_{-\infty}^{+\infty} dp_y \int_0^{+\infty} dp_z \frac{\nu' F(\alpha, \beta)}{\nu(E/c)\,\partial\nu/\partial p_x} \cdot \exp\left\{ -\frac{(p_x^2 + p_y^2 + p_z^2)}{2m_e kT} \right\}. \tag{51}$$

In Eq. (50), N is the number of free electrons per unit volume and k is Boltzmann's constant.

All the properties of the scattered light were now contained in Eq. (50), which had been derived from the equations of statistical mechanics. For the discussion of the possible shift in the frequency of light (emitted from the interior of the sun) by the free electrons in the solar atmosphere, Eq. (50) had a particularly convenient structure: the contributions to the ν-integral on the right-hand side were additive and Dirac was able to discuss the question of the frequency shift for each component ν of the incident radiation separately. There still remained the mathematical problem of evaluating the function $\psi(\nu, \nu')$, Eq. (51), which contains the function $F(\alpha, \beta)$ that determines the intensity of the scattered light of frequency ν'. Two possibilities were available for calculating this intensity: either using classical electrodynamics or the light-quantum theory. Since the practical results would be the same in both cases, Dirac opted for the classical solution, which yielded the function

$$F_0(\alpha, \beta) = \frac{e^4}{h^2c^2\alpha}\left(1 - \frac{\beta}{\alpha^2} + \frac{\beta}{2\alpha^4}\right) \tag{52}$$

for electrons having zero three-momentum. Since the average velocity of the electrons in normal stellar atmospheres, including the sun's, is small compared to the velocity of light, Dirac replaced $F(\alpha, \beta)$ in Eq. (51) by $F_0(\alpha, \beta)$ and also expanded the integrand in powers of p_y/m_ec and p_z/m_ec. He then integrated term by term to determine $\psi(\nu, \nu')$ as a series of ascending powers of the small dimensionless quantity, $(kT/m_ec^2)^{1/2}$, whose value for $T = 6000°\text{C}$ is about 10^{-3}. Retaining only the first orders in $(kT/m_ec^2)^{1/2}$, he obtained the result

$$\psi(\nu, \nu') = \sqrt{\frac{2m_ekT}{1 - \cos\theta'}}\ \frac{F_0}{\nu}\ \exp\left\{-\frac{(\nu' - \nu)}{a_\nu^2}\right\}, \tag{53}$$

with

$$a_\nu^2 = \frac{4kT}{m_ec^2}\nu^2(1 - \cos\theta'), \tag{54}$$

provided the incident frequencies are of the order of kT/h. Due to Eq. (53), where θ' is the angle between the incident and the scattered radiation, the thermally distributed electrons produce a symmetrical broadening of a spectral line of frequency ν, with half-width a_ν. For a temperature of 6000°C and the scattering angle of 90°, Dirac found that a_ν/ν is about 2×10^{-3}, or an infinitely sharp spectral line in the visible region will assume a half-width of 10 Å.

So far, the scattering of light by electrons did not give rise to any asymmetry structure. However, by carrying through his evaluation of the $\psi(\nu, \nu')$-integral to the second order in the parameter $(kT/m_ec^2)^{1/2}$, Dirac obtained a shift, $\delta\nu$, of the centre of gravity of the incident line given by the equation

$$\frac{\delta\nu}{\nu} = \left(\frac{4kT}{m_ec^2} - \frac{h\nu}{m_ec^2}\right)(1 - \cos\theta'). \tag{55}$$

The second term in this expression represents just the Compton effect due to scattering by electrons, which is of the order of 0.024 Å for the radiation of wavelength 5000 Å. The additional term, which is positive and describes Milne's blue-shift, cancels the Compton effect more or less completely since for frequencies of visible light and the temperature of the solar atmosphere (chromosphere) the quantity $h\nu$ is of the same order of magnitude as kT. Now the observed lines of the sun are absorption lines, whose intensity or blackness is determined by the amount of the absorbing material that lies along the path between the point where the radiation is created and the surface. The free electrons in the solar atmosphere add to the absorption, but their effect is comparatively negligible; Dirac estimated the absorption coefficient from the electrons to be a million times smaller than the absorption coefficient of the solar matter. 'The observed wings of the solar lines [which show a definite red-shift],' he concluded, 'can therefore hardly be accounted for by this effect' (Dirac, 1925b, p. 831). Thus, what remained as the theoretical explanation for the observed red-shift was the one due to general relativity.

The work on the astrophysical problem was only marginally of interest to Dirac, although later he would occasionally return to such questions.[151] The next paper he wrote dealt again with the principle of adiabatic invariance in quantum theory; it was received by the *Proceedings of the Cambridge Philosophical Society* on 5 November 1925 (Dirac, 1925c). In this paper he applied adiabatic transformations to atomic systems upon which a magnetic field, whose strength is varied infinitely slowly, is imposed. A difficulty arises in such a problem: systems subject to velocity-dependent forces might not always satisfy Hamilton's equations of motion when these forces are changed.[152] This question had really not been discussed in the literature until fall 1925 (although adiabatically changing magnetic fields had been applied frequently to atomic systems), and Dirac decided to investigate it carefully.

The Cartesian momenta of an electron, of charge $-e$ and mass m_e, moving within an atom with the velocity $(\dot{x}, \dot{y}, \dot{z})$ under the action of a *constant* magnetic field of strength $\mathbf{H}$, are given by the expressions

$$p_x = m_e\dot{x} - \frac{e}{c}A_x, \qquad p_y = m_e\dot{y} - \frac{e}{c}A_y, \qquad p_z = m_e\dot{z} - \frac{e}{c}A_z, \tag{56}$$

where A_x, A_y and A_z are the components of $\mathbf{A}$, the vector potential at the position of the electron (curl $\mathbf{A} = \mathbf{H}$). The dynamical variables satisfy the equations of motion with the Hamiltonian H,

$$H = H_0 + \frac{e}{c}(A_x\dot{x} + A_y\dot{y} + A_z\dot{z}), \tag{57}$$

where H_0 is the Hamiltonian of the atom in the absence of the magnetic field and terms of the second order in the A's are neglected.[153] The magnetic term, which is

[151] Dirac acknowledged Milne's supervision of this work at the end of his paper (Dirac, 1925b, p. 832). It was communicated by Milne to the Royal Astronomical Society and published in its *Monthly Notices*. Later, Dirac helped S. Chandrasekhar in the calculation of stellar absorption and opacity coefficients on the basis of Fermi statistics for electrons and atomic nuclei (see Chandrasekhar, 1931).

[152] If there are no velocity-dependent forces, the validity of the Hamiltonian equations of motion is never in doubt. Then there always exist dynamical variables which do not depend explicitly on the adiabatic parameter a, and all other sets of dynamical variables are obtained by contact transformations. In the case of a velocity-dependent force no set of variables can be found which is independent of the parameter a, and this fact may give rise to complications.

[153] The full Hamiltonian contains terms of the second order in the vector potential, which are $(e^2/2m_ec^2)(A_x^2 + A_y^2 + A_z^2)$.

the second term in Eq. (57), can be written as $(e/2m_ec)|\mathbf{H}|p_\phi$, where $|\mathbf{H}|$ denotes the absolute value of the field strength and p_ϕ the component of the angular momentum of the electron in the direction of the magnetic field. Since this term is a constant of motion, the quantum integrals of the atom in a constant magnetic field are given by the same functions of the position and momentum coordinates as for the field-free atom, the momenta being defined by Eqs. (56). Thus, one may hope that the adiabatic principle will hold if the external magnetic field is turned on infinitely slowly; however, the crucial question still remains whether or not the canonical equations of motion are valid at any instant during the adiabatic change.

On first inspection the answer seems to be negative because from Eqs. (56) one finds, for example, that the equation of motion for p_x contains a term proportional to the time derivative of the vector potential, which *does not* occur in the canonical equation of motion for x. However, a closer examination reveals that a changing magnetic field creates an electric field, and the latter in turn accelerates the electron. Consequently the true equations of motion, at any instant of the adiabatic transformation of the system, assume the canonical form with the Hamiltonian H' given by

$$H' = H_0 + \frac{e}{c}\left(A_x\dot{x} + A_y\dot{y} + A_z\dot{z}\right) - e\phi, \tag{58}$$

where ϕ is the scalar potential. Since the order of magnitude of ϕ is determined by the inverse of the time taken to establish the magnetic field, the deviation of H' from H, Eq. (57), is zero in the adiabatic limit. Hence the adiabatic principle is applicable to atomic systems in magnetic fields, provided the magnetic field is turned on infinitely slowly.

Towards the end of his note on the adiabatic principle Dirac established another point which Sommerfeld, in the fourth edition of *Atombau und Spektrallinien* (Sommerfeld, 1924d, p. 403), had simply stated without argument or proof: that the instantaneous velocity of the electron in an atom will not be affected if a magnetic field were established suddenly. This would seem to be rather strange, for the fast turning on of a magnetic field creates a strong electric field. Nevertheless, Dirac showed in detail that the magnetic field is established by means of the electromagnetic waves arriving at the position of the electron with equal amplitudes from all directions in a plane perpendicular to the magnetic field vector.[154]

The note on the adiabatic hypothesis for atomic systems in the presence of a magnetic field was suggestive of important features characteristic of Dirac's

[154]Dirac proceeded as follows. Suppose that a magnetic field of amplitude C, say in the z-direction, is created by a plane wave advancing in the x-direction, then simultaneously an electric field of amplitude $E_y = -C$ arises, which would accelerate the electron in the y-direction. A second electromagnetic wave of the same amplitude propagating in the negative x-direction should arrive with such a phase that it causes, at the position of the electron, a magnetic field in the z-direction and an electric field in the y-direction, both with amplitude $+C$. The electric field in the y-direction is then cancelled up to the retardation effect. This retardation effect can be made very small if one assumes finally that electromagnetic waves arrive from all directions in the xy-plane at the electron and establish the constant magnetic field of strength $|\mathbf{H}| = 2C$.

work. One of them was the profound confidence he had in the validity of the Hamiltonian methods in quantum theory. This confidence was the main reason why, as late as fall 1925, he again returned to the consideration of the adiabatic principle, a topic which was no longer fashionable.[155] Another feature of Dirac's work was a continuity of themes: that is, he returned again and again to topics he had once taken up, and he would continually seek to improve or extend the mathematical formulation. This constant improvement eventually led him to increasingly more perfect general equations, until he could regard the final scheme as 'beautiful.' The pursuit of mathematical beauty in physical theory became the guideline of his scientific work, and he summarized his credo in the following words:

> I feel that a theory, if it is correct, will be a beautiful theory, because you want the principle of beauty when you are establishing fundamental laws. Since one is working from a mathematical basis, one is guided very largely by the requirement of mathematical beauty. If the equations of physics are not mathematically beautiful that denotes an imperfection, and it means that the theory is at fault and needs improvement. There are occasions when mathematical beauty should take priority over [temporary] agreement with experiment.
>
> Mathematical beauty appeals to one's emotions, and the need for it is accepted as an article of faith; there is no logical reason behind it. It just seems that God constructed the Universe on the basis of beautiful mathematics and we have found that the assumption that basic ideas should be expressible in terms of beautiful mathematics is a profitable assumption to make.
>
> In an approximate theory the mathematics may not be beautiful. Newton's theory has some beauty, but Einstein's theory of gravitation has greater beauty. Pure mathematicians, even if they are not physicists at all, and not in the least concerned with gravitation are still interested in Einstein's equations because they find the equations beautiful. The whole idea of Minkowski space and its equations is a beautiful thing because it is connected with the Lorentz group. Also nonrelativistic quantum mechanics is a beautiful theory because it is complete.
>
> I would consider the theory of complex variables a very beautiful theory because of the great power that one has with Cauchy integrals. The same I feel with projective geometry, but not with some other branches of mathematics, such as the theory of sets and topology.
>
> A beautiful theory has universality and power to predict, to interpret, to set up examples and to work with them. Once you have the fundamental laws, and you want to apply them, you don't need the principle of beauty any more, because in treating practical problems one has to take into account many details and things become messy anyway. (Dirac, Conversations, see Mehra, 1972, p. 59, footnote 78)

The continuity of thought and method in Dirac's style of attacking the problems of atomic theory led him to important results. He thus gradually

[155] In comparison to the many papers dealing with the applications of the correspondence principle, only a few contributions had been made to the adiabatic principle since the time it was established by Paul Ehrenfest (1913c). Among these were the papers and thesis of J. M. Burgers (1918), but hardly anybody sought to improve the mathematical formulation of the adiabatic principle afterwards.

deepened his understanding of many questions. On the other hand, the fact that he stuck faithfully to his methods prevented him from appreciating or inventing approaches that were radically different from his own.[156] For instance, his deep faith in the action-angle scheme of classical mechanics posed a hindrance, for as he admitted later: 'I now see that it was a mistake; [by] just thinking of action and angle variables one would never have gotten on the new mechanics. So without Heisenberg and Schrödinger, I would never have done it myself' (Dirac, Conversations, p. 41). However, as soon as he recognized the validity of the new quantum-mechanical ideas, he built an entire scheme upon them.

[156]For example, Dirac was not able to follow Bohr's method of approaching the problems and difficulties of quantum theory: that is, to consider the apparently most contradictory aspects of atomic phenomena, to arrive at certain physical conclusions about them, and to wait for the appropriate mathematical formulation to come afterwards.

Chapter IV
The Reformulation of Dynamical Laws

IV.1 Key to the Quantum Mystery

From fall 1922 to spring 1925 there grew the realization on the Continent—in Copenhagen, Göttingen, Munich, Leyden and Berlin—that quantum theory was faced with fundamental difficulties. These difficulties were not quite as clearly perceived in England as on the Continent, not even in Cambridge where all new developments were followed with the greatest interest. However, in early 1925 this situation changed considerably, at least in the inner circle of theoretical physicists in Cambridge. It began with Ralph Fowler's visit to Bohr's Institute in Copenhagen, where he spent more than two months from the end of January to the middle of April. He brought back with him to Cambridge an intimate knowledge of the latest results in dispersion theory and the theory of complex spectra, which he immediately passed on to his students in a course of lectures entitled 'Recent Developments.'[157] In these lectures he focused very much on the successes of what he called the 'refined' application of the correspondence principle and not so much on the difficulties of quantum theory. He did so because he knew that Niels Bohr, the foremost spokesman of these difficulties, intended to visit Cambridge soon and wished to speak about them himself.

In spring 1925 Bohr had received an invitation from Rutherford to speak in Cambridge on recent developments in theoretical atomic physics. He had accepted it in the hope not only of seeing Rutherford again and presenting his talk, but of having the occasion for 'still more private discussions about our present theoretical troubles, which are of alarming character indeed' (Bohr to Rutherford, 18 April 1925). From these discussions with Rutherford and Fowler he hoped to gather some hints about how to proceed in a situation that had become even more desperate after the experiment, made by Walther Bothe and Hans Geiger, had confirmed Einstein's light-quantum point of view and disproved the radiation theory of Bohr–Kramers–Slater, which had been the very foundation of a successful treatment of dispersion. In a long letter, sent on the same day as he received the depressing news about the experiment from Geiger, another earlier collaborator of Rutherford's, Bohr explained his worries in some detail to Fowler. He concluded: 'I am looking forward to discussing all these things with you in Cambridge and I intend in my lecture there to make a clean breast of my

[157] See footnote 141 for the contents of Fowler's lectures on 'Recent Developments.'

bad conscience as to our present theoretical troubles' (Bohr to Fowler, 21 April 1925).

Although Niels Bohr had exchanged letters with Rutherford during all the years since he left Manchester from Copenhagen in 1916 and had also visited him in Cambridge in fall 1923, this was the first time that he desired help from the great experimentalist. Available to Bohr at that time in Copenhagen were several experts of atomic and quantum theory—Hendrik Kramers, Ralph Kronig and Wolfgang Pauli; and yet, the problems seemed so hard to him that he seized the opportunity of discussing them with someone whom he trusted and respected profoundly, even though he was not a theoretician. Moreover, Fowler, whose stay at Copenhagen Bohr had 'enjoyed immensely' (Bohr to Rutherford, 18 April 1925), had reported to him about the experimental and theoretical work going on in Cambridge with great enthusiasm; Bohr wanted to learn more about it, hoping that it might aid in showing the way out of his troubles. It was therefore natural that he turned towards Rutherford and Cambridge to give a full account of the 'outstanding difficulties which offer such uncomforting indications of essential deficiencies in our usual description of natural phenomena' (Bohr to Rutherford, 28 April 1925).

The topics, which Bohr covered in his lecture on 'Problems of the Quantum Theory' in Cambridge in May 1925, had already been announced in his letter of 21 April to Fowler and discussed in a paper entitled '*Über die Wirkung von Atomen bei Stößen*' ('On the Action of Atoms in Collisions'), submitted to *Zeitschrift für Physik* a few weeks previously (Bohr, 1925a).[158] In the latter, Bohr had proposed to organize atomic phenomena into two separate types: First, those for which there exists an inverse reaction, and in which energy and momentum are conserved. He called these 'reciprocal interactions' ('*reziproke Wechselwirkungen*,' Bohr, 1925a, p. 153). An example is the excitation of atoms by electron impact, whose inverse reaction is the so-called collision of the 'second kind.' Second, there are processes, such as the dispersion of light by atoms, where the coupling between the emission and absorption processes seems to be less tight; hence energy and momentum might be conserved only in the statistical average, and the interaction assumes an essentially nonmechanical character, which Bohr termed 'irreciprocal' ('*irreziprok*,' Bohr, 1925a, p. 153). A great many processes could thus be classified in one of the two categories, and their properties described at least qualitatively. For example, why do the equations of classical mechanics account quite well for the observed ionization of gases by collisions with fast moving charged particles? The reason, according to Bohr's scheme, was that the process was 'reciprocal.' On the other hand, in collision processes between mass-particles, deviations from the classical description had to be expected if the conditions of 'reciprocity' were not satisfied. It was on this basis

[158] Bohr attended a *Soirée* of the Royal Society in London on 13 May 1925. His lecture on 'Problems of the Quantum Theory' in Cambridge was scheduled on 15 May, and on 18 May he was invited to speak before the Kapitza Club.

that Bohr explained why Fowler had theoretically obtained, in the case of α-particles penetrating through matter, a too low rate of capture of electrons by α-particles in comparison with the experimental results (Fowler, 1924b).

This seemingly satisfactory theoretical picture of atomic phenomena, as Bohr had conceived it, lost its foundation by the outcome of the Bothe–Geiger experiment because the latter proved the validity of *exact* energy and momentum conservation in individual atomic processes, including those of radiation. Hence, the same kind of description, as in Bohr's 'reciprocal' processes, had to apply in cases where radiation was absorbed and emitted, and the separation of interactions into two different types, which seemed to provide the clue to the theoretical understanding of atomic phenomena, had to be abandoned. As a result, the successes of the dispersion-theoretic approach could not be understood any more. 'In fact,' Bohr confessed to Fowler, 'the possibility of describing the experiments without radical departure from the ordinary space-time description is so remote that we may just as well surrender at once and prepare ourselves for a coupling between the changes of state in distant atoms of the kind involved in light-quantum theory' (Bohr to Fowler, 21 April 1925).

Nothing could be more desperate for Bohr than to admit the essential correctness of the light-quantum hypothesis. His talk at Cambridge depicted quantum theory as passing through its deepest crisis, and Paul Dirac listened to it carefully.

The fact that collision processes involving emission and absorption of radiation should obey the same rules as the processes without radiation, and that energy and momentum were always strictly conserved, did not surprise Dirac at all. In his paper on detailed balancing he had assumed throughout the conservation of energy and momentum (Dirac, 1924d), and, in general, he harboured no prejudice against Einstein's conception of the light-quanta. Nevertheless Bohr, whom he heard for the first time in May 1925, impressed him with his presentation of the difficulties of quantum theory.[159] Dirac knew from Fowler's lectures and his own reading that Bohr's collaborators, especially Heisenberg and Kronig, had been very successful in explaining the properties of complex spectra of atoms by using the methods of dispersion theory. From the point of view of the available dynamical and mechanical methods, which could not even be applied to the helium atom, these successes appeared like a miracle. Now Bohr had reminded his Cambridge audience that the dispersion-theoretic approach, which had been so successful, needed a new foundation because the one he had been advocating was wrong. That is, the duality in the theoretical description could not be kept, as there were no fundamentally different types of atomic interactions, especially no interactions in which the laws of conservation of energy and momentum failed. Hence the need had arisen to construct a fundamentally new,

[159]In conversations with me Dirac mentioned that Bohr 'came and lectured sometimes' and that he was 'very happy to meet Bohr.' Since Dirac was not in Cambridge at the time of Bohr's previous visit in September 1923, he first saw Bohr in May 1925. (J. Mehra)

energy-momentum-conserving quantum theory that would retain the good results of dispersion theory (including the successful treatment of 'irreciprocal' processes) as well as the good results of the classical theory of 'reciprocal processes.' This task of discovering a universal quantum theory attracted Dirac, but he soon found that it was very difficult to make progress.

Two and a half months after Bohr's talk another messenger from Copenhagen and Göttingen, Werner Heisenberg, addressed the Kapitza Club in Cambridge. In his talk on 28 July 1925 Heisenberg summarized the theoretical results on term spectroscopy and anomalous Zeeman effects of spectral lines, and emphasized the need to search for the correct dynamical theory. In private conversations with Fowler, his host, Heisenberg mentioned his most recent work in which he had proposed new, general quantum-mechanical relations from which some of the spectroscopic results could be derived. Fowler, who for some time had become interested in everything Heisenberg wrote, requested the latter to send him a copy of the proof-sheets as soon as it was available, and Heisenberg did so. Shortly after the middle of August 1925 Heisenberg's paper on the quantum-theoretical re-interpretation of kinematic and mechanical relations arrived in Cambridge, and Fowler handed it over to Dirac for a closer study of its contents. This was more than a month before the paper appeared in print (Heisenberg, 1925c).

Heisenberg's work did not appear to be easy reading to Dirac, for it discussed problems in a language to which he was not accustomed. Heisenberg had introduced his paper with some philosophical remarks emphasizing the necessity of using only observable quantities in the quantum theory to be constructed. Although the philosophy did not seem to be unreasonable, it was difficult to give it a unique mathematical formulation.[160] The only help in this direction came from Heisenberg's remark at the end of the introduction, in which he stated: 'One can regard the frequency condition and dispersion theory of Kramers [Kramers, 1924a] together with its extensions in recent papers [Born, 1924b; Kramers and Heisenberg, 1925; Born and Jordan, 1925a] as the first important steps towards such a quantum mechanics' (Heisenberg, 1925c, p. 880). At first sight, this hint did not assist Dirac very much; he was aware of the results contained in these papers, but he also knew perfectly well that dispersion theory did not yet provide any dynamical scheme for treating the problems of atomic physics. Indeed, the dispersion-theoretic equations had nothing to do with the only dynamical scheme that he thought really made sense: the Hamiltonian scheme. The lack of any reference to this scheme in Heisenberg's paper disturbed Dirac and prevented him in the beginning from finding anything worthwhile in it. Moreover, the principal example, which Heisenberg had treated with the help

[160] As we have mentioned earlier (Chapter I), Dirac was not attracted by philosophical arguments, especially when they did not lead immediately to clear mathematical equations. Moreover, the question, as to which quantities were observable in atomic systems and which not, had by no means an immediate answer either in 1925 or later. The selection of observable quantities always involved some arbitrariness.

of his new methods, was the anharmonic oscillator; it held no particular appeal for Dirac, who at the time was interested in the theory of the helium atom. However, when after some days he returned to Heisenberg's paper he discovered that it contained more than just another correspondence approach to dispersion problems.[161] It became clear to Dirac that it indeed provided, as the author had claimed, 'some new quantum mechanical relations' (Heisenberg, 1925c, p. 880).

In analyzing Heisenberg's paper with greater care Dirac noticed that the main point was contained in a new mathematical definition of the dynamical variables. Heisenberg had represented $x(n,t)$, a variable of a periodic quantum system in the stationary state n, by what he called a 'quantum-theoretical series,' that is,

$$x(n,t) = \sum_{\alpha=-\infty}^{+\infty} x(n,n-\alpha)\exp\{i\omega(n,n-\alpha)t\}, \tag{59}$$

where $x(n,n-\alpha)$ is the coefficient associated with $\omega(n,n-\alpha)$, the quantum frequency (multiplied by 2π) of the system. For two such quantities, $x(n,t)$ and $y(n,t)$, he had proposed the multiplication rule,

$$\begin{aligned} x(n,t)y(n,t) = \sum_{\beta=-\infty}^{+\infty} \sum_{\alpha=-\infty}^{+\infty} & x(n,n-\alpha)y(n-\alpha,n-\alpha-\beta)\cdot \\ & \exp\{\cdot i\omega(n,n-\alpha-\beta)t\}, \end{aligned} \tag{60}$$

with

$$\omega(n,n-\alpha-\beta) = \omega(n,n-\alpha) + \omega(n-\alpha,n-\alpha-\beta). \tag{61}$$

By identifying the complete set of quantities, $x(n,t)$, for n running over all stationary states of the system, with the dynamical variable $x(t)$, Heisenberg had concluded from Eq. (60) that: 'Whereas in classical theory $x(t)y(t)$ is always equal to $y(t)x(t)$, this is not necessarily the case in quantum theory' (Heisenberg, 1925c, p. 884). Somewhat frightened by this situation, which had never before arisen in the mathematical description of natural processes, Heisenberg then treated the example of the anharmonic oscillator in which only products of the same dynamical quantity, the position variable q, seemed to occur.

In order to integrate the equation of motion of the anharmonic oscillator, Heisenberg had used the re-interpreted quantum condition,

$$h = 4\pi m_0 \sum_{\alpha=0}^{\infty} \{|q(n,n+\alpha)|^2\omega(n+\alpha,n) - |q(n,n-\alpha)|^2\omega(n,n-\alpha)\}, \tag{62}$$

where m_0 is the oscillating mass. Heisenberg noticed that Eq. (62) had an

[161] This fact is not quite obvious from the paper itself. Apart from the calculation with the anharmonic oscillator, Heisenberg made use of his method of quantum-theoretical re-interpretation only to point out the consistency of Kramers' dispersion formula and to rederive the intensity formulae of R. Kronig (1925a) and H. Hönl (1925) for complex multiplets of spectral lines.

important consequence: If the coefficients $q(n, n \pm \alpha)$ were identified with $A(n, n \pm \alpha)$, the emission and absorption amplitudes of Kramers' dispersion formula for P_{qu}, the electric moment of an atom (electron charge $-e$, frequencies $\omega(n, n \pm \alpha)$), induced by an incident electromagnetic wave of amplitude $E_0 \cos\{\omega_0 t\}$, that is,

$$P_{qu} = e^2 E_0 \cos\{\omega_0 t\} \frac{4\pi}{h} \sum_{\alpha=0}^{\infty} \left[\frac{|A(n, n+\alpha)|^2 \omega(n+\alpha, n)}{\omega^2(n+\alpha, n) - \omega_0^2} - \frac{|A(n, n-\alpha)|^2 \omega(n, n-\alpha)}{\omega^2(n, n-\alpha) - \omega_0^2} \right], \tag{63}$$

then the limit of P_{qu} for very high frequency of the incident radiation could be calculated. For $\omega_0 \gg |\omega(n, n \pm \alpha)|$, one obtains with the help of Eq. (62) the result

$$P_{qu} = -e^2 \frac{E_0 \cos\{\omega_0 t\}}{\omega_0^2 m_e}, \tag{64}$$

which is completely identical with the classical expression for the scattering moment of the free electron of mass m_e. (See our discussion in Volume 1, Section VI.1.)

With respect to discovering a consistent quantum-theoretical dynamics the derivation of Eq. (64) was an important result. It showed, for the first time, that there existed an analogy between classical dispersion theory, a consistent dynamical scheme, and the dispersion-theoretic equations of Kramers and others that had no dynamical foundation before. Thus, Heisenberg's new scheme seemed to fulfill what Bohr had demanded in his talk of May 1925 in Cambridge from a correct quantum theory of atomic phenomena: it showed a systematic way of uniting good classical results (such as the scattering formula, Eq. (64)) and good quantum results (such as Kramers' formula, Eq. (63)). The only mathematical way of achieving this goal was the introduction of the quantum-theoretical Fourier series, Eq. (59), which obeyed the multiplication rule given by Eq. (60). In order to determine whether the new formalism really provided the key to the solution of the quantum difficulties that Bohr had discussed, Dirac decided to investigate how the application of Heisenberg's multiplication rule would work in the dispersion theory of Kramers and Heisenberg, which he had learned from Fowler's lectures.[162]

The starting point of the quantum dispersion-theoretic approach was the classical expression of P_{cl}, the electric moment of an electron bound in atom,

[162]In the first part of his lectures on 'Recent Developments,' Fowler had discussed the paper of Heisenberg and Kramers in detail (see footnote 141). At that time the paper had just appeared in print (Kramers and Heisenberg, 1925), and Fowler requested Bohr to send him several copies for his use (Fowler to Bohr, 4 May 1925).

which is induced by the incident radiation (electric field $E_0 \cos\{\omega_0 t\}$, as in Eqs. (63) and (64), of frequency ω_0),

$$P_{\mathrm{cl}} = \frac{\pi}{2} \operatorname{Re} \sum_{\alpha} \sum_{\beta} \left[\left(\beta \frac{\partial A_\alpha}{\partial J} \right) \exp\{i(\alpha\omega)t\} \left(\frac{E_0 A_\beta}{\beta\omega + \omega_0} \right) \exp\{i(\beta\omega + \omega_0)t\} \right]$$
$$- \frac{\pi}{2} \operatorname{Re} \sum_{\alpha} \sum_{\beta} \left[A_\alpha \exp\{i(\alpha\omega)t\} \alpha \frac{\partial}{\partial J} \left(\frac{E_0 A_\beta}{\beta\omega + \omega_0} \right) \exp\{i(\beta\omega + \omega_0)t\} \right], \tag{65}$$

where Re denotes the real part of the expression following it, and α and β assume all integral values (Kramers and Heisenberg, 1925, p. 688, Eq. (15)). The A_α's are the Fourier coefficients of P_0,

$$P_0 = \frac{1}{2} \sum_{\alpha} A_\alpha \exp\{i(\alpha\omega)t\}, \tag{66}$$

the electric moment of the atom in the absence of external radiation, corresponding to the harmonic frequencies $(\alpha\omega)$. In a system of f degrees of freedom with fundamental frequencies $\omega_1, \ldots, \omega_f$, the bracket $(\alpha\omega)$ denotes $\sum_{\alpha_k} \alpha_k \omega_k$, the sum of terms, with integer α_k. Also, in Eq. (65), the differential expressions $\partial(\ldots)/\partial J$ denote $\sum_k \partial(\ldots)/\partial J_k$, the sum of the differential quotients with respect to the f action variables J_k. From the previous work on dispersion theory Dirac was aware of the fact that no classical expression could be 'translated' or re-interpreted successfully quantum-theoretically unless it could be cast in a form similar to the right-hand side of Eq. (65), or, in the general form

$$z_\gamma = \sum_{\alpha} \sum_{\substack{\beta \\ (\alpha+\beta=\gamma)}} \left\{ \beta \frac{\partial x_\alpha}{\partial J} y_\beta - \alpha \frac{\partial y_\beta}{\partial J} x_\alpha \right\}, \tag{67}$$

where x_α and y_β are the Fourier coefficients of physical quantities—as, for example, the A_α's in Eq. (65)—depending only on J_k, the action variables of the periodic system. Since J, the action variable, assumes only discrete values in quantum theory,

$$J = J_0 \pm \alpha h \qquad (\alpha = 0, 1, 2, \ldots), \tag{68}$$

the quantum translation of the classical expression (67), for a system of one degree of freedom, had to be written as

$$z_\gamma(J) = \frac{1}{h} \sum_{\alpha} \sum_{\substack{\beta \\ \alpha+\beta=\gamma}} \left[\{x_\alpha(J) - x_\alpha(J - \beta h)\} y_\beta(J) \right.$$
$$\left. - x_\alpha(J)\{y_\beta(J) - y_\beta(J - \alpha h)\} \right], \tag{69}$$

where the values of the action variables for the Fourier coefficients x_α and y_β are indicated explicitly. Now, in terms of Heisenberg's quantum-theoretical reinterpretation, the expressions $x_\alpha(J)$, $x_\alpha(J-\beta h)$, $y_\beta(J)$, $y_\beta(J-\alpha h)$ and $z_\gamma(J)$ could simply be replaced by $x(n,n-\alpha)$, $x(n-\beta,n-\alpha-\beta)$, $y(n,n-\beta)$, $y(n-\alpha,n-\alpha-\beta)$ and $z(n,n-\gamma)$, respectively. Hence, Eq. (69) reads

$$hz(n,n-\gamma) = \sum_{\alpha} \sum_{\substack{\beta \\ (\alpha+\beta=\gamma)}} \big[x(n,n-\alpha)y(n-\alpha,n-\alpha-\beta) - y(n,n-\beta)x(n-\beta,n-\alpha-\beta)\big] \tag{70}$$

$$= \sum_{\alpha} \big[x(n,n-\alpha)y(n-\alpha,n-\gamma) - y(n,n-\alpha)x(n-\alpha,n-\gamma)\big],$$

where the sum over α extends to all integers. Due to Eq. (60), the result on the right-hand side yields $hz(n,n-\gamma)$, the coefficient of the quantum Fourier series for $hz(n,t)$; hence $h \cdot z(t)$, the dynamical variable associated with it, is just the difference of the 'Heisenberg products' $x(t)y(t)$ and $y(t)x(t)$.

Thus, from his analysis Dirac concluded that *all* the expressions obtained hitherto from the so-called dispersion-theoretic approach to quantum systems could be written in the form of differences of products of 'quantum variables,' as given in Eq. (70), in which Heisenberg's multiplication rule, Eq. (60), had to be applied.[163]

Ironically, Heisenberg had not consciously made any use of the property of noncommutativity of variables in quantum mechanics; rather he had made

[163]The considerations connecting Eq. (67) to Eq. (70) are to be found on a sheet which Dirac kept together with his manuscript on Bose–Einstein statistics (AHQP, Microfilm No. 36, Section 9). There he wrote:

'$$\frac{\partial x_\tau}{\partial J}\tau' y_{\tau'} - \frac{\partial y_{\tau'}}{\partial J}\tau\, x_\tau = \tau'\frac{\partial x_\tau}{\partial J}\, y_{\tau'} - \tau\frac{\partial y_{\tau'}}{\partial J}\, x_\tau$$

$$= \{x_\tau(J+\tau' h) - x_\tau(J)\}\, y_{\tau'}(J) - x_\tau(J)\{y_{\tau'}(J+\tau h) - \cdots\}$$

$$= \{x(n,n-\tau) - x(n-\tau',n-\tau-\tau')\}\, y(n,n-\tau')$$

$$- x(n,n-\tau)\{y(n,n-\tau') - y(n-\tau,n-\tau-\tau')\}\, x_\tau$$

$$\rightarrow x(n,n-\tau).$$'

The inverse process of the quantum-theoretical reformulation of the classical expression, i.e., the translation of the quantum difference expression into a differential expression in the limit of high quantum numbers n, is found in Dirac's published paper, and we shall discuss it in the next section.

In our discussion, we have supplied the factor h where it was missing, and have replaced $\tau'(\partial x_\tau(J)/\partial J)y_{\tau'}(J)$ (of Dirac's notation) by $(1/h)\{x_\tau(J) - x_\tau(J-\tau' h)\}\, y_{\tau'}(J)$, etc. However, these changes do not affect the result obtained by Dirac in his note. The latter has to be regarded as evidence of Dirac's earlier step towards his first paper on quantum mechanics (Dirac, 1925d).

attempts at avoiding it. What Dirac revealed by his analysis was that Heisenberg had not really been able to throw noncommuting products out of the theory, even though he had chosen the example of the anharmonic oscillator, in whose equation of motion just the commuting products of the same dynamical variable, the position coordinate q, occurred. But, in integrating the equation of motion, Heisenberg had also applied the quantum condition, Eq. (62), which, as Dirac discovered, could be written in the form

$$1 = \frac{2\pi i}{h} m_0 \sum_{\alpha=-\infty}^{+\infty} \{\dot{q}(n,n+\alpha)q(n+\alpha,n) - q(n,n+\alpha)\dot{q}(n,n+\alpha)\}. \quad (71)$$

If the quantity $m_0\dot{q}$ is identified with p, the momentum variable of the oscillator, as in classical mechanics, and the multiplication law, Eq. (60), is used, Eq. (71) becomes

$$\{pq - qp\}(n,n) = \frac{h}{2\pi i}, \quad (72)$$

which states that all the constant terms in the Fourier series of the difference $qp - pq$ are different from zero. Hence, the variables q and p, denoting the position and momentum of the one-dimensional oscillator, respectively, do not commute.

Equation (72) was not contained in Heisenberg's paper, and Dirac commented later on: 'Heisenberg was in that way [i.e., by the quantum condition (62)] very reluctantly led to non-commutative algebra. He was very reluctant because it was so foreign to all ideas of physicists at that time, and when it first turned up he thought there must be something wrong with his theory and tried to correct it but he was just forced to accept it' (Dirac, 1972, p. 152). In comparison with Heisenberg, Dirac had a great psychological advantage: he had no prejudice against noncommuting quantities for several reasons.[164] First, he already knew an example of such noncommuting quantities in mathematics, the quaternions,

[164] Dirac said: 'Now when Heisenberg noticed that [his physical quantities did not satisfy the commutative law of multiplication] he was really scared. It was such a foreign idea. Physicists from the earliest times had always thought of the variables that they were using as quantities satisfying the ordinary laws of algebra. It was quite inconceivable that two physical things when multiplied in one order should not give the same result as when multiplied in the other order. It was thus most disturbing to Heisenberg. He was afraid this was a fundamental blemish in his theory and that probably the whole beautiful idea would have to be given up.

'. . . I saw that the noncommutation was really the dominant characteristic of Heisenberg's theory . . . So I was led to concentrate on the idea of noncommutation and to see how the ordinary dynamics which people had been using until then should be modified to include it.

'At this stage, I had an advantage over Heisenberg because I did not have his fears. I was not afraid of Heisenberg's theory collapsing. It would not have affected me as it would have affected Heisenberg . . . .

'I think it is a general rule that the originator of a new idea is not the most suitable person to develop it because his fears of something going wrong are really too strong and prevent his looking at the method from a purely detached point of view in the way that he ought to.' (Dirac, 1971, pp. 20–24).

and although he did not think of them when investigating the content of Heisenberg's paper he did assume—contrary to most physicists in those days—that noncommuting quantities were possible.[165] Second, Dirac thought about the problems of dispersion theory in summer 1925 for the first time with a fresh mind and was not concerned with technical details. He had heard from Niels Bohr that some unusual methods might be called for in obtaining a satisfactory theory, and now he came to the conclusion that noncommutativity was the crucial unusual feature revealed by quantum phenomena. To be sure, noncommuting quantities had not been used in quantum theory earlier, but they offered the only hope of establishing a *unified description* of atomic processes—of what Bohr had called 'reciprocal' and 'irreciprocal' interactions.

In Dirac's mind one of the quantum phenomena, which could not be accounted for by conventional mechanical description, was the observation that in the Stern–Gerlach experiment angular momentum, the special orientation of atoms in space, was transferred to the atoms of the deflected beam almost instantaneously, whereas classical physics predicted a time of the order of 10^9 s for it to happen.[166] Another fundamental problem was the manner in which energy accumulates in a photoelectron, that is, an electron before it is kicked out of an atom by incident radiation, especially when the intensity of the latter is extremely small.[167] How would the property of noncommutativity resolve these difficulties? While Dirac did not immediately undertake to work on these problems, he was very much concerned with a third difficulty of the old quantum theory: namely, to account for the observed line spectrum of the neutral helium atom.[168] He

[165] The first rewriting of Eq. (62) as Eq. (72) was done by M. Born and P. Jordan (1925b, p. 870, Eq. (37)) in their paper published in September 1925. Dirac's discovery of the noncommutativity of the dynamical variables q and p (in the beginning of September 1925) was independent of Born's earlier discovery (in July 1925). While Born noticed Eq. (72) to be the consequence of the fact that q and p are matrices, Dirac's conclusion was derived from a general analysis of the structure of formulae obtained in the so-called dispersion-theoretic approach.

Only after introducing his own noncommuting quantities, the q-numbers, did Dirac remember the quaternions as providing the most general algebra. He concluded that quaternions might be a perfect tool for quantum theory; however, he did not make any progress in quantum mechanics by using quaternions.

[166] The classical estimate was presented in the paper of A. Einstein and P. Ehrenfest (1922, p. 32) and had been a riddle ever since. The difficulty had also been discussed by N. Bohr in his review paper on the fundamental postulates (Bohr, 1923a), where he had argued that the orientation of the momenta, parallel and antiparallel to the direction of the inhomogeneous magnetic field, was already present in the atom and could, therefore, be stimulated by the external field nearly instantaneously (Bohr, 1923a, p. 149, footnote 1). However, Bohr had not given a proper mechanical and mathematical description of the phenomenon, and he still worried about it in 1925.

[167] A. Sommerfeld had discussed this difficulty in some detail in the fourth edition of his *Atombau und Spektrallinien*, (Sommerfeld, 1924d, pp. 58–59). He argued that it demonstrated the necessity of using both the light-quantum picture and the wave picture in order to describe radiation phenomena. He concluded that: 'Here modern physics stands between irreconcilable opposites and must frankly confess "*non liquet*"' (Sommerfeld, 1924d, p. 59).

[168] Dirac mentioned all three problems in discussions with me as the outstanding problems of atomic theory that bothered him. He would elaborate on the problem of the accumulation of energy in the photoelectron in a paper dealing with the emission and absorption of radiation (Dirac, 1927b, p. 265). (J. Mehra)

recalled later:

> The young people in those days were trying to account for the spectrum of helium by setting up a theory of interaction of Bohr orbits, and there is no doubt that they would have continued along these lines if it had not been for Heisenberg and Schrödinger. They would have probably continued for decades on those lines working out the interactions between Bohr orbits, with a large number of people working on the problem and continually modifying their assumptions and comparing the results of their calculations with experimental progress which had been made. People would have found that some assumptions gave results in tolerable agreement with experiment, and they would have become accepted more or less, and people would have then gone on to build up more elaborate theories dealing in greater detail with this interaction . . . But you can see that such work would have been very unlikely to lead to any really important developments . . . However, the one fundamental idea which was introduced by Heisenberg and Schrödinger was that one must work with non-commutative algebra, . . . [and these] people studying [interacting] Bohr orbits . . . would have never thought of introducing that kind of mathematics into their work. (Dirac, 1972, pp. 148–150)

Dirac had himself been working along lines which he admitted would have led nowhere. In Heisenberg's introduction of noncommuting dynamical variables he saw the way out of the previous difficulties, for the equations containing them (e.g., Eq. (72)) 'apply as well to non-periodic systems, of which none of the constituent particles go off to infinity, such as a general atom' (Dirac, 1925d, p. 653). In the helium atom, where he expected some kind of aperiodic motion to be present, one would not have to assign a single quantum number n to each stationary state; hence the difficulties of associating a nonintegral quantum number with a particular state of the atom, would not exist. However, even here 'the quantum variables would still have harmonic components, each related to two n's [i.e., two indices], and Heisenberg's multiplication could be carried out exactly as before' (Dirac, 1925d, p. 653).

Thus, the search for the solution of the helium problem guided Dirac's first work in quantum mechanics to a large extent. Though his conception of this problem turned out to be wrong and he did not succeed in achieving his goal, he was right in his first impression about the crucial role of noncommutativity. As he said later: 'I suddenly saw that Heisenberg's idea provided the key to the whole mystery' (Dirac, Conversations, p. 14).

IV.2 The Dynamical Significance of Noncommutativity

Once Dirac had discovered that Heisenberg's work contained the essential step that would help in overcoming the outstanding difficulties of atomic theory, he proceeded to answer a question that seemed to him fundamental for formulating a complete dynamical theory. How were the mathematical expressions that occurred in Heisenberg's scheme, in particular the differences of products of variables of the type given by the right-hand side of Eq. (70), related to the concepts of classical mechanics? 'At that time,' Dirac explained later, 'I was expecting some kind of connection between the new mechanics and Hamiltonian dynamics (as Hamiltonian dynamics was used so much with Sommerfeld's development of the Bohr theory) and it seemed to me that this connection should show up with large quantum numbers' (Dirac to van der Waerden, 28 August 1961).

The requirement that, in the limit of large quantum numbers, classical and quantum expressions describing the same physical situation should become identical had been one of the fundamental assumptions of Bohr's atomic theory from its very beginning in 1913. No one working in quantum theory had ever doubted the validity of the correspondence principle. Thus, it would appear to be quite a natural step to consider the high quantum number limit of the difference, $xy - yx$, where x and y are quantum dynamical variables in the sense of Heisenberg, and to look for its counterpart in classical theory. However, with Dirac the psychological situation was entirely different. He had not been one of Bohr's boys, like Hendrik Kramers, who played the correspondence game with great virtuosity and derived many results from it. Dirac had thus far refused to make use of correspondence arguments, for, in his opinion, they did not yield a unique prescription for writing down well-defined equations. Even after reading the papers of Kramers (1924a, b) and Kramers and Heisenberg (1925) on dispersion theory he was unwilling to proceed along similar lines to treat other problems of atomic theory. It would have been against his grain to employ the correspondence principle to obtain such results as those of Heisenberg (1925a, b) and Kronig (1925a, b) on the intensity of complex multiplet lines of atoms subject to magnetic fields. In contrast to these, Heisenberg's work on the quantum-theoretical re-interpretation of kinematics and mechanics appeared to be on a different level, for it provided clear prescriptions for formulating mathematical equations. Thus, his first task was to find the connection between Heisenberg's new mechanics and classical mechanics, and, in particular, to search for the classical expression corresponding to the difference, $xy - yx$. And that search would involve no mathematical ambiguity, as all steps for determining the high quantum number limit of 'Heisenberg products' were well defined.

Dirac considered a multiply periodic system of f degrees of freedom. The variables of this system, x, y, etc., are represented by quantum-theoretical, f-fold Fourier sums; for instance, the variable x is given by the harmonic expansion

$$x(n,t) = \sum_{\alpha_1=-\infty}^{+\infty} \cdots \sum_{\alpha_f=-\infty}^{\infty} x(n_1, \ldots, n_f; n_1 - \alpha_1, \ldots, n_f - \alpha_f) \cdot \exp\left\{ i \sum_{k=1}^{f} \omega(n_k, n_k - \alpha_k) t \right\}, \tag{73}$$

where n stands for a set of indices, n_k, $k = 1, \ldots, f$, having discrete values. For a given set of such numbers the sum on the right-hand side of Eq. (73) consists of terms, which Dirac called 'harmonic components' or just 'components' (Dirac, 1925d, p. 643), each component belonging to one particular transition with a definite quantum frequency, given by $\sum_{k=1}^{f} \omega(n_k, n_k - \alpha_k)$, with $\alpha_k = 0, \pm 1, \pm 2, \ldots$, etc. He now considered that the system is in a stationary state, such that all n_k's take on large values and only the transitions to neighbouring states, $n_k - \alpha_k$, with $|\alpha_k| \ll n_k$, are important. Then $x(n_1, \ldots, n_f; n_1 - \alpha_1, \ldots, n_f - \alpha_f)$, the coefficients of the expansion—which Dirac called 'amplitudes' (Dirac, 1925d, p. 643), abbreviating the term 'Fourier amplitude' of classical theory or 'transition amplitude' of quantum theory—would vary slowly with the values of the indices n_k, and Dirac felt justified in replacing n_k's by suffixes, κ_k, that are varying quantities. Thus, he rewrote the amplitudes as

$$\begin{aligned} x(n, n-\alpha) &= x(n_1, \ldots, n_f; n_1 - \alpha_1, \ldots, n_f - \alpha_f) \\ &= x_{\alpha_1, \ldots, \alpha_f; \kappa_1, \ldots, \kappa_f} = x_{\alpha\kappa}. \end{aligned} \tag{74}$$

Since, with the above approximations, the differences between the values of n_k and $n_k + \alpha_k$ do not matter, the values of the suffixes κ_k determine completely the states of the quantum system. In the Bohr–Sommerfeld semiclassical theory, the states of a multiply periodic system are defined by the values of the action variable, $J_k = (n_k + \text{const.})h$, h being Planck's constant. In the limit of high quantum numbers the classical and the quantum-mechanical descriptions should become identical, that is, the quantum-mechanical variables should approach the 'corresponding' classical ones; hence Dirac was allowed to identify κ_k, the suffixes of the quantum variables, with the classical action variable J_k. He could thereby establish the correspondence between $x(n, n-\alpha)$, the quantum-mechanical amplitude, and $x_{\alpha J}$, the classical Fourier coefficient.

With the help of these considerations Dirac evaluated the difference $xy - yx$ in the correspondence limit; that is, he evaluated the expression

$$\begin{aligned} (xy - yx)(n,t) = \sum_{\alpha,\beta} \{ & x(n, n-\alpha) y(n-\alpha, n-\alpha-\beta) \\ & - y(n, n-\beta) x(n-\beta, n-\alpha-\beta) \} \cdot \\ & \exp\{ i\omega(n, n-\alpha-\beta)t \}, \end{aligned} \tag{75}$$

in the limit $|\alpha|, |\beta| \ll n$, where n represents the set of indices, $n_1, \ldots, n_f$. By replacing κ_k in Eq. (74) by the action variable J_k he now rewrote the terms on the right-hand side of Eq. (75) as

$$\begin{aligned}&\{x(n,n-\alpha)y(n-\alpha,n-\alpha-\beta)-y(n,n-\beta)x(n-\beta,n-\alpha-\beta)\}\cdot\\&\quad\exp\{i\omega(n,n-\alpha-\beta)t\}\\&=h\sum_{k=1}^{f}\left(\beta_k\frac{\partial x_{\alpha J}}{\partial J_k}y_{\beta J}-\alpha_k\frac{\partial y_{\beta J}}{\partial J_k}x_{\alpha J}\right)\exp\{i\omega(n,n-\alpha-\beta)t\},\end{aligned}\tag{76}$$

thus inverting the steps that had led him previously from Eq. (67) to Eq. (70).[169]

Equation (76) could not as yet be compared to an appropriate classical expression, even if one were to identify the quantum frequency, $\omega(n, n-\alpha-\beta)$, with the classical frequency, $(\alpha+\beta)\omega$, in the given state—which is permitted in the approximation considered—for it also contained the numbers α_k and β_k. However, the factors $\beta_k y_{\beta J}$ and $\alpha_k x_{\alpha J}$ could be rewritten with the help of the equations

$$\frac{\partial}{\partial w_k}(y_{\beta J}\exp\{i(\beta\omega)t\})=2\pi i\beta_k y_{\beta J}\exp\{i(\beta\omega)t\}\tag{77a}$$

and

$$\frac{\partial}{\partial w_k}(x_{\alpha J}\exp\{i(\alpha\omega)t\})=2\pi i\alpha_k x_{\alpha J}\exp\{i(\alpha\omega)t\},\tag{77b}$$

where w_k $(=(\omega_k/2\pi)t)$ is the angle variable conjugate to J_k. Thus, Dirac finally obtained the result

$$\begin{aligned}(xy-yx)(n,t)=\frac{ih}{2\pi}\sum_{k=1}^{f}\sum_{\alpha,\beta}\Bigg[&\frac{\partial}{\partial w_k}(x_{\alpha J}\exp\{i(\alpha\omega)t\})\cdot\frac{\partial}{\partial J_k}(y_{\beta J}\exp\{i(\beta\omega)t\})\\&-\frac{\partial}{\partial J_k}(x_{\alpha J}\exp\{i(\alpha\omega)t\})\cdot\frac{\partial}{\partial w_k}(y_{\beta J}\exp\{i(\beta\omega)t\})\Bigg],\end{aligned}\tag{78}$$

which is valid in the correspondence limit. In this limit the left-hand side represents $xy-yx$, the product difference of the quantum variables in the state in which n_k's assume high numerical values. On the right-hand side of Eq. (78), the quantities within parentheses, which are differentiated with respect to the classical action and angle variables, represent the classical variables in the state that 'corresponds' to the quantum-mechanical stationary state mentioned above.

[169] In replacing the factor $y(n-\alpha, n-\alpha-\beta)$ by $y_{\beta J}$, Dirac neglected the difference between the action variable J in the states n and $n-\alpha$; he did the same when he wrote $x(n-\beta, n-\alpha-\beta)$ as $x_{\alpha J}$. He also assumed that the difference, $x(n, n-\alpha)-x(n-\beta, n-\alpha-\beta)$ and $y(n, n-\beta)-y(n-\alpha, n-\alpha-\beta)$, could be approximated by the differential quotient with respect to the action variable J; thus, for example, $x(n, n-\alpha)-x(n-\beta, n-\alpha-\beta)=h\beta(\partial x_{\alpha J}/\partial J)$, etc. Note that in Eq. (76) the suffixes α and J stand for f suffixes $\alpha_1, \ldots, \alpha_f$ and $J_1, \ldots, J_f$, respectively.

Hence Eq. (78) holds for all states that satisfy the conditions governing the correspondence limit, and Dirac concluded that the difference of products of the quantum variables 'corresponds' to a classical expression containing the differences of products of two derivatives. Thus, he arrived at his general 'correspondence' relation,

$$xy - yx \rightarrow \frac{ih}{2\pi} \sum_{k=1}^{f} \left(\frac{\partial x}{\partial w_k} \frac{\partial y}{\partial J_k} - \frac{\partial y}{\partial J_k} \frac{\partial x}{\partial J_k} \right). \tag{79}$$

The analysis of Heisenberg's paper, which occupied Dirac for about two to three weeks, was thus completed. When, on a Sunday in September 1925, he found the differential expression (79), it occurred to him that, apart from the factor $ih/2\pi$, it might be the so-called Poisson bracket of classical mechanics, although he was not sure about it.[170] The next morning he confirmed his guess by looking up the exact definition of Poisson brackets in the library. The expression within parentheses in Eq. (79) was indeed the Poisson bracket of the classical variables x and y, with the action and angle variables, J_k and w_k, representing the fundamental dynamical variables describing the multiply periodic system.[171] He discovered that by this recognition everything fell into place. Dirac concluded that the correspondence relation, Eq. (79), would enable him to cast Heisenberg's quantum-mechanical relations into the Hamiltonian framework. All he had to do was to translate into quantum language the results obtained in classical dynamics a century before.

The brackets in question had been introduced in mechanics by Siméon Denis Poisson in a paper read at the *Institut de France* on 16 October 1809.[172] Poisson

[170] In his Varenna lectures, 1972, Dirac said:

> I was in Bristol at the time I started on Heisenberg's theory. I had returned home for the last part of the summer vacation, and I went back to Cambridge at the beginning of October, 1925, and resumed my previous style of life, intensive thinking on the problems during the week and relaxing on Sunday, going for a long walk in the country alone. It was during one of the Sunday walks in October, when I was thinking very much about this $uv - vu$, in spite of my intention to relax, that I thought about Poisson brackets. (Dirac, 1977, pp. 121–122)

This statement suggests that the Sunday in question was either 4 or 11 October 1925. Both dates, however, would not have left Dirac much time to continue his analysis of Heisenberg's paper and to develop his own mathematical formulation of the theory. Earlier, Dirac always maintained to me (J. Mehra) that it was in September when the idea about the Poisson brackets first came to him. Because of the great amount of work which he still had to perform until the beginning of November when the first paper was submitted (Dirac, 1925d), a date in the second half of September, probably Sunday, 20 September, is more likely. (See Mehra, 1972, p. 31.)

[171] The book which Dirac consulted was Whittaker's *Analytical Dynamics*. (See also Dirac, 1977, p. 122.)

[172] S. D. Poisson was born on 21 June 1781 at Pithiviers in the department of Loiret, France. He first studied medicine and chemistry, and finally mathematics. In 1798 he entered the *École Polytechnique* at Paris, where he attracted the attention of Joseph Louis Lagrange and Pierre Simon de Laplace. At the *Polytechnique* he was made an examiner and deputy professor in 1802, and full professor in 1806. In 1808 he became astronomer to the *Bureau des Longitudes*, and a year later he was appointed first professor of pure mechanics at the newly instituted *Faculté des Sciences* of the Sorbonne.

Poisson worked in pure and applied mathematics; he published about 300 memoirs and several books, including a two-volume standard textbook on mechanics, entitled *Traité de mecanique* (1811),

attempted to calculate the motion of a planet around the sun in the presence of the perturbing gravitational interaction of the other $(n-1)$ planets (Poisson, 1809). He approximated this astronomical n-body problem by assuming the sun to be fixed in space and including the perturbing gravitational action of the other planets in a potential, Ω, which is a function of time and the space-point of the planet under investigation. The entire n-body system possesses $3n$ degrees of freedom, and its motion is described by as many Lagrangian equations, that is,

$$\frac{d}{dt}\left(\frac{\partial L}{\partial \dot{q}_k}\right) - \frac{\partial L}{\partial q_k} = -\frac{\partial \Omega}{\partial q_k}, \qquad k = 1, \ldots, 3n, \tag{80}$$

for the $3n$ position coordinates, q_k, and their time derivatives, $\dot{q}_k$. The function L is the Lagrangian of the unperturbed planet–sun system in the fixed gravitational potential of the sun.

Poisson proposed to solve these equations in the following way. First he wrote down c_ρ, the known $6n$ first integrals of the unperturbed motion, that is, the constants obtained by integrating the $3n$ second-order equations (80) with the perturbing potential Ω equal to zero. In this case, c_ρ may be considered as functions of the position and momentum coordinates, q_k and p_k $(= \partial L/\partial \dot{q}_k)$, as well as of the time t, such that their overall time dependence vanishes.[173] However, under the action of a finite perturbation potential, Ω, the c_ρ will vary slowly with time in accordance with the equations[174]

$$\frac{dc_\rho}{dt} = (c_\rho, c_1)\frac{\partial \Omega}{\partial c_1} + \cdots + (c_\rho, c_{6n})\frac{\partial \Omega}{\partial c_{6n}}, \qquad \rho = 1, \ldots, 6n, \tag{81}$$

---

and books on the theory of waves, capillarity, mathematical theory of heat and the application of the theory of probability to court judgments. In pure mathematics he contributed, in particular, to the theory of definite integrals, Fourier series and probability (Poisson distribution). His researches in mechanics were inspired by the ideas of his teachers Lagrange and Laplace, and he devoted great attention to general questions, such as the introduction of the momentum coordinate p instead of Lagrange's velocity coordinate $\dot{q}$. He worked on all parts of the mathematical physics of his day: on potential theory (Poisson's equation), capillarity, elasticity of materials (Poisson's constant), heat, acoustics, and especially on electrostatics and magnetism, contributing to each field he touched on.

Poisson was elected a member of the Academy of Sciences, Paris, in 1812, and was made a peer in 1837. He died in Paris on 25 April 1840, having been occupied up to his death almost entirely with teaching and research.

[173] In astronomy, for example, the six elliptic elements of a planet, i.e., the major axis, the eccentricity, the two angles fixing the plane, the length of the perihelion and the average period during which the planet passes once through the perihelion position, represent the first integrals c_ρ.

[174] To derive Eqs. (81), one observes that Eq. (80) can be written as

$$\frac{dp_k}{dt} = \frac{\partial p_k}{\partial t} - \frac{\partial \Omega}{\partial q_k}, \qquad k = 1, \ldots, 3n.$$

Inserting this result into the expressions for the total time derivative of the c_ρ,

$$\frac{dc_\rho}{dt} = \sum_{k=1}^{3n} \frac{\partial c_\rho}{\partial q_k}\frac{dq_k}{dt} + \sum_{k=1}^{3n} \frac{\partial c_\rho}{\partial p_k}\frac{dp_k}{dt} + \frac{\partial c_\rho}{\partial t}, \qquad \rho = 1, \ldots, 6n,$$

and keeping in mind that the left-hand sides of these equations are zero for the unperturbed motion

in which the brackets are defined as

$$(c_\rho, c_\sigma) = \sum_{k=1}^{3n} \left(\frac{\partial c_\rho}{\partial q_k} \frac{\partial c_\sigma}{\partial p_k} - \frac{\partial c_\sigma}{\partial q_k} \frac{\partial c_\rho}{\partial p_k} \right). \tag{82}$$

Equations (81) yield $6n$ partial differential equations of the first order; their solutions describe the motion of the n-body system, because from c_ρ one can calculate, for instance, the $6n$ position and momentum coordinates, q_k and p_k.

The advantage of using Eqs. (81) lies in a result that Poisson obtained after a lengthy calculation: that the bracket expressions, defined by Eqs. (82), are independent of time. Poisson had noted this observation more or less as a curiosity which helps in evaluating the perturbation equations (81). More than thirty years later, in a letter to the *Académie des Sciences*, Paris, Carl Gustav Jacob Jacobi described it as 'the most profound discovery of Monsieur Poisson' and 'the most important theorem of dynamics' ('*la plus profonde découverte de M. Poisson*' and '*le théorème . . . le plus important de la mécanique*,' Jacobi, 1841, p. 529).

Jacobi had come across Poisson's bracket expressions in connection with the investigation of the complete solution of the Hamiltonian equations of motion for a n-body system, that is, of the equations,

$$\frac{\partial q_k}{\partial t} = \frac{\partial H}{\partial p_k}, \qquad \frac{\partial p_k}{\partial t} = - \frac{\partial H}{\partial q_k}. \tag{83}$$

Assuming that f_1 and f_2 are constants of the motion, he had found that f_3, the Poisson bracket of f_1 and f_2, must be a constant of the motion as well; hence, from two integrals of a mechanical system one can always derive a third. 'After I had discovered this theorem, I reported to the academies at Berlin and Paris about it as if it were an entirely new discovery,' remarked Jacobi in his lectures

---

(i.e., when one replaces each $\dot{q}_k$ by $\partial q_k/\partial t$ and each $\dot{p}_k$ by $\partial p_k/\partial t$), one obtains the equations

$$\frac{dc_\rho}{dt} = - \sum_{k=1}^{3n} \frac{\partial c_\rho}{\partial p_k} \frac{\partial \Omega}{\partial q_k} = - \sum_{k=1}^{3n} \frac{\partial c_\rho}{\partial p_k} \sum_{\sigma=1}^{6n} \frac{\partial \Omega}{\partial c_\sigma} \frac{\partial c_\sigma}{\partial q_k}.$$

Since Ω does not depend on the momenta p_k, the equations

$$\sum_{\sigma=1}^{6n} \frac{\partial \Omega}{\partial c_\sigma} \left(\sum_{k=1}^{3n} \frac{\partial c_\rho}{\partial q_k} \frac{\partial c_\sigma}{\partial p_k} \right) = 0, \qquad \rho = 1, \ldots, 6n$$

hold, which may be added to the right-hand side of the expressions for dc_ρ/dt to yield finally the equations

$$\frac{dc_\rho}{dt} = - \sum_{\sigma=1}^{6n} \frac{\partial \Omega}{\partial c_\sigma} \sum_{k=1}^{3n} \left(\frac{\partial c_\rho}{\partial p_k} \frac{\partial c_\sigma}{\partial q_k} - \frac{\partial c_\rho}{\partial q_k} \frac{\partial c_\sigma}{\partial p_k} \right), \qquad \rho = 1, \ldots, 6n.$$

These last equations are identical to Eqs. (81).

on dynamics in winter 1842–1843. 'However, soon thereafter I realized that this theorem had already been discovered and buried for thirty years, for one had not suspected its true significance and had just used it as an auxiliary theorem for a totally different problem' (Jacobi, 1866, p. 269). Thus, Jacobi had proved independently what he called 'Poisson's theorem'—that is, the time independence of the Poisson bracket of two constants of motion of a mechanical system—with the help of a 'Jacobi identity,'

$$(F_1, F_2, F_3) = ((F_1, F_2), F_3) + ((F_2, F_3), F_1) + ((F_3, F_1), F_2) = 0, \tag{84}$$

which is valid for any arbitrary functions, F_1, F_2, F_3, of q_k and p_k, the dynamical variables of the system under investigation. Jacobi noticed that the time derivative of every such function, F, becomes identical to its Poisson bracket with H, the Hamiltonian, that is,

$$\frac{dF}{dt} = (F, H); \tag{85}$$

moreover, if both F_1 and F_2 do not depend on time, and $F_3 = H$, the constancy of the Poisson bracket, (F_1, F_2), follows immediately from Eq. (84).

In classical mechanics there occur many examples in which the Poisson brackets of two functions of the dynamical variables assume constant numerical values independent of the properties of the system under investigation. An example of this is the bracket expressions belonging to a given fundamental set of dynamical variables, say the position coordinates q_k and their conjugate momentum coordinates p_k, that is,

$$(q_k, p_l) = \delta_{kl}, \qquad (q_k, q_l) = (p_k, p_l) = 0, \tag{86}$$

where δ_{kl} is the Kronecker symbol ($\delta_{kl} = 1$ for $k = l$ and zero otherwise). Jacobi discovered that *all* sets of dynamical variables, Q_k and P_k, which are obtained from q_k and p_k by a transformation that leaves the Hamiltonian equations of motion (83) unaltered, satisfy the equation (Jacobi, 1838)

$$(Q_k, P_l) = \delta_{kl}, \qquad (Q_k, Q_l) = (P_k, P_l) = 0. \tag{87}$$

Moreover, one can prove that the Poisson brackets of any two functions, F_1 and F_2, remain invariant under all such transformations; that is, the equation

$$\sum_{k=1}^{3n} \left(\frac{\partial F_1}{\partial Q_k} \frac{\partial F_2}{\partial P_k} - \frac{\partial F_1}{\partial P_k} \frac{\partial F_2}{\partial Q_k} \right) = \sum_{k=1}^{3n} \left(\frac{\partial F_1}{\partial q_k} \frac{\partial F_2}{\partial p_k} - \frac{\partial F_1}{\partial p_k} \frac{\partial F_2}{\partial q_k} \right) \tag{88}$$

holds for any arbitrary system of $3n$ degrees of freedom. Since the transformation to a new set of variables plays an important role in the solution of dynamical problems, Poisson brackets have been used extensively with great success.

Shortly after 1870 Sophus Lie discovered that Poisson brackets could be particularly useful in describing the contact transformations of mechanical systems (see Lie, 1875, 1888). These transformations are given by the equations

$$Q_k = q_k + \Delta\epsilon \frac{\partial \phi}{\partial p_k}, \qquad P_k = p_k - \Delta\epsilon \frac{\partial \phi}{\partial q_k}, \tag{89}$$

where $\phi = \phi(q_1, \ldots, q_{3n}; p_1, \ldots, p_{3n})$ is a constant of the motion and $\Delta\epsilon$ represents an infinitesimal change of ϵ, the parameter of the mechanical system that happens to be canonically conjugate to ϕ. An arbitrary function F is then altered by an amount ΔF, where

$$\Delta F = \Delta\epsilon\,(F, \phi). \tag{90}$$

As an example, one may choose H, the Hamiltonian of a conservative system, as the constant of motion, then Eq. (90) assumes the special form

$$\Delta F = \Delta t\,(F, H), \tag{91}$$

which is identical with Eq. (85), the equation of motion for the function F. Thus, according to Lie, it is possible to interpret the evolution of a system in time—that is, the change of all its variables—as a succession of infinitesimal contact transformations generated by the transformation function $H\Delta t$.

In Whittaker's *Analytical Dynamics*, which Dirac consulted about the definition of Poisson brackets, the use of the latter in classical mechanics was treated in some detail. For example, in Chapter 11, dealing with the transformation theory of dynamics, the validity of Eqs. (87) as a sufficient condition for a contact transformation of the set of variables q_k, p_k into the set Q_k, P_k was emphasized (Whittaker, 1917, §131). Whittaker had reproduced, in the text as well as in the problems accompanying it, many formulae in which the equations of Hamiltonian dynamics were written with the help of Poisson brackets. Dirac was quite familiar with the contents of Whittaker's book, then the standard treatise on theoretical mechanics in England. That is why he recalled the notion of the Poisson bracket when, on discovering the differential expression (79), he pondered about its significance. The rereading of the appropriate sections of Whittaker's book did not take him very long, but it proved to be most useful for his further work on quantum mechanics.

Since the difference of 'Heisenberg products,' $xy - yx$, apart from the factor $ih/2\pi$, corresponded to the Poisson bracket (x, y) in classical theory, Dirac could now attempt to write down the equations of quantum mechanics in the Hamiltonian scheme essentially by replacing the Poisson brackets in classical equations—such as Eqs. (84)–(87)—by the differences of 'Heisenberg products.' This procedure, he claimed, also exhausted the content of the correspondence principle. It did so not by applying vague limiting procedures or performing delicate manipulations of classical expressions, different for each atomic problem, but by simply using the prescription of Eq. (79). As Dirac noted:

> If a series of classical operations leads to the equation $0 = 0$, the corresponding series of quantum operations must also lead to the equation $0 = 0$, and not to $h = 0$, since there is no way of obtaining a quantity that does not vanish by a quantum operation with quantum variables such that the corresponding classical operation with the corresponding classical variables gives a quantity that does vanish. The possibility . . . of deducing by quantum operations the inconsistency $h = 0$ thus cannot occur. *The correspondence between the quantum and classical theories lies not so much in the limiting agreement when $h \to 0$ as in the fact that the mathematical operations on the two theories obey in many cases the same laws.* (Dirac, 1925d, p. 649)

Although the goal was clearly defined, namely, to search for the quantum-mechanical counterpart of every fundamental classical mechanical equation containing Poisson brackets, the actual task was by no means easy. Dirac was aware that at every stage of the translation procedure he would have to examine in detail whether the noncommutativity of the dynamical variables would render a given classical relation meaningless in quantum theory. In some way Dirac's difficulties were similar to those that Heisenberg encountered when he tried to prove energy conservation in the quantum-theoretical scheme for the anharmonic oscillator.[175] However, his confidence in the vision of a Hamiltonian scheme describing atomic systems would overcome all difficulties. Within a few weeks of his discovery of the dynamical significance of noncommutativity, Dirac completed the formulation of 'The Fundamental Equations of Quantum Mechanics' (Dirac, 1925d).[176]

[175] As we have discussed in Volume 2, Section IV.4, Heisenberg could not derive energy conservation along the conventional line of argumentation from the fact that the anharmonic oscillator formally obeyed the same equation of motion as in classical theory. The reason was that the conclusions drawn on the basis of the classical procedure were valid only if the dynamical variables would commute.

[176] Although the discovery of the correspondence of the expression $xy - yx$ to the classical Poisson bracket was a brilliant result, Dirac was not the only person in fall 1925 to hit upon it. In late October of the same year Hendrik Kramers visited Wolfgang Pauli in Hamburg and became involved, just as Dirac had been a few weeks earlier, in an investigation of the quantum condition (62). He sought to interpret the physical significance of this equation by referring to the properties of an electron dispersing light in an atom. At the end of his conclusions he added that Pauli had drawn his attention to the fact that the difference, $pq - qp$, in Eq. (72), which could be derived from Eq. (62), corresponded to the Poisson bracket of momentum and position variables, p and q, in classical theory (Kramers, 1925c, p. 376).

Kramers and Pauli were aware not only of Heisenberg's paper on the quantum-theoretical re-interpretation of kinematics (Heisenberg, 1925c), but also of the following work of Born and Jordan (1925b) and of Born, Heisenberg and Jordan (1926) on matrix mechanics. The development of matrix mechanics had the same starting point as Dirac's, that is, the recognition of the property of noncommutativity in the product of quantum dynamical variables. Yet, in spite of the fact that Pauli and Kramers, unlike Dirac, were acquainted with the latest status of quantum mechanics in Göttingen and had independently come across the relation (79), giving the analogy with the Poisson bracket, they did not make use of these ideas to derive further conclusions. Of course, by the time they recognized the analogy they had no chance to publish anything earlier than Dirac; however, not knowing about the work in Cambridge, Kramers would have certainly indicated an interest in pursuing this correspondence analogy further, if he had had any. The fact remains that in this respect Pauli and Kramers were far behind Dirac; they were still too much imbued with the correspondence principle philosophy which had dominated the work of Niels Bohr and his collaborators for several

IV.3 Steps towards a Quantum Algebra

From his investigations Dirac concluded that most of the quantum-mechanical equations, obtained hitherto, exhibited the same formal structure as the corresponding classical equations. In particular, the sum of two variables in a classical equation was replaced by a sum of two analogous quantum variables in the quantum equation, and a classical product was replaced by a quantum-mechanical product. The mathematical operations in corresponding equations were sometimes the same, sometimes not. For example, the quantum-mechanical process of addition obeyed exactly the same law as the addition in classical theory: that is, the sum of two position coordinates, x_1 and x_2, could be obtained by simply adding the components, $x_1(nm)$ and $x_2(nm)$, which multiply the same frequency function $\exp\{i\omega(nm)t\}$ in the Fourier expansions for x_1 and x_2, as in Eq. (73). The main difference between classical and quantum mechanics had thus far been brought in by the new multiplication law. Dirac now asked himself whether it was possible to build systematically a complete scheme of quantum mechanics by following the suggestion that 'it is not the equations of classical mechanics that are in any way at fault, but that the mathematical operations by which physical results are deduced from them [in atomic physics] require modification' (Dirac, 1925d, p. 642). Guided by this idea he began to study the necessary modifications of the well-known mathematical operations of classical theory.

It is interesting that in searching for a consistent mathematical framework in which, in particular, the multiplication law of quantum variables could be incorporated, Dirac did not turn to the theory of matrices or quaternions, although he had come across the latter in his mathematical studies at Bristol. It is true that quaternions were not in fashion at that time, neither in mathematics nor in physics, and nobody had used them until that time in any of the fields Dirac was interested in. It was different with matrices, however; they had been used at many places in relativity theory as tensors of second rank. But Dirac did not think of the mathematical tools of relativity when he investigated the mathematical laws obeyed by quantum variables, and the reason is not difficult to understand. In none of the books on relativity available to him, especially not in Eddington's *Mathematical Theory of Relativity* (Eddington, 1923b), had any attention been drawn to the fact that the product of two tensors does not commute.[177] The noncommutativity of multiplication was, however, the principal

years. Thus, even after accepting the mathematics of noncommuting variables in quantum theory they expected very slow progress towards a satisfactory general theory. Unlike Dirac, both Kramers and Pauli lacked real confidence in the fundamental validity of the mathematical structure of classical mechanics for quantum systems.

[177]We have made note of the same fact earlier in connection with Max Born's analysis of Heisenberg's discovery of the noncommutativity of the product of quantum dynamical variables. Although Born was familiar with matrices, especially with their use in relativity theory, when he confronted them in Heisenberg's work his mind went back to the lectures of Rosanes that he had attended as a young student in Breslau. Thus, neither Born's nor Dirac's formulations of quantum variables were influenced by prior experience of tensor calculus in relativity theory.

feature of quantum mechanics, and since it did not play a role in any calculus he was familiar with, Dirac felt strongly the necessity of *creating a new calculus*. The laws of this calculus would completely govern the behaviour of dynamical variables. Thus, in Dirac's case, the formulation of the mathematical rules of calculation—'quantum algebra' as he called it[178]—and the derivation of the fundamental equations of quantum mechanics became part of the same goal.

In the equations of classical mechanics, which Dirac had to recast in quantum theory, there occurred not only the operations of addition and multiplication, but also those that defined functions such as the inverse, $1/x$, or the square root, $\sqrt{x}$, of a given variable x. Such functions also had to make sense in quantum mechanics, and Dirac proposed to define them with the help of the equations

$$\frac{1}{x}\cdot x = 1 \qquad \text{or} \qquad x\cdot\frac{1}{x} = 1 \tag{92a}$$

and

$$\sqrt{x}\cdot\sqrt{x} = x, \tag{92b}$$

respectively. That is, when the right-hand sides of Eqs. (92a) are applied to an arbitrary quantum variable y, they reproduce y itself; and when y is multiplied twice from the same side with $\sqrt{x}$, the result, in agreement with Eq. (92b), is identical to multiplying y by x.[179]

In his earlier study of Heaviside's calculus, Dirac had come across a similar method of defining new variables in an indirect way, that is, by substituting them in the equations for known variables and requiring that the known results should follow.[180] Dirac's acquaintance with Heaviside's methods made it easier for him to develop his mathematical tools for quantum mechanics. However, an even greater influence in this development must be attributed to the symbolic calculus that had been used in geometry, for it helped Dirac in discovering the nature of the most important operation that occurs in dynamical equations: the differentiation of a quantum variable with respect to a parameter, which may be the time or another variable.

In classical theory, the process of differentiation of a dynamical variable is a special analytic operation: if the dynamical variable x depends on a parameter v, one finds from this dependence the differential quotient, dx/dv, provided the

[178]The notion of 'quantum algebra' first occurred as the title of Section 2 in Dirac's paper on the fundamental equations of quantum mechanics (Dirac, 1925c, p. 642). Later on he wrote a complete paper 'On Quantum Algebra' (Dirac, 1926e).

[179]Dirac was aware of the fact that the use of $\sqrt{x}$ in multiply periodic systems might lead to difficulties. 'In particular,' he mentioned in his published paper, 'one may have to introduce sub-harmonics, i.e., new intermediate frequency levels, in order to express $\sqrt{x}$' (Dirac, 1925d, p. 645). To avoid this difficulty, however, he proposed to rationalize each equation of classical theory before interpreting it as a quantum-mechanical equation.

[180]Heaviside had introduced symbolic quantities, i.e., differential and integral operators, and defined their properties by observing their effect on certain functions occurring in the description of electrical circuits.

function $x(v)$ exhibits a certain analytic property or smoothness. For the purpose of discussing the analogous operation in quantum mechanics, which he called 'quantum differentiation,' Dirac could not refer to any such analytic property, the reason being that he did not yet see how to define the functional dependence of the quantum variable x on a parameter v. However, he took note of the elementary fact that, in classical dynamics, if x is a dynamical variable of the system, so is the differential quotient dx/dv. He assumed that the same situation should exist in quantum mechanics. Thus, for instance, in the case of a multiply periodic system of f degrees of freedom both quantities, x and dx/dv, can be represented by expansions of the type given in Eq. (73). Hence both of them are quantum variables having the harmonic components, $x(nm)\exp\{i\omega(nm)t\}$ and $(dx/dv)(nm)\exp\{i\omega(nm)t\}$, respectively; n, m denote f pairs of indices, each assuming all integral values (m_k replacing the $n_k - \alpha_k$ of Eq. (73)), the associated frequencies being given by $\omega(nm) = \sum_{k=1}^{f} \omega(n_k, m_k)$. 'Quantum differentiation' (Dirac, 1925d, p. 645) then had to provide a unique prescription for obtaining $(dx/dv)(nm)$, the amplitudes of the variable dx/dv, from $x(nm)$, the amplitudes of the variable x. Dirac conceived the problem of discovering the mathematical structure of the equations relating $(dx/dv)(nm)$ and $x(nm)$ as a geometrical problem. 'For getting ideas I worked geometrically,' he explained. 'Once the ideas are established, one can put them in algebraic form and proceed to deduce consequences' (Dirac, Conversations, p. 60).

In geometry, objects like points, lines or figures of a higher dimension are often described by certain symbols, and their relations by equations containing operations between these symbols. The particular operations that transform a given object—say, a triangle into another triangle—were known to be linear transformations. Dirac visualized the process of 'quantum differentiation' of a quantum variable x with respect to a parameter v as such a geometrical transformation; he required that there be a linear relation between the components of the geometrical object associated with dx/dv and the components of the geometrical object associated with x. Taking the harmonic components, $(dx/dv)(nm)\exp\{i\omega(nm)t\}$ and $x(nm)\exp\{i\omega(nm)t\}$, as the geometrical components in question, Dirac wrote the equation

$$\frac{dx}{dv}(nm) = \sum_{n'm'} a(nm; n'm')\, x(n'm'), \tag{93}$$

with the help of which the amplitudes $dx(nm)/dv$ can be obtained from the amplitudes $x(nm)$ by a linear transformation with the 'coefficients' $a(nm; n'm')$. The double sum over the indices n' and m' on the right-hand side of Eq. (93) extends, in principle, over all integers from $-\infty$ to $+\infty$; hence the transformation in question does not belong to the conventional three-dimensional space of human perception.[181] This fact, however, did not bother Dirac at all, for he was

[181] The space of quantum mechanics is different from the spaces considered in relativity theory and in the calculus of quaternions, both of which have four dimensions. Therefore, it is hardly surprising that neither the tensor calculus of relativity nor the quaternions served Dirac as the starting point of his investigations of the mathematical structure of quantum mechanics.

familiar with higher-dimensional spaces and knew that mathematicians, since mid-nineteenth century, had considered geometries in spaces of arbitrarily many dimensions. Thus, the form of Eq. (93) seemed to him a legitimate consequence of mathematical theory. He also felt justified in making use of other results, such as those of Hermann Graßmann's in his 'calculus of extension' ('*Ausdehnungslehre*,' Graßmann, 1844, 1862), in order to derive further consequences about the mathematical structure of 'quantum differentiation.'[182]

The essential point, which interested Dirac, was the following. Graßmann had described geometrical objects in terms of what he called 'orders' ['*Stufen*'], denoting, in particular, the quantities of 'first order' by the symbols a, b, etc. He represented a and b by the equations

$$\begin{aligned} a &= \alpha_1 e_1 + \alpha_2 e_2 + \cdots + \alpha_n e_n, \\ b &= \beta_1 e_1 + \beta_2 e_2 + \cdots + \beta_n e_n, \end{aligned} \tag{94}$$

where α_i and β_i may be considered as the components of a vector in an n-dimensional affine space, with e_i being the basis elements or unit vectors ($i = 1, \ldots, n$).

Graßmann introduced the operations of addition and multiplication for the two quantities, a and b, as follows. The sum of a and b was defined simply by adding the components, α_i and β_i, belonging to the same basis element e_i. Postulating that the product, ab, should depend linearly on the components of the factors, and that its definition should not depend on the choice of the basis vectors, Graßmann found two different types of products: One, the 'algebraic' product, satisfied the law of ordinary multiplication in arithmetic; however, the other product, which he called 'combinatorial' or 'external' product, was defined by the equation

$$a \cdot b = \sum_{i,j} \alpha_i \beta_i [e_i e_j], \tag{95a}$$

with

$$[e_i e_j] = \begin{cases} -[e_j e_i] & \text{for} \quad i \neq j \\ 0 & \text{for} \quad i = j, \end{cases} \tag{95b}$$

and it provided an example of a noncommutative operation.[183]

[182] Hermann Günther Graßmann (born 15 April 1809 in Stettin, Prussia) studied theology and philology in Berlin and then turned, without any formal training in the field, to mathematics. He became a high school teacher of mathematics in Berlin and Stettin. He died in Stettin on 26 September 1877, having contributed to mathematics, physics and Sanskrit philology. His principal mathematical work was '*Ausdehnungslehre*' (Graßmann, 1844), in which he created a new algebra of n-dimensional space. The importance of Graßmann's mathematical results was recognized only gradually, especially after his death. In Germany his ideas were propagated, in particular by Felix Klein.

[183] Graßmann arrived at the noncommutative product at about the same time as Hamilton introduced the quaternions, whose product is also not commutative. Later, Graßmann became interested in higher complex numbers (of which the quaternions were the first example), and he proposed sixteen different kinds of multiplication rules for n-dimensional objects a and b, leading to as many types of hyper-complex numbers (Graßmann, 1855).

Dirac had learned about the symbolic calculus, such as Graßmann's, in which the product of two quantities is not necessarily commutative, at Henry Baker's tea parties.[184] He also learned there that the symbolic operations of addition and multiplication could be given a geometrical interpretation in projective spaces of arbitrarily many dimensions. In particular, all operations, by which a symbol was transformed into another symbol of the same kind, were to be associated with rotations of the geometrical object represented by the symbol. It occurred to Dirac that he should identify the quantum variables with similar symbols and visualize the operation of 'quantum differentiation' as a rotation in a projective space, the latter having infinitely many dimensions. Then, in a symbolic calculus of the kind proposed by Graßmann, dx/dv could be expressed by means of an operation containing only the sums and products of noncommuting symbols—that is, *'quantum differentiation' would become an algebraic operation applied to quantum variables*. The detailed structure of 'quantum differentiation' would have to follow, of course, from the particular requirements or laws imposed upon it by its detailed correspondence to classical differentiation.

In searching for these laws Dirac had to start from those properties of classical differentiation that define it as a linear operation, for he was convinced from his geometrical visualization that quantum differentiation was also a linear procedure. He therefore considered as the fundamental laws of differentiation the two equations

$$\frac{d}{dv}(x+y) = \frac{dx}{dv} + \frac{dy}{dv} \tag{96}$$

and

$$\frac{d}{dv}(x \cdot y) = \frac{dx}{dv} \cdot y + x \cdot \frac{dy}{dv}, \tag{97}$$

which express the differential quotients of the sum and the product, respectively, of two variables, x and y, in terms of the differential quotients of the individual variables and the basic algebraic operations of addition and multiplication. Dirac assumed these laws to be valid also when x and y are quantum variables. However, in quantum mechanics, unlike the classical theory, one has to keep the order of the factors also in the product on the right-hand side of Eq. (97).

Quantum differentiation automatically satisfies the first law, Eq. (96), for with the help of Eq. (93) one can show that

$$\begin{aligned} \frac{d(x+y)}{dv}(nm) &= \sum_{n'm'} a(nm; n'm')\{x(n'm') + y(n'm')\} \\ &= \sum_{n'm'} a(nm; n'm')x(n'm') + \sum_{n'm'} a(nm; n'm')y(n'm'), \end{aligned} \tag{98}$$

[184]Dirac did not know the details of Graßmann's work, nor were they presented in Baker's *Principles of Geometry*. But he had learned that Graßmann had invented a symbolism for the treatment of problems in projective geometry and that this symbolism allowed one to consider noncommutative multiplication in a consistent way (Dirac, Conversations, p. 70).

which is identical with the law of addition for the amplitudes. However, the second law, Eq. (97), holds only if the transformation coefficients, $a(nm; n'm')$, are restricted in a particular way. In order to determine these restrictions, Dirac calculated, again with the help of Eq. (93), the identical amplitudes (denoted by the indices n, m) of the two expressions, $d(x \cdot y)/dv$ and $(dx/dv) \cdot y + x \cdot (dy/dv)$, respectively, and deduced from Eq. (97) the relation

$$\begin{aligned} &\sum_{n'm'r} a(nm; n'm')x(n'r)y(rm') \\ &\quad = \sum_{rn'r'} a(nr; n'r')x(n'r')y(rm) + \sum_{rr'm} x(nr)a(rm; r'm')y(r'm'), \end{aligned} \tag{99}$$

which holds for the set of indices $n = (n_1, \ldots, n_f)$ and $m = (m_1, \ldots, m_f)$, all assuming integral values. Since Eq. (99) must be true for any arbitrary variables, x and y, it follows that the coefficients multiplying the same product amplitude, $x(n'r)y(rm')$, on both sides must be equal; thus, one obtains

$$\delta_{rr'}a(nm; n'm') = \delta_{mm'}a(nr'; n'r) + \delta_{nn'}a(rm; r'm'), \tag{100}$$

where the Kronecker symbol $\delta_{rr'}$ is equal to 1 for $r = r'$ and 0 otherwise. By exploiting Eq. (100) Dirac arrived at a set of equations between the different transformation coefficients, from which he found that all $a(nm; n'm')$ were zero except those which had at least two equal sets of indices. He then rewrote these nonzero coefficients in the following manner:

$$\begin{aligned} a(nr'; nr) &= -a(rm; r'm) = a(rr') \qquad \text{for } r \neq r', \\ a(rr'; rr') &= a(r'r') - a(rr). \end{aligned} \tag{101}$$

That is, he introduced new coefficients, $a(rr')$, depending only on two sets of indices, $r = (r_1, \ldots, r_f)$ and $r' = (r'_1, \ldots, r'_f)$, for a system of f degrees of freedom.

Using the new coefficients, defined by Eq. (101), Dirac simplified Eq. (93) for the amplitude of the differential quotient of any quantum variable, obtaining

$$\frac{dx}{dv}(nm) = \sum_r \{x(nr)a(rm) - a(nr)x(rm)\}. \tag{102}$$

On the basis of Eq. (102) he made a bold assumption: If he were to interpret the coefficients $a(nm)$ as the amplitudes of a new quantum variable, a, the equation for 'quantum differentiation' would take the particular form

$$\frac{dx}{dv} = xa - ax. \tag{103}$$

That is, the differential quotient of an arbitrary variable x with respect to any parameter v can be expressed by a difference of products of x with another quantum variable a. Although Eq. (103) followed from Eq. (102) only by formal analogy with the law of multiplication of quantum amplitudes, Dirac felt certain that it represented a fundamental law of quantum mechanics.[185] To support this point of view he considered a special case of Eq. (103) and showed that it yielded a well-known equation of the new theory. He assumed, in particular, that all amplitudes of the harmonic expansion of a are zero, except the ones denoted by $a(nn)$; that is, he assumed that the variable a is a constant of motion of the system under investigation. He further identified the amplitudes $a(nn)$ with $-i\Omega(n)$, where $\Omega(n)$ are the frequency levels (multiplied by 2π) of the system. Then Eq. (102) led to the equation

$$\frac{dx}{dv}(nm) = -i\{\Omega(m) - \Omega(n)\}x(nm). \tag{104}$$

Since the difference within the curly brackets is equal to $\omega(nm)$, the transition frequency of the system from the state denoted by the indices n to the state denoted by the indices m, the left-hand side of Eq. (104), for physical reasons, must be put equal to $(dx/dt)(nm)$, the amplitude of the differential quotient of the variable x with respect to time. Dirac felt confident that in the case of other differential quotients as well the quantity a could be identified with a known variable of the system.[186]

With the discovery of the correspondence between $xy - yx$, the difference of products of the quantum variables x and y, and (x, y), the classical Poisson bracket expression, on the one hand, and the correspondence between classical differentiation and a difference of products of quantum variables, on the other, Dirac had arrived at an important conclusion. He had demonstrated that the differential operations of classical dynamics could be replaced, in quantum mechanics, by algebraic operations. Thus, the equations of quantum mechanics, unlike those of classical theory, would not fall into two classes: differential equations and purely algebraic equations. Rather all equations in the new theory would be of *one* type. They would be algebraic equations in which two fundamental operations occurred: addition and multiplication of quantum variables. Once it was known how these operations were to be performed, the equations were well defined. Dirac's next task now simply became: to formulate the fundamental equations of quantum mechanics corresponding to the Hamiltonian scheme of classical mechanics.

[185] From Eq. (103) it follows that a, like the quantum variable x, can be expressed by an expansion. Or, in geometrical interpretation, the object a is represented by its homogeneous coordinates, $a(nm)$, in a similar way as the object x. However, from this it does not yet follow that the two objects have to be the same. For example, in plane projective geometry both objects, the 'point' and the 'straight line,' are represented by three homogeneous coordinates. The essential thing that makes the 'line' different from the 'point' is that the basis elements of the two objects are different. In the case of the quantities, x and a, which enter into Eq. (103), the basis elements are the functions, $\exp\{i\omega(nm)t\}$, and they are identical in both cases. Hence a can be interpreted as a dynamical variable in the same way as x.

[186] For example, when a denotes a constant of motion other than the energy states of the system, say $(2\pi/ih)\phi$, then one obtains from Eq. (103) $dx/dv = (2\pi/ih)(x\phi - \phi x)$.

IV.4 The Hamiltonian Scheme of Quantum Mechanics

With his investigations of the equations governing the dynamical behaviour of systems in quantum mechanics Dirac found himself in a situation he had not encountered before. In his previous work he had concentrated on improving particular points of already *existing* theories; he had drawn new or deeper conclusions by means of a more consistent and rigorous application of fundamental concepts—concepts which he learned about from the literature or from his teachers. Now, for the first time, he pursued an idea entirely his own: this was *his* idea that on the basis of a correspondence equation, relating a quantum-theoretical product difference to a classical Poisson bracket, a complete theory of quantum mechanics could be constructed. True, Dirac had arrived at this idea in his usual manner by improving upon someone else's results: his investigation had begun with his desire to give an unambiguous and well-defined mathematical expression to Heisenberg's observation that 'quantum-theoretical magnitudes' do not commute. However, in pursuing this task Dirac set up an original scheme of atomic mechanics that had not occurred before in the literature. This fact was not lost upon Ralph Fowler, to whom Dirac showed his results and who advised him to publish them immediately in a paper.

In his paper on 'The Fundamental Equations of Quantum Mechanics' Dirac summarized the results of his analysis of Heisenberg's paper and his own discoveries, and he used both to write down the equations of what corresponded in quantum mechanics to the Hamilton–Jacobi theory of the so-called multiply periodic systems in classical mechanics (Dirac, 1925d). As the starting point of the new theory Dirac considered the question whether, for any quantum system, it was possible to introduce quantities analogous to the dynamical variables that describe the corresponding classical system. He declared that this was indeed possible, that the 'quantum-theoretical magnitudes,' x, y, etc., which Heisenberg had defined by means of expansions of the form given in Eq. (73) (see Heisenberg, 1925c, p. 882), were just the 'quantum variables' (Dirac, 1925d, p. 647) of the multiply periodic system. However, in order to serve the same purpose as the classical variables the new variables had to satisfy certain conditions. The first requirement would evidently be that the 'quantum variables' describe a quantum system as completely as the dynamical variables describe a classical system, for only then could they enter the equations of motion whose integration would yield the possible solutions that determine the state of the system. By comparing the analogous classical and quantum-mechanical situations Dirac showed that quantum variables did indeed satisfy this requirement.

In classical theory any variable x of a multiply periodic system of f degrees of freedom can be written as

$$x = \sum_{\alpha_1 = -\infty}^{+\infty} \cdots \sum_{\alpha_f = -\infty}^{+\infty} x(\alpha_1, \ldots, \alpha_f)_{J_1, \ldots, J_f} \exp\left\{ i \sum_{k=1}^{f} (\alpha_k \omega_k)_{J_1, \ldots, J_f} t \right\}. \quad (105)$$

The $x(\alpha_1, \ldots, \alpha_f)_{J_1, \ldots, J_f}$ are the Fourier coefficients of the expansion in terms of harmonic components, each belonging to a particular frequency which is a sum of the higher harmonics, $\alpha_k \omega_k$ ($\alpha_k = 0, \pm 1, \pm 2$, etc.). Since x satisfies an equation of motion, the Fourier coefficients and the ground frequencies, ω_k, depend on the values assumed by J_k, the action variables of the system. The same values of J_k also determine the state of the system completely. This can be seen by introducing, besides the action variables, the conjugate angle variables, w_k,

$$\dot{w}_k = \frac{\partial H}{\partial J_k} = \frac{\omega_k}{2\pi}, \qquad k = 1, \ldots, f, \tag{106}$$

where H is the Hamiltonian of the system. The angle variables, w_k $(= (\omega_k / 2\pi)t)$, are thus obtained directly from the ground frequencies (which themselves depend only on J_k). Moreover, the action and the angle variables together constitute a complete set of dynamical variables of the multiply periodic system; hence any other variable, say x, can be expressed as a function of them, that is, $x = x(J, w)$.

Turning to the analogous multiply periodic system in quantum mechanics, Dirac realized that the expansion of an arbitrary quantum variable x, which corresponds to the f-fold Fourier sum in Eq. (105), is given by the right-hand side of Eq. (73). In order for x to satisfy its proper equation of motion, the indices $n_1, \ldots, n_f$ have to assume adequate, discrete values. Thus, he arrived at the conclusion that it is the indices n_k which replace the classical action variables J_k in describing the quantum system, and that a particular state of the latter is defined by the values which the set of indices assumes. However, the indices n_k could not be identified immediately with 'quantum' action variables because the operations between quantum variables, in general, exhibited features that were alien to classical theory. For example, in order to calculate $xy(n, n - \alpha - \beta)$, the amplitudes of the product of two variables x and y, where $\alpha + \beta = 0, \pm 1, \pm 2$, etc. and the indices n_k assume a given set of numbers, one has to know—due to the product rule, Eq. (60)—the amplitudes $x(n, n - \alpha)$ and $y(n - \alpha - \beta)$, where $\alpha = 0, \pm 1, \pm 2$, etc. While the first set of amplitudes, $x(n, n - \alpha)$, represents the variable x when the system is in the state denoted by the indices n_k—that is, *one* particular solution of the equation of motion—the second set of amplitudes, $y(n - \alpha, n - \alpha - \beta)$, represents the variable y when the system is in *all* possible states. Therefore, if products of dynamical variables occur in the equations used for obtaining the stationary states of the system under investigation, one cannot, in general, obtain a single, complete solution by taking from these equations only the components of the dynamical variables—$x(n, n - \alpha) \exp\{i\omega(n, n - \alpha)t\}$, $y(n, n - \alpha) \exp\{i\omega(n, n - \alpha)t\}$, etc.—for which the indices n_k assume a fixed set of values. 'The quantum solutions,' noted Dirac, 'are all interlocked, and must be considered as a single whole' (Dirac, 1925d, p. 643). From this interlocking of the solutions of a quantum system, which expresses at the same time an interlocking of the stationary states, it follows that the quantum variables exhibit properties that are not matched by those of the corresponding classical variables.

In order to establish the fundamental dynamical equations of quantum mechanics Dirac made use of his key discovery: the fact that, in the limit of high 'quantum numbers,' the classical Poisson bracket of two variables, (x, y), corresponds to the quantum-theoretical expression, $xy - yx$, multiplied by the factor $2\pi/ih$. He even concluded that: 'No classical expression involving differential coefficients can be interpreted on quantum theory unless it can be put into this form [of Poisson brackets]' (Dirac, 1925d, p. 650). This restriction did not seem to be serious, however, for he knew from classical dynamics that differential quotients, dx/dv, could be replaced by Poisson brackets of the variable x with ϕ, an integral of motion that is associated with the parameter v (Whittaker, 1904, 1917, Section 144; see our Eq. (90), Section IV.2). If the parameter v is the time, the integral of motion will be the Hamiltonian. Hence Dirac took it for certain that the equations of motion—namely, the equations for the time derivatives of the variables as well as all other fundamental equations of classical dynamics—can be written in such a way that they contain differential quotients only in combinations that form Poisson brackets. Having arrived at this manner of formulating the classical equations he could recast them as quantum-mechanical equations by using 'the fundamental assumption that *the difference between the Heisenberg products of two quantum quantities* [*variables*] *is equal to* $ih/2\pi$ *times their Poisson bracket expression*' (Dirac, 1925d, p. 648). In applying this assumption he had to require, of course, that the order of factors within a product occurring in a classical Poisson bracket was fixed in a certain way, say as xy; that is, the quantum expression obtained by taking the Poisson bracket between xy and z would, in general, be different from the expression obtained by taking the Poisson bracket between yx and z. Dirac called the 'quantum Poisson bracket expression' that Poisson bracket in which the order of factors in a product of variables was fixed, and denoted it by square brackets (Dirac, 1925d, p. 650). It is this bracket that occurs in the equation

$$xy - yx = \frac{ih}{2\pi}[x, y] \tag{107}$$

for the quantum-mechanical product differences.[187]

Equation (107) provided the necessary tool for obtaining the fundamental equations of quantum mechanics in a *consistent* way. Dirac noticed several mathematical reasons to confirm this consistency. First, the bracket expression on the right-hand side of Eq. (107) is antisymmetric in the variables x and y

[187] Dirac used the same nomenclature as Whittaker's *Analytical Dynamics* (1904, 1917, Section 130). Whittaker had called (x, y), the classical Poisson brackets, the 'Poisson bracket expressions,' the same as Jacobi who had used the term '*Klammerausdrücke*' (Jacobi, 1866). The first to use the term 'Poisson brackets' was Sophus Lie (Lie, 1888, p. v). Dirac denoted the classical and the quantum-theoretical Poisson bracket as $[x, y]$, but he understood the latter always in the sense of Eq. (107).

because of the corresponding property of the classical Poisson bracket, that is,

$$[x, y] = -[y, x]. \tag{108}$$

Second, the 'quantum Poisson bracket' obeys the laws of composition given by

$$[x + y, z] = [x, z] + [y, z] \tag{109}$$

and

$$[xy, z] = [x, z]y + x[y, z], \tag{110}$$

which have the same structure as the laws, Eqs. (96) and (97), obeyed by classical and quantum differentiations; thus, one would not run into difficulties with respect to the order of factors in products of quantum variables. Finally, Dirac obtained a third consistency check for Eq. (107) by calculating the quantum-mechanical expression corresponding to the Jacobi bracket expression,

$$[x, y, z] = [[x, y], z] + [[y, z], x] + [[z, x], y]. \tag{111}$$

In classical theory this expression is zero, and the result is called Jacobi's identity. If one substitutes the quantum-mechanical product differences on the right-hand side of Eq. (111), that is,

$$\begin{aligned}[x, y, z] = \left(\frac{ih}{2\pi}\right)^2 \{(xy - yx)z - z(xy - yx) + (yz - zy)x \\ - x(yz - zy) + (zx - xz)y - y(zx - xz)\},\end{aligned} \tag{112}$$

one notices immediately that it is zero.[188]

After convincing himself that Eq. (107) did not give rise to mathematical difficulties, Dirac sought to apply it to obtain quantum-mechanical equations from analogous equations in classical theory. First of all, he evaluated the quantum Poisson brackets for q_k and p_k, the fundamental set of variables of a given system of f degrees of freedom, and obtained the results

$$q_k p_l - p_l q_k = \frac{ih}{2\pi}[q_k, p_l] = \frac{ih}{2\pi}\delta_{kl} \tag{113a}$$

[188] In Whittaker's book the proof of Jacobi's identity was indicated in *Example 1* following Section 130, where the Poisson bracket was introduced. Whittaker had not associated the name 'Jacobi's identity' with the equation $((x, y), z) + ((y, z), x) + ((z, x), y) = 0$; hence Dirac did not refer to it by name.

The demonstration of the quantum-mechanical Jacobi identity was the first of many examples in which the quantum results were immediately evident, while the derivation of the corresponding classical results was fairly intricate.

and

$$q_k q_l - q_l q_k = 0, \qquad p_k p_l - p_l p_k = 0, \qquad k, l = 1, \ldots, f. \tag{113b}$$

These equations were generalizations of Eq. (72), which he had derived from Heisenberg's 'quantum condition,' Eq. (62), and he referred to them also as 'quantum conditions' (Dirac, 1925d, p. 650) even though they contained much more information than the original equation. It follows from Eq. (113a) that the time-dependent terms in the harmonic expansion of $qp - pq$ are zero because the classical Poisson bracket of any two variables is a constant of motion. More explicitly, the amplitudes, $(qp - pq)(nm)$, will be zero unless $n = m$.

In quantum mechanics the quantum conditions play an important role, unlike the analogous equations in classical theory that are empty tautologies. Indeed, Eqs. (113) are automatically satisfied in classical mechanics when the definition of the Poisson brackets is taken into account. The quantum conditions, on the other hand, lead to theoretical and practical consequences. For example, with their help, one can change the order of factors, q_k and p_k, in the products, which may simplify a given calculation with quantum variables, provided the latter can be expressed in terms of such products. The quantum conditions were also found to close a gap that had remained thus far in Dirac's arguments: they would help in determining the values of the indices n occurring in the amplitudes of the quantum variables, and, thus, they would contribute fundamentally to the calculation of the stationary states of the system under investigation. However, before treating this question, Dirac turned his attention to the equations that determine the dynamical behaviour of quantum systems.

In classical mechanics the equations of motion are the Hamiltonian equations for the dynamical variables. If H—a function of q_k and p_k, the position coordinates and their canonically conjugate momenta, respectively—is the Hamiltonian of the system, then the equations for q_k and p_k can be formulated with the help of Poisson brackets as [189]

$$\frac{dq_k}{dt} = (q_k, H), \qquad \frac{dp_k}{dt} = (p_k, H). \tag{114}$$

The analogous quantum-mechanical equations are easily obtained, for Dirac noticed that Eqs. (114) 'will be true on the quantum theory for systems for which the orders of the factors of products occurring in the equations of motion [that is, in the Hamiltonian, in particular] are unimportant' (Dirac, 1925d, p. 651).

[189] Although Eqs. (114) did not occur explicitly in Whittaker's *Analytical Dynamics*, they could be deduced from the discussion in Section 133 and 134, where Whittaker pointed out that one 'can interpret these equations [i.e., the Hamiltonian equations of motion for the q_k and the p_k] as implying that the transformation from the values of the variables at time t to their values at time $t + dt$ is an infinitesimal contact-transformation' (Whittaker, 1904, pp. 292–293). In mathematical terms, the resulting equations are, $dq_k = (q_k, H)dt$ and $dp_k = (p_k, H)dt$, because the contact transformation in question is expressed as the Poisson brackets of the variables, q_k and p_k, with the Hamiltonian H. The above equations are identical to Eqs. (114).

Indeed, in all these cases one may replace the classical round brackets by the square brackets of Eq. (107). For example, if H is a function of the q_k's only, then the time derivatives dq_k/dt are zero in quantum theory just as in classical theory, and the time derivatives dp_k/dt are obtained by evaluating the quantum Poisson brackets of the p_k's with the q-factors occurring in the Hamiltonian with the help of Eq. (113a). Similarly one proceeds in the case when H is a function of the momentum variables only; then p_k's are constants of motion and can be taken as 'uniformizing variables.' Finally, when H is a so-called *separable* Hamiltonian—that is, it is a sum of two terms, H_1 and H_2, such that H_1 depends only on the position variables and H_2 only on the momentum variables—the evaluation of the equations of motion proceeds along similar lines.

Dirac extended his scheme immediately to systems described by *nonseparable* Hamiltonians as well, noting that the Hamiltonian equations of motion 'may be taken to be true for systems in which these orders are important if one can decide upon the orders of the factors [q and p] in H' (Dirac, 1925d, p. 651). The reason is that Eqs. (109), (110) and (113) give an unambiguous prescription for calculating the 'quantum Poisson bracket' of an arbitrary product of factors q_k and p_l with any of the dynamical variables of the system. With the help of the same equations Dirac also derived the equation of motion for an arbitrary variable x, which is an algebraic function of the q's and p's, namely,

$$\frac{dx}{dt} = [x, H]. \tag{115}$$

Taken together with Eq. (107), this equation represents the fundamental quantum-mechanical equation of motion.

On comparing Eq. (115) with the analogous classical equation, Eq. (85), one sees immediately that both have identical structure. Dirac discovered that the same situation occurs for all dynamical equations, which, in classical theory, can be written with the help of Poisson brackets. For example, if C is a constant of motion of a classical system, its Poisson bracket with the Hamiltonian H is zero; an identical condition, that is,

$$[C, H] = 0, \tag{116}$$

must hold in quantum mechanics in order that the quantum variable C be a constant of motion of the quantum system whose Hamiltonian is H. From Eqs. (107) and (116) it is quite easy to check whether a given variable is a constant of motion in quantum mechanics, for one has only to calculate a difference of quantum products. Dirac also encountered no difficulty in proving the quantum-mechanical analogue of what Jacobi had called 'Poisson's theorem': indeed, $[C_1, C_2]$, the quantum Poisson bracket of two constants of motion, each of which satisfies Eq. (116), is again a constant of motion, as it follows from the quantum-mechanical Jacobi identity that the bracket $[[C_1, C_2], H]$ is also zero.

The recognition, that quantum mechanics could be founded on the basis of classical mechanics and a systematic application of the fundamental assumption contained in Eq. (107) enabled Dirac to establish, *without going into any detailed calculations of specific examples*, a complete set of quantum-mechanical equations that would account for any system whose classical analogue could be treated with the help of the Hamilton–Jacobi scheme. This fundamental set of equations includes, besides Eq. (107) and the quantum conditions, Eqs. (113), also the Hamiltonian equations defining the canonical variables, i.e., the equations corresponding to the classical Eqs. (114). Thus Dirac required that the quantum Poisson bracket of the canonical variables, Q_k and P_k, satisfy the equations

$$[Q_k, P_l] = \delta_{kl}, \qquad [Q_k, Q_l] = 0, \qquad [P_k, P_l] = 0. \tag{117}$$

With the help of this theoretical scheme Dirac expected to be able to treat quantum systems that are of practical interest in atomic physics. And, provided the application of the equations could be made in an unambiguous way, he expected to be able to calculate results in agreement with the observed data. However, before proceeding to discuss any particular system, Dirac summarized the consequences which appeared to follow from the mathematical structure of the fundamental equations of quantum mechanics. These were concerned with the general postulates, which Bohr had introduced more than a decade before, to account for the behaviour of atoms: the postulates of the existence of stationary states and that of the frequency condition, expressing the frequency of the emitted or absorbed radiation in terms of the energy differences of the states involved.

The Bohr–Sommerfeld semiclassical theory of atomic mechanics had made use of the action and angle variables, J_k and w_k. If the atomic system was assumed to be multiply periodic, the action variables became constants of motion; once the Hamiltonian had been expressed as depending on these constants alone, the energy states of the system were obtained directly, for one had just to insert integral multiples of Planck's constant h for J_k. The first question to consider in quantum mechanics was, naturally, whether action and angle variables made any sense in the new theory, and, if so, how would they be defined. Dirac foresaw no problem concerning their existence in quantum mechanics: he simply postulated that any multiply periodic system of f degrees of freedom *can* be described by f pairs of quantum action and angle variables, J_k and w_k.[190] After all, the angle variables w_k had already appeared implicitly in Eq.

[190] Dirac's attitude towards the use of action and angle variables in quantum mechanics was necessarily very different from the one taken by Born, Jordan, Heisenberg and Pauli in matrix mechanics. The Göttingen physicists and their followers encountered serious difficulties when they sought to represent the angle variables by matrices because the angles are nonperiodic coordinates. However, Dirac had no such problem, for he was convinced from the very beginning that his fundamental equations would account for both periodic and nonperiodic systems and could contain both periodic and nonperiodic variables. He therefore admitted the angle variables, in particular, as perfectly valid objects in his theory. (We shall return to a detailed discussion of action and angle variables in Sections V.2 and V.3.)

(73), the Fourier expansion of the quantum variables describing multiply periodic systems, if there the product $\omega_k t$ in the periodic exponential were identified with w_k. However, in order to define the action variables, Dirac referred to an indirect method, which was familiar from the corresponding classical situation. For instance, in the discussion of the three-body problem it had been found convenient to introduce, besides the action and angle variables, another complete set of dynamical variables, ξ_k and η_k, defined by the equations

$$\xi_k = \sqrt{\frac{J_k}{2\pi}}\ \exp\{2\pi i w_k\}, \qquad \eta_k = -i\sqrt{\frac{J_k}{2\pi}}\ \exp\{-2\pi i w_k\}. \tag{118}$$

ξ_k and η_k are special periodic variables whose harmonic expansions, according to Eq. (105), consist only of one particular frequency term, namely, the one with $\alpha_k = +1$ or -1, respectively. Dirac assumed that 'there will be a corresponding set of canonical variables on the quantum theory, each containing only one kind of component, so that $\xi_k(nm) = 0$ except when $m_k = n_k - 1$ and $m_l = n_l$ $(l \neq k)$, and $\eta_k(nm) = 0$ except when $m_k = n_k + 1$ and $m_l = n_l$ $(l \neq k)$,' and he considered 'the existence of such variables as the condition for the system to be multiply periodic on the quantum theory' (Dirac, 1925d. p. 651). All further consequences followed from this definition.[191]

The essential property of the dynamical variables ξ_k and η_k in quantum mechanics is that the products, $\xi_k \eta_k$ and $\eta_k \xi_k$, are found to be constants of motion of the multiply periodic system under investigation. This means that the amplitudes of the harmonic expansions of $\xi_k \eta_k$ and $\eta_k \xi_k$ are nonzero only when the two sets of indices, n_k and m_k, assume identical values, $n_k = m_k$. Moreover,

[191]The set of variables, defined by Eqs. (118), is related to what Max Born called 'Poincaré's "orthogonal" canonical coordinates' ('*die Poincaréschen "rechtwinkligen" kanonischen Koordinaten*' (Born, 1925, p. 317)), $\hat{\xi}_k$ and $\hat{\eta}_k$, which are defined as (see Poincaré, 1893, Section 142)

$$\hat{\xi}_k = \sqrt{\frac{J_k}{2\pi}}\ \{\sin 2\pi w_k\}, \qquad \hat{\eta}_k = \sqrt{\frac{J_k}{2\pi}}\ \{\cos 2\pi w_k\}.$$

In classical theory it is not difficult to prove that ξ_k and η_k, defined by Eqs. (118), are canonical variables because they satisfy the equation

$$(\xi_k, \eta_k) = \frac{\partial \xi_k}{\partial w_k}\frac{\partial \eta_k}{\partial J_k} - \frac{\partial \xi_k}{\partial J_k}\frac{\partial \eta_k}{\partial w_k} = 1,$$

with $(\xi_k, \xi_k) = 0$ and $(\eta_k, \eta_k) = 0$. However, this result cannot be taken over into quantum mechanics because the order of w_k and J_k cannot be changed when evaluating the 'quantum Poisson bracket.' The difficulty can be seen by evaluating $[\xi_k, \eta_k]$, which becomes identical to $1 + \frac{1}{2}(\sqrt{J_k} \cdot \exp\{2\pi i w_k\} - \exp\{2\pi i w_k\}\sqrt{J_k})1/\sqrt{J_k}\exp\{-2\pi i w_k\} - \frac{1}{2}1/\sqrt{J_k}(\sqrt{J_k}\exp\{2\pi i w_k\} - \exp\{2\pi i w_k\}\sqrt{J_k}) \cdot \exp\{-2\pi i w_k\}$. Only if one is allowed to commute the factor $J_k^{-1/2}$ in the second term with the round bracket in front of it, will the second term cancel the first one. Thus, ξ_k and η_k will be canonical variables provided this condition is satisfied. For this reason Dirac preferred to use the existence of *canonical* variables ξ_k and η_k as a *definition* of multiply periodic systems in quantum mechanics.

the product coefficients satisfy the relations

$$\xi_k\eta_k(nn) = \eta_k\xi_k(mm), \tag{119}$$

where n's denote the sets of f indices, $n_1, \ldots, n_f$, and m's the sets of f indices, $m_1, \ldots, m_f$, such that all indices of the set m assume values identical with those of the set n, except m_k, whose value is equal to $n_k - 1$.[192] In classical theory, the products $\xi_k\eta_k$ (or $\eta_k\xi_k$), multiplied by the factor $2\pi i$, are identical to the action variables J_k. This definition cannot be taken over into quantum mechanics because the factors in the quantum product do not commute. However, with the analogy to the classical expression in the back of his mind, Dirac argued that the relation between the quantum action variables J_k and the products of ξ_k and η_k might be given by either of the equations

$$\xi_k\eta_k = \frac{J_k}{2\pi i} \tag{120a}$$

and

$$\frac{1}{2}(\xi_k\eta_k + \eta_k\xi_k) = \frac{J_k}{2\pi i}, \tag{120b}$$

or by any other equation having the property that its left-hand side, when expressed in terms of classical variables, is identical to $\xi_k\eta_k$. 'A detailed investigation of any dynamical system is necessary in order to decide what it is,' he concluded (Dirac, 1925d, p. 652).[193]

To test his assumptions and methods developed so far in connection with the action-angle variables, Dirac showed that he was able to treat all the examples discussed earlier by Heisenberg. In the case of the one-dimensional harmonic oscillator (with mass m_0 and frequency constant ω_0) he had only to identify the amplitudes of his variables ξ and η with the transition amplitudes (i.e., the

[192] Evidently, the amplitudes of the products are obtained in the way shown for the product $\xi_k\eta_k$. The sum over the indices n in the defining equation for the amplitude $\xi_k\eta_k(nn)$, that is, $\xi_k\eta_k(nn) = \sum_m \xi_k(nm)\eta_k(mn)$, reduces, because of Eqs. (118), to one term, namely, the one in which the indices satisfy the equations $m_1 = n_1, \ldots, m_{k-1} = n_{k-1}$, $m_k = n_k - 1$, and $m_{k+1} = n_{k+1}, \ldots, m_f = n_f$. In the same way one proves that there is no term in the sums for the amplitudes $\xi_k\eta_k(nn')$ or $\eta_k\xi_k(mm')$, if $n \neq n'$ or $m \neq m'$.

[193] In case Eq. (120b) gives the relation between the action variables J_k and the variables ξ_k and η_k, Dirac introduced still another set of variables, ξ'_k and η'_k, that is,

$$\xi'_k = \frac{1}{\sqrt{2}}(\xi_k + i\eta_k), \qquad \eta'_k = \frac{1}{\sqrt{2}}(\xi_k - i\eta_k),$$

in terms of which he expressed J_k by the simple relation $J_k = \pi(\xi'^2_k + \eta'^2_k)$. The variables ξ'_k and η'_k, which correspond to Poincaré's coordinates for the case of the three-body system of classical mechanics (Poincaré, 1893, Section 142), have two amplitudes when the set of indices, $n_1, \ldots, n_f$, assumes any given set of f integers, namely, those with $m_1 = n_1, \ldots, m_{k-1} = n_{k-1}$, $m_k = n_k \pm 1$, $m_{k+1} = n_{k+1}, \ldots, m_f = n_f$ (Dirac, 1925d, p. 652).

amplitudes of the position variable q) with the help of the equations

$$\begin{aligned} \xi(n,n-1) &= \left(\frac{m_0\omega_0}{2}\right)^{1/2} q(n,n-1), \\ \eta(n-1,n) &= -i\left(\frac{m_0\omega_0}{2}\right)^{1/2} q(n-1,n). \end{aligned} \tag{121}$$

The action variable, J, could then be represented in terms of these ξ and η with the help of Eq. (120b).[194]

In Heisenberg's second example, the rigid rotator, the situation was quite different, although the action variable of the corresponding classical system looks very similar to that of the one-dimensional harmonic oscillator. In fact, it is proportional to the square of an amplitude, the distance of the rotating mass from the axis of rotation. Since this distance is fixed throughout the motion, that is, it is a constant of motion, the action variable cannot be expressed as a rational function of the periodic variables ξ and η, defined by Eq. (118), and the same should be true in quantum mechanics. However, no argument spoke against constructing the quantum action variable J in a more complicated manner from the products $\xi\eta$ and $\eta\xi$. The same method could be applied to atomic systems having several degrees of freedom.

After demonstrating how the action variables of periodic systems could be obtained, at least in principle, Dirac proceeded to define their stationary states. First he noticed that the harmonic expansion of any constant of motion, C, due to Eq. (73), consists only of one component, $C(nn)$, for each set of indices n_k. 'It thus becomes convenient,' he argued, 'to suppose each set of n's to be associated with a definite state of the atom, as on Bohr's theory, so that each $C(nn)$ belongs to a state in precisely the same way in which *every* quantity occurring in classical theory belongs to a certain configuration' (Dirac, 1925d, p. 652). Thus, if for a given system one has established a relation between quantum variables, all of which are constants of motion, one can derive from it a set of equations, each containing only the amplitudes, $C(nn)$, belonging to one definite state. And each of these relations 'will be the same as the classical theory relation, on the

[194] Heisenberg had not presented the quantum-theoretical reformulation of the classical action variable J, which, in the case of the one-dimensional harmonic oscillator in the state n, is given by the expression

$$\frac{\pi m_0}{2}\left\{|q_{-1}(n)|^2\omega_0 + |q_{+1}(n)|^2\omega_0\right\}.$$

(See Heisenberg, 1925c, p. 885, Eq. (14), for $\alpha = \pm 1$ and $\omega_n = \omega_0$.) However, Dirac had no difficulty in finding the analogous quantum expression, which is evidently,

$$\frac{\pi}{2} m_0\omega_0\{q(n,n-1)q(n-1,n) + q(n,n+1)q(n+1,n)\},$$

and thus equal to the amplitude $\pi i\{\xi\,\eta(nn) + \eta\xi(nn)\}$.

assumption that the classical laws hold for the description of stationary states,' Dirac concluded, noting in particular that 'the energy will be the same function of the J's as on the classical theory' (Dirac, 1925d, p. 652). Dirac expected that it was possible to express the Hamiltonian of any multiply periodic quantum system of f degrees of freedom as a function of the quantum action variables J_k ($k = 1, \ldots, f$) alone, in exact analogy to the classical situation. Thus, he had arrived at his first goal: a justification of Bohr's postulate concerning the mechanical nature of stationary states.

With this definition of the stationary states of a quantum system it became straightforward to derive Bohr's second postulate, the frequency condition. Dirac took Eq. (115), the equation of motion for x, an arbitrary variable of the system, and replaced the quantum Poisson bracket of x with the Hamiltonian H by the difference of products with the help of Eq. (107). He then obtained the following equation for the amplitudes of the time derivative of x,

$$\frac{dx}{dt}(nm) = \frac{2\pi}{ih}\{x(nm)H(mm) - H(nn)x(nm)\}, \tag{122}$$

which is valid for all sets of index pairs (nm). $H(nn)$ and $H(mm)$ are two different amplitudes of the Hamiltonian that correspond to the states defined by the particular values of the sets of indices, n_k and m_k, respectively. Now the left-hand side of Eq. (122), in agreement with the harmonic expansion for x, must be equal to the amplitude $x(nm)$ multiplied by the factor $i\omega(nm)$, where $\omega(nm)$ is 2π times the transition frequency $\nu(nm)$. Hence Eq. (122) is satisfied only if the frequency condition,

$$\nu(nm) = \frac{\omega(nm)}{2\pi} = \frac{1}{h}\{H(nn) - H(mm)\}, \tag{123}$$

holds for any values of the two sets of indices n_k and m_k.

The derivation of Bohr's fundamental postulates, both of which he knew to be alien to classical mechanics, convinced Dirac that the principal difficulties of atomic theory had now been removed. He was even able to go a step beyond the results of the Bohr–Sommerfeld theory, for he showed how half quantum numbers, which had been employed occasionally in order to explain certain spectroscopic data, could be obtained from the new theory. This point was connected with a general question which Dirac had not discussed so far: namely, the assumption that the values of the indices n_k and m_k, denoting the amplitudes of quantum variables of multiply periodic systems, are integral numbers. This assumption had thus far not really been justified, as the indices could take on arbitrary discrete sequences of different values. However, Dirac went on to prove the assumption by taking note of a procedure in Heisenberg's paper. From the quantum condition, Eq. (62), and the additional physical assumption of the existence of a lowest state in the system, Heisenberg had concluded in the case of a one-dimensional oscillator that the indices n and m, denoting the transition amplitudes $q(nm)$, assume integral values, $n, m = 0, 1, 2$, etc. (Heisenberg, 1925c, pp. 888–889, Eq. (20) and the following equation). Dirac found a way of generalizing this derivation for the case of a multiply periodic system of several degrees of freedom. He took the amplitudes of the product differences, $\xi_k\eta_k - \eta_k\xi_k$, which, due to Eq. (113a), are equal to $ih/2\pi$. Postulating further the existence of a lowest state of the system, and defining this state by the condition that all indices n_k assume the value zero, he obtained the equation

$$\xi_k\eta_k(nn) = -n_k\frac{ih}{2\pi}, \qquad k = 1, \ldots, f. \tag{124}$$

That is, the values of n_k, an index in the amplitude of the quantum product $\xi_k\eta_k$, become associated with the value of the amplitude itself. The action variables J_k, apart from a factor h (Planck's constant), were the 'quantum numbers' of the Bohr–Sommerfeld theory. Provided the relation between J_k and the products $\xi_k\eta_k$ and $\eta_k\xi_k$ was known, Dirac could evaluate J_k with the help of Eq. (124). For example, in the case that this relation is given by Eq. (120a), he found each J_k to become equal to $n_k h$. The action values thus assume the values 0, h, $2h$, etc., which are integral multiples of Planck's constant, in agreement with the quantum conditions of the Bohr–Sommerfeld theory. However, for systems in which the J_k's are obtained from Eq. (120b), their values will be half-integral multiples of Planck's constant, that is, $\frac{1}{2}h$, $\frac{3}{2}h$, $\frac{5}{2}h$, etc. Half-integral quantum numbers could thus really occur in multiply periodic quantum systems.

These preliminary results, which followed from the new dynamical theory, especially from the fundamental noncommutativity of quantum variables, allowed Dirac to conclude that he was on the right track. It was important for him that he had not obtained these results from specific, suitably chosen examples, but deduced them directly from general equations which were applicable to all systems in atomic physics. The power of Dirac's methods was recognized immediately by others when they got to know his work. Next to Fowler, the first person to know about it was Heisenberg, whom Dirac sent a manuscript-copy of the paper that was to appear within a few weeks. Heisenberg answered Dirac on the same day he received his paper, and remarked: 'I have read your extraordinarily beautiful paper on quantum mechanics with the greatest interest, and there can be no doubt that all your results are correct as far as one believes at all in the newly proposed theory' (Heisenberg to Dirac, 20 November 1925).[195] He especially liked Dirac's proof of the conservation of energy and the derivation of the frequency condition, and he frankly admitted that the paper from Cambridge was 'also really better written and more concentrated than our attempts here [in

[195] Heisenberg recalled that the manuscript he received 'was handwritten in Dirac's very nice way of writing' (Heisenberg, Conversations, p. 274). After studying the manuscript in detail, Heisenberg wrote to Dirac (23 November, 1 December 1925; on 1 December, a letter and a postcard) in which he raised questions concerning the derivation of specific results. Although Dirac's replies to Heisenberg have been lost, his response can be determined from his following published papers. We shall discuss some of the points in the following sections.

Göttingen]' (Heisenberg to Dirac, 20 November 1925). Dirac was very happy about Heisenberg's reaction to his paper. He was not particularly disturbed when he learned that meanwhile another theoretical scheme of quantum mechanics had been developed in Göttingen (by Born and Jordan, and Born, Heisenberg and Jordan), from which many of his results had also been derived. Being aware of the active competition from other places, Dirac began to develop and perfect the mathematical details of his theory and to calculate with its help the properties of atomic systems, including those that could not be considered as being multiply periodic.

Chapter V
q-Numbers at Work

During the months following the submission of his paper on the fundamental equations of quantum mechanics to the *Proceedings of the Royal Society of London* (Dirac, 1925d), Dirac devoted himself to perfecting his scheme and its applications to specific atomic problems. He obtained many results, which he published in a rapid sequence of four papers, completed between January and July 1926. In the first paper of this series he developed the mathematical calculus of what he called 'q-numbers'; he applied this calculus to establish systematically the quantum-mechanical theory of action-angle variables, which he tested right away in a 'preliminary investigation of the hydrogen atom' (Dirac, 1926a). In the following paper Dirac gave the general theory of many-electron atoms, showing how multiply periodic systems could be treated with the help of q-numbers (Dirac, 1926b). In a third paper he tried to extend the formulation of the laws of quantum mechanics in such a way that they could be applied to systems whose Hamiltonian contained the time explicity or in which the relativistic features were important (Dirac, 1926c). Finally, after submitting his work on 'Quantum Mechanics' as his doctoral dissertation in May 1926 (Dirac, 1926d), he concluded the set of papers with a note on 'quantum algebra,' in which he discussed the definition and properties of the most general q-number functions and completed, to a certain extent, the mathematical methods underlying the calculation with quantum variables (Dirac, 1926e).

In these papers Dirac demonstrated in some detail how his theory, based on q-number calculus, was able to solve the problems of atomic physics. His work was performed and published simultaneously with that of Continental physicists on matrix mechanics, its extensions and modifications. Though at times Dirac referred to certain specific results obtained by Born, Heisenberg, Jordan and Pauli, the methods of matrix mechanics never served as an example or a guide for his own work. In the papers, which he published in the first half of 1926, Dirac entirely followed his own ideas; these ideas had already brought him, before he saw any of the papers on matrix mechanics, to the formulation of the fundamental equations of quantum mechanics in fall 1925. For help in developing his ideas Dirac relied on the resources he had used earlier: Henry Frederick Baker's *Principles of Geometry* and Edmund Taylor Whittaker's *Analytical Dynamics*. Sticking closely to these resources saved him much time, which was of the essence in the early days of quantum mechanics when everybody involved worked with great enthusiasm, vigour and speed.

V.1 On q- and c-Numbers

In order to formulate the laws of quantum mechanics one had to introduce quantities whose multiplication was not commutative. For two such quantities, z_1 and z_2, one obtained in general the inequality

$$z_1 z_2 \neq z_2 z_1. \tag{125}$$

Dirac had recognized this noncommutativity to be the essential property, which distinguished quantum variables from classical variables, as the latter were always described by ordinary numbers obeying the usual laws of arithmetic. However, in spite of the unusual properties of the symbols occurring in the new theory, Dirac had been able to show that one could build a 'workable theory' (Dirac, 1926a, p. 561) of the dynamical quantum variables, which contained no internal contradictions. In continuing his investigations Dirac became involved in a detailed study of the new mathematical symbols and the laws governing them. Apart from the noncommutativity which he had observed, the quantum variables satisfied all other laws of arithmetical numbers; that is, a quantum-theoretical symbol z operated in most respects like an ordinary number. This brought Dirac to consider quantum variables as being represented by a new kind of numbers. 'The fact that the variables used for describing a dynamical system do not satisfy the commutative law means, of course, that they are not numbers in the sense of the word previously used in mathematics,' he wrote in the introduction to his second paper on quantum mechanics. 'To distinguish the two kinds of numbers, we shall call the quantum variables q-numbers and the numbers of classical mathematics which satisfy the commutative law c-numbers, while the word number alone will be used to denote either a q-number or a c-number' (Dirac, 1926a, p. 562).[196]

Dirac was happy about the idea of treating the numbers used in quantum and classical mechanics, respectively, on an equal footing. As he emphasized later on, 'I suppose that it was the main point in my early work, that I did appreciate that there would be a close analogy between the q-numbers and ordinary numbers' (Dirac, Conversations, p. 60). To some extent this analogy was suggested by the physical correspondence between dynamical quantum variables and classical variables. But more than the physical correspondence, the analogy arose from basic mathematical arguments: Dirac's introduction of q-numbers was stimulated directly by his experiences in geometry, especially the ones he gathered at Henry Frederick Baker's tea parties. Dirac's formulation of the first section in his second paper on quantum mechanics, entitled 'The Algebraic Laws Governing

[196]The notion of q-numbers was not mentioned in the earlier paper (Dirac, 1925d), where Dirac spoke only about 'quantum algebra' or the algebra of quantum variables. q-numbers first occurred in Section 1 of Dirac's second paper on quantum mechanics, and they were introduced in the sentences quoted above. 'The reason for the notation q-number and c-number was rather obvious. q stands for quantum or queer. c stands for classical or commuting' (Dirac, Private Communication).

Dynamical Variables' and dealing with the mathematical laws obeyed by q-numbers, owed much to the third section in Baker's *Principles of Geometry* (Vol. I, Baker, 1922), where the algebraic laws of the symbols employed in projective geometry were presented. In fact, by writing down this section of his paper Dirac demonstrated in detail how much he had benefited from Baker's geometrical tea parties, for he was able to apply what he had learned there in a completely different context, that is, to the new theory of quantum mechanics.[197] There were two principal lessons that he took over from geometry: First, that there existed a complete set of laws for noncommuting 'numbers,' similar to the one for ordinary, commuting numbers; second, since the symbols used in conventional geometry did commute, Dirac concluded that 'at present one can form no picture of what a q-number is like' (Dirac, 1926a, p. 562).[198]

Ironically, the last observation seemed to remove the very basis of the principle, which had guided Dirac in discovering the laws of quantum mechanics, namely, their geometrical visualization. However, the situation was a bit more complex, for a complete geometrical visualization was not actually needed in the process of creating the mathematical symbols that should enter into quantum algebra. The geometrical analogy had served its purpose by leading Dirac to the idea that quantum variables might be thought of as symbols similar to the ones used in geometry. It did not, therefore, matter at all whether or not the symbols thus emerging would obey *all* the laws that the symbols of conventional geometry did. But it was important that in all geometries, the conventional *and* the unconventional ones, the operations of linear transformation could be performed, and that was what Dirac needed. 'Once the ideas are established,' he remarked later on, 'one can put them in algebraic form and one can proceed to deduce their consequences. That's just a question of algebra' (Dirac, Conversations, p. 60). In other words, the geometrical visualization had served its purpose, now one could forget about it.

Although some examples of noncommuting symbols were known at that time —Dirac had come across quaternions, and Baker had given the example of two by two matrices in his book (Baker, 1922, pp. 67–68)—people were not familiar with such cases. Thus, the q-numbers describing the quantum variables were indeed queer objects. For instance, Dirac noticed that 'one cannot say that a q-number is greater or less than another' (Dirac, 1926a, p. 562). However, the quantum dynamical variables had to have the following properties: 'If z_1 and z_2 are two q-numbers or one q-number and one c-number, there exist the numbers $z_1 + z_2, z_1 z_2, z_2 z_1$, which will in general be q-numbers but may be c-numbers'; moreover, the q-numbers 'satisfy all the ordinary laws of algebra, excluding the

[197] As we have mentioned earlier (in Section III), the participants at Henry Baker's tea parties, including Dirac, sometimes discussed the physical problems of relativity theory. However, nobody in 1923 and 1924 thought of applying geometrical ideas to the problems of quantum theory.

[198] By 'conventional geometry' in this context, as explained in some detail in Baker's *Principles of Geometry*, was meant that geometry in which the so-called 'theorem of Pappus' did hold. We shall discuss this point later in this section.

commutative law of multiplication' (Dirac, 1926a, p. 562). Hence Dirac was perfectly justified in taking over the 'laws of operations of the symbols' from Baker (1922, pp. 63–64): the commutative law of addition for arbitrary q-numbers, z_1 and z_2, that is,

$$z_1 + z_2 = z_2 + z_1; \tag{126}$$

the associative law of addition for arbitrary z_1, z_2, z_3,

$$(z_1 + z_2) + z_3 = z_1 + (z_2 + z_3); \tag{127a}$$

the associative law of multiplication,

$$(z_1 z_2) z_3 = z_1 (z_2 z_3); \tag{127b}$$

and the distributive law of multiplication, which is contained in the two equations,

$$z_1(z_2 + z_3) = z_1 z_2 + z_1 z_3 \tag{128a}$$

and

$$(z_1 + z_2) z_3 = z_1 z_3 + z_2 z_3. \tag{128b}$$

The similarity of these passages in Dirac's paper to the relevant ones in Baker's *Principles of Geometry* went far, and one discovers that even the phrasing was almost identical.[199] Clearly, Dirac was not concerned with the problem of constructing the mathematical laws or axioms of q-numbers; those he took over as known from geometry. His problem was different: namely, to investigate those

[199]For example, Eqs. (126), (127a, b) and (128a, b) could be found in Baker's book. It is interesting to compare Dirac (1926a, Section 1) with Baker. Baker had written:

> The symbols which we first introduce are, speaking in general terms, subject to all the laws of ordinary algebra, except the commutative law of multiplication. They are not necessarily, and will not be finally, arranged in order of magnitude, so that, in their entirety, they are wider than the numbers of arithmetic. In order to emphasize this, we give here a formal statement of the properties assumed by them, and some examples.
>
> We use the symbol = to mean "may be replaced by." Thus if a, b be two of the symbols the statement $a = b$ is the same as $b = a$.
>
> From any two symbols, say a and b, another symbol can be formed, represented by $a + b$, or by $b + a$, independent of the order in which a, b are taken. Further, if $a = b$ and $x = y$, then $a + x = b + y$. Also if a, b, c be any three of the symbols, the symbol thus formed from a and $b + c$ is equivalent with that so formed from $a + b$ and c; that is,
>
> $$a + (b + c) = (a + b) + c.$$
>
> We shall speak of the process of forming $a + b$ as *addition*, and of $a + b$ as the *sum* of a and b; and we may say that the addition of two symbols is *commutative*, and the addition of three symbols is *associative*. (Baker, 1922, p. 63)

Similarly, Baker introduced what he called the 'multiplication' of two symbols a, b, noting that ab may be equal to ba, 'but this is not to be assumed in general' (Baker, 1922, p. 64). Finally, if a, b, c are three symbols, he wrote down the equations $(ab)c = a(bc)$ and $(a + b)c = ac + bc$, $c(a + b) = ca + cb$, that is, the multiplication is both associative and distributive (Baker, 1922, pp. 64–65).

special features and properties of q-numbers that would assist in defining quantum variables because these properties would help him in performing calculations with the dynamical equations of quantum mechanics.

The symbols used in geometry also included the special symbol 0, which was called 'singular' and replaced the zero of conventional arithmetic, while the symbol 1 was used to represent unity (Baker, 1922, p. 65).[200] Dirac took over both into quantum mechanics in order to give the equations

$$z_1 z_2 = 0, \qquad z_2 z_1 = 0, \tag{129}$$

and

$$x_1 z = 1, \qquad z x_2 = 1, \tag{130}$$

a proper meaning. Equations (129) state that one of the factors, z_1 or z_2, is singular, or that both are singular. In case that the q-number z is not singular, Eqs. (130) define a *unique* inverse of z, that is, both solutions, x_1 and x_2, become identical with z^{-1}, the 'inverse' of z.[201]

[200] Baker (1922, p. 65) noted the properties of the symbol 0 as being given by the equations

$$a + 0 = 0 + a, \qquad a \cdot 0 = 0 \cdot a = 0,$$

and those of the symbol 1 by

$$a \cdot 1 = 1 \cdot a = a.$$

[201] Again, Dirac followed closely the presentation in Baker's book. He stated in his paper:

One may define further numbers, x say, by means of equations involving x and the z's, such as $x^2 = z$, which defines $z^{1/2}$, or $xz = 1$, which defines z^{-1}. There may be more than one value of x satisfying such an equation [i.e., $xz = z'$], but this is not so for the equation $xz = 1$, since if $x_1 z = 1$ and $x_2 z = 1$ then $(x_1 - x_2)z = 0$, which gives $x_1 = x_2$ provided $z \neq 0$. (Dirac, 1926a, p. 562)

Baker, on the other hand, had written about the same definitions as follows:

To every symbol, a, which is not singular, there corresponds another symbol, denoted by a^{-1}, having the property that

$$a^{-1} \cdot a = a \cdot a^{-1} = 1,$$

and such that, if $a = b$, then $a^{-1} = b^{-1}$. Thus

$$ab^{-1} \cdot b = a \cdot b^{-1}b = a \cdot 1 = a, \qquad a \cdot a^{-1}b = aa^{-1} \cdot b = 1 \cdot b = b,$$

while, if a, b be symbols such that $ab = 1$, and a be not singular,

$$a^{-1} = a^{-1} \cdot 1 = a^{-1}(ab) = a^{-1}a \cdot b = b,$$

from which, if also b is not singular,

$$b^{-1} = 1 \cdot b^{-1} = aa^{-1} \cdot b^{-1} = ab \cdot b^{-1} = a \cdot bb^{-1} = a.$$

Further, if $x = ab$, then $x^{-1} = b^{-1}a^{-1}$; for, if $y = b^{-1}a^{-1}$,

$$x \cdot b^{-1} = a, \qquad xb^{-1}a^{-1} = 1, \qquad xy = 1, \qquad y = x^{-1}.$$

Also, if a is not singular, there is one and only one symbol, x, such that $ax = b$, and one and only one symbol y such that $ya = b$, where b is an arbitrary symbol. For $ax = b$, $ya = b$ lead, respectively, to

$$x = a^{-1}a \cdot x = a^{-1} \cdot ax = a^{-1}b, \qquad y = y \cdot aa^{-1} = ya \cdot a^{-1} = ba^{-1}.$$

This involves that the statement $ax = az$ requires $x = z$, the symbol a not being singular. (Baker, 1922, p. 65)

With Eqs. (125)–(130) the mathematical foundation of the q-number calculus was laid. Now, for instance, it became straightforward to write down the inverse of $z_1 z_2$, the product of two q-numbers z_1 and z_2, as

$$\frac{1}{z_1 z_2} = \frac{1}{z_2} \cdot \frac{1}{z_1}, \tag{131}$$

or to obtain the differential quotient of the inverse of a q-number with the equation

$$\frac{d}{dt}\left(\frac{1}{z}\right) = -\frac{1}{z} \cdot \frac{dz}{dt} \cdot \frac{1}{z}, \tag{132}$$

keeping in mind the noncommutativity property from Eq. (125).[202] Moreover, Dirac noticed that the expansions for the expression, $(1 + z)^n$, where n is an integer, and for the function, $\exp\{z\}$, in power series of the q-number z, were formally the same as in ordinary algebra. The usual multiplication law of exponentials, however, is not valid because $\exp\{z_1 z_2\}$ is equal to the product of $\exp\{z_1\}$ and $\exp\{z_2\}$ only if the q-numbers z_1 and z_2 commute. In general, $f(z)$, any function of q-number z, can be defined by means of an algebraic relation expressing $f(z)$ in terms of z. It is also possible to define $\partial f(z)/\partial z$, the differential quotient of such a function, by an algebraic relation; this relation is found to be the same as when z is a c-number, provided the relation between $f(z)$ and z contains no q-number that does not commute with z and $f(z)$. For example, if $f(z)$ is the power z^n, with n a c-number, then df/dz, its differential quotient, will be given by nz^{n-1}.

Whenever, in his early papers on quantum mechanics, Dirac discussed functions he thought of those that have a very specific dependence on their variables. ‘One has been obliged,’ he wrote, ‘to restrict the word function to apply only to analytic functions specified by algebraic equations involving the operations of addition, multiplication and reciprocation . . . , and one has usually contented oneself with taking a theorem involving arbitrary functions to be true generally when one has proved it true for power series’ (Dirac, 1926e, p. 412). Although this restriction did not hinder the immediate application of the q-number methods to calculate the properties of atomic systems, Dirac felt that there remained a gap that had to be filled. He, therefore, took up this question in a paper, entitled ‘On Quantum Algebra,’ which Ralph Fowler communicated to

[202] Equation (131) is proven to be true by observing that

$$z_1 z_2\left(\frac{1}{z_2} \cdot \frac{1}{z_1}\right) = z_1\left(z_2 \cdot \frac{1}{z_2}\right)\frac{1}{z_1} = z_1 \cdot 1 \cdot \frac{1}{z_1} = 1;$$

and Eq. (132) follows by multiplying the identity

$$0 = \frac{d}{dt}\left(z \cdot \frac{1}{z}\right) = \frac{d}{dt}\left(\frac{1}{z}\right) \cdot z + \frac{1}{z} \cdot \dot{z},$$

from both sides by $1/z$.

the *Proceedings of the Cambridge Philosophical Society* in July 1926. In this paper he proposed 'a general definition of a function which appears to enable one to establish theorems involving arbitrary functions with rigour' (Dirac, 1926e, p. 412).[203] The practical consequences of the new definition of functions in quantum mechanics were not important insofar as the results derived from it had been taken to be obvious. Yet, this paper on q-numbers, in spite of the often highly technical and sometimes artificial definitions and derivations, provided more than just a short mathematical appendix to Dirac's dynamical quantum theory; rather it gave the final conclusions from the analogy of q-number methods to the symbolic calculus used in geometry, and in particular it presented the geometrical visualization of q-numbers. Indeed, this work of Dirac's led to fundamental insights into quantum mechanics.

Dirac based the new definition of q-number functions on the elementary observation that every number, which commutes with a given q-number x, also commutes with any power of x, say x^2. Powers of a variable represent the type of algebraic functions that had already been adequately defined in quantum mechanics, but the property just stated was shared by a much wider class of objects depending on the q-number x. Dirac, therefore, proposed to take sets of two quantum numbers, x and X, such that any number which commutes with x also commutes with X, and then defined X to be a function of a single variable, x. 'A function of a q-number variable,' he noted, 'is thus specified by a pair of q-numbers that satisfy a certain condition and is quite different from a function of a real c-number variable, for the specification of which an infinite number of pairs of numbers are required, which need not satisfy any conditions' (Dirac, 1926e, p. 413).

The restriction imposed upon q-number functions by the new definition relates the values of the variables and of the functions in a specific manner. Let x and X be a pair of q-numbers belonging to a given function, and let y be a q-number obtained from x through the equation

$$y = bxb^{-1}, \tag{133}$$

where b is an arbitrary, nonsingular q-number. Then the q-number Y, which is given by the equation

$$Y = bXb^{-1}, \tag{134}$$

is the same function of y that X is of x. All pairs, (y, Y), (z, Z), etc., constructed from an original pair, (x, X), with the help of Eqs. (133) and (134), where different, nonsingular q-numbers b enter, represent a q-number function of a

[203]The same restriction in the definition of functions, which occurred in Dirac's early papers, occurred also in matrix mechanics (see, e.g., Born, Heisenberg and Jordan, 1926). Dirac tried to fill, in his last paper on quantum algebra, a gap which had existed so far in all approaches to quantum mechanics.

single variable.[204] Dirac called the specific pairs, (x, X), (y, Y), etc., the 'particular values' of the independent variables and the function, respectively (Dirac, 1926e, p. 413). It should be noted that in quantum mechanics, where the functions and the independent variables become quantum variables, these 'values' can be identified with the values of the quantum variables in stationary states of the system under investigation only when the latter are solutions of the equations of motion; Dirac always assumed this to be the case.

The functions thus defined possess the property that any of them must commute with any other function of the same variable. As a result, a c-number may be regarded as a particular case of a function of any q-number variable since it commutes with all functions of arbitrary variables. It was more important, however, that Dirac could use the same idea to construct functions of several variables: that is, he proposed to 'define X to be a function of the variables $x_1, x_2, \ldots, x_n$ only, if X commutes with every number that commutes with $x_1, \ldots, x_n$' (Dirac, 1926e, p. 415). Again there are restrictions on the functions defined in this way as compared to the case of real c-number functions of several variables. In particular, the value assumed by each variable cannot vary independently of the others, but the values of *all* variables have to change according to the same scheme, given by the set of equations

$$y_1 = bx_1b^{-1}, \ldots, y_n = bx_nb^{-1}, \tag{135}$$

with the same q-number b for all variables. Only when all $x_1, \ldots, x_n$ commute

[204] Evidently, with the help of Eqs. (133) and (134) one can generate all pairs (y, Y) defining a function from a basic pair, say (x, X), provided the following properties, which Dirac listed in his paper, are satisfied:

(i) All values of b that satisfy [(133)] must give the same value for Y defined by [(134)].
(ii) Y defined by [(134)] must be a function of the single variable y, i.e., any number that commutes with y must also commute with Y.
(iii) Functions of a single variable can be formed by the addition or multiplication of functions of that variable.
(iv) A function of a function of a certain variable must be a function of that variable. (Dirac, 1926e, p. 413)

Dirac proved the validity of conditions (i) to (iv) with the help of the symbolic calculus, which he learned from Baker's *Principles of Geometry*. For example, in order to show that (i) holds, he assumed that two roots b and b' of Eq. (133) exist; thus,

$$y = bxb^{-1} = b'xb'^{-1}.$$

Multiplying from the left by b^{-1} and from the right by b', he obtained

$$xb^{-1}b' = b^{-1}b'x,$$

noting that $b^{-1}b'$ commutes with x. Since, by definition, $b^{-1}b'$ also commutes with X, i.e.,

$$b^{-1}b'X = Xb^{-1}b',$$

the equation, obtained by multiplying it by b from the left and b'^{-1} from the right, that is,

$$b'Xb'^{-1} = bXb^{-1} = Y,$$

is also valid. Hence condition (i) is satisfied.

with one another is it possible to admit independent changes as in c-number theory.[205]

In spite of these restrictions, Dirac did enlarge the class of q-number functions with the new definition. However, they would be useful in the dynamical calculations of quantum mechanics only if the two operations of differentiation and integration could be defined on the q-number functions as well. In the case of differentiation, he first turned to Q, a function of a single (q-number) variable q. In addition, he introduced the particular q-number p, which satisfied the equation

$$qp - pq = \frac{ih}{2\pi}, \tag{136}$$

and constructed a new q-number Q' as

$$Q' = \frac{2\pi}{ih}(Qp - pQ). \tag{137}$$

Finally, he proposed to 'define Q' to be the differential coefficient of Q with respect to q, and write it dQ/dq' (Dirac, 1926e, p. 415).

The new definition of the differentiation of a function satisfied all conditions which the internal consistency of the quantum-mechanical scheme required. For example, the ordinary rules for the differential coefficient of a sum, of a product, and of a function were found to hold. Moreover, Dirac demonstrated, in every necessary detail, that dQ/dq is a function of q only; that is, it is of the same type as Q and is defined uniquely by Eqs. (136) and (137). From these properties Dirac concluded immediately that 'the differential coefficients of simple algebraic functions have their ordinary values' (Dirac, 1926e, p. 417). He had thus established the consistency of his new definition of dQ/dq with the earlier definition of differential coefficients in quantum mechanics. He also saw no difficulty in applying the same idea for introducing the concept of a partial differential coefficient, and concluded: 'If Q is a function of the variables $q, q_1, q_2, \ldots, q_n$ that commute with one another, to define $\partial Q/\partial q$ one requires a p that satisfies [Eq. (136)] and that commutes with $q_1, q_2, \ldots, q_n$. One can then define $\partial Q/\partial q$ by the left-hand side of [Eq. (137)]' (Dirac, 1926e, p. 418).

[205] To demonstrate the necessity of conditions (135), Dirac considered the example of two variables, x_1 and x_2, satisfying the equation

$$x_1x_2 - x_2x_1 = 1.$$

Hence, x_1x_2 and $1 + x_2x_1$ represent identical functions. If x_1 and x_2 are varied independently, that is,

$$y_1 = bx_1b^{-1}, \qquad y_2 = b'x_2b'^{-1},$$

one obtains

$$y_1y_2 = bx_1b^{-1}b'x_2b'^{-1} \qquad \text{and} \qquad 1 + y_2y_1 = 1 + b'x_2b'^{-1}bx_1b^{-1}.$$

The two right-hand sides are equal only if $b = b'$.

The above definitions, especially those for functions of several variables and their partial differential coefficients, contained severe restrictions from the mathematical point of view. Yet, for most functions occurring in the description of quantum systems, these restrictions are automatically satisfied. Dirac knew about this fact, and, although he did not go into it in his paper, it is not difficult to show it on the basis of the information available to him at the time. First, one may discuss the conditions imposed by Eqs. (135) on the set of variables of a function of several variables. Such functions do represent the properties of a system of f degrees of freedom if they depend on q_k and p_k, $k = 1, \ldots, f$, that is, a complete set of canonical variables of the system. Now in classical theory, all changes of q_k and p_k can be expressed by contact transformations. For example, when one studies the evolution of the system in time, the values of the variables at time $t + \Delta t$, that is, $q_k(t + \Delta t)$ and $p_k(t + \Delta t)$, are obtained from $q_k(t)$ and $p_k(t)$, the values of the variables at an infinitesimal moment earlier (i.e., at time t), by the equations

$$q_k(t + \Delta t) = q_k(t) + (q_k(t), H)\Delta t \tag{138a}$$

and

$$p_k(t + \Delta t) = p_k(t) + (p_k(t), H)\Delta t, \tag{138b}$$

where H is the Hamiltonian of the system. In quantum mechanics, according to Dirac's prescription, one had just to replace the classical Poisson bracket by the quantum Poisson bracket in order to arrive at the following equation,

$$q_k(t + \Delta t) = \exp\left\{ - \frac{2\pi}{ih} H\Delta t \right\} q_k(t) \exp\left\{ + \frac{2\pi}{ih} H\Delta t \right\}, \tag{139}$$

expressing the value of $q_k(t + \Delta t)$ in terms of the value of $q_k(t)$. Equation (139) is analogous to Eq. (138a), provided one takes, in the expansion of the functions $\exp\{\pm(2\pi/ih)H\Delta t\}$ in powers of the variable $(2\pi/ih)H\Delta t$, only terms up to the first order, which is allowed because Δt is infinitesimally small. Equation (139) represents a special case of contact transformations in quantum mechanics. Dirac discovered that a general canonical transformation, which transforms p_k and q_k, a given fundamental set of variables of the system, into another set of canonical variables, Q_k and P_k, is indeed always given by the equations

$$Q_k = bq_k b^{-1} \quad \text{and} \quad P_k = bp_k b^{-1}, \tag{140}$$

where b is some suitable q-number representing a dynamical variable of the system.[206] Since the transformation of a system to uniformizing variables is a particular contact transformation, which determines the solution of the equations of motion, Eqs. (140) play an important role in quantum mechanics. They place

[206] Equations (140) did not occur in Dirac's first paper on quantum mechanics, but were presented, without any 'derivation,' in his second paper (Dirac, 1926a, p. 564, last equation). In the

restrictions on what Dirac called the 'values' of the quantum variables. However, by comparing Eqs. (140) with Eqs. (135) it is easily seen that these *dynamical restrictions* and the *formal mathematical restrictions*, imposed upon the variables of a q-number function of several variables, are identical. Hence the functions employed in quantum mechanics, which depend upon the fundamental variables q_k and p_k, are q-number functions in the sense considered by Dirac.

Dirac noticed that the situation is less satisfactory when one turns to the question whether the partial differential quotients (coefficients) of quantum variables also obey the definition of the partial differential coefficients of q-numbers. One can, of course, make use of Eqs. (136) and (137) to obtain $\partial F/\partial q_k$ and $\partial F/\partial p_k$, the differential quotients of a function F that depends upon the variables q_k and p_k of a system, that is,

$$\frac{\partial F}{\partial q_k} = \frac{2\pi}{ih}(F p_k - p_k F), \qquad \frac{\partial F}{\partial p_k} = \frac{2\pi}{ih}(F q_k - q_k F). \tag{141}$$

These definitions lead to no problem, as long as the function F depends only on the position variables, or only on the momentum variables. But a difficulty may arise, if canonically conjugate pairs, say q_k and p_k, are among the variables of F. Since p_k does not commute with q_k, the differential quotients $\partial F/\partial q_k$ and $\partial F/\partial p_k$ are not given in a unique manner by Dirac's definition, except when the function F is *separable* in the variables q_k and p_k. One must, therefore, extend the definition of the differential coefficient of q-number functions in order to complete the mathematical description of quantum-mechanical systems.[207]

Besides differentiation, the process of integrating a function over a certain domain of its variables plays a fundamental role in mechanics. Integration can simply be defined as the inverse of differentiation. Thus, to obtain the integral of $Q'(q)$, which is a q-number function of the variable q, one first writes it as a quantum Poisson bracket of another function $Q(q)$ with the q-number p, the bracket being defined with the help of Eq. (136). In order that Q be an integral of Q', the condition must hold that 'the only functions whose differential coeffi-

---

above example one finds that

$$q_k(t + \Delta t) = b q_k(t) b^{-1} \qquad \text{with} \qquad b = \exp\left\{-\frac{2\pi i}{h} H \Delta t\right\}.$$

Hence, it is not difficult to convince oneself that the most general transformations which leave the Poisson brackets invariant are just the ones defined by Eqs. (140). Born, Heisenberg and Jordan had also taken Eqs. (140), in their matrix language, of course, as representing the most general canonical transformation of the dynamical variables (see Born, Heisenberg and Jordan, 1926, Chapter 1, p. 565, Eq. (17)). Dirac saw the three-man paper shortly before submitting his second paper; there he noticed the agreement of his definition with that of the Göttingen physicists.

[207] For instance, such an extension of the definition of the partial derivatives must be made already when F is the product qp, where q is the position and p the momentum variable of a system of one degree of freedom. The reason can be seen from the following observation: Since q and p satisfy the equation $qp = pq + ih/2\pi$, then qp and $pq + ih/2\pi$ must yield the same function of q and p. But it becomes impossible to define, say the partial derivative with respect to p in a consistent way because one arrives at two different results, q and $q + ih/2\pi$, respectively.

cients are zero shall be constants (i.e., c-numbers)' (Dirac, 1926e, p. 418). In the case of functions of a single variable q, this condition means that all q-numbers, which commute with p and also with every q-number that commutes with q, must be c-numbers. Only together with this condition would the process of integration lead to the usual results when applied to simple algebraic functions. Since Dirac saw no possibility of deriving this condition from the laws of q-numbers, he made it an independent axiom of his calculus. He expressed it in the most general form as: 'If a number commutes with every number that commutes with the q-numbers $q, q_1, q_2, \ldots, q_n$ and with one number that does not commute with q but commutes with $q_1, q_2, \ldots, q_n$, then it commutes with every number that commutes with $q_1, q_2, \ldots, q_n$' (Dirac, 1926e, p. 418). Clearly, Dirac's axiom—and this is the one and only axiom he introduced besides the laws of q-numbers taken from the symbolic calculus in geometry—leads, in the special case of one variable, to the condition which makes the integration of a function unique.

Although the above axiom was recognized to be necessary, Dirac still had to consider the question of whether it contradicted the other laws of the q-number calculus or the known laws of quantum mechanics. For this purpose he stated the geometrical interpretation of the axiom as follows: 'Each q-number may be represented by a point in space of an infinite number of dimensions in such a way that the condition that a q-number shall commute with certain given q-numbers is that it shall lie in a certain flat space' (Dirac, 1926e, p. 418). He knew well the meaning of this geometrical statement and its consequences from the discussion of the so-called 'Pappus' theorem' in Baker's *Principles of Geometry*, where it had been shown that '*the assumption of Pappus' theorem, and the restriction of the algebraic symbols to such as are commutative in multiplication are, subject to the other assumptions made, equivalent*' (Baker, 1922, p. 78).[208] Originally, Pappus' theorem was a typical theorem of geometry in a plane, but mathematicians of the nineteenth century had generalized it to apply also to flat linear spaces of higher dimensions.[209] By identifying Baker's symbols with

[208] Pappus' theorem was considered in a paper which M. H. Newman of St. John's College—whom Dirac knew—submitted in June 1925 to the *Proceedings of the Cambridge Philosophical Society* (Newman, 1925). In this paper Newman simplified a proof of the theorem due to Hilbert, based on the assumption of the axiom of Archimedes.

[209] In a plane, Pappus' theorem claims that three points, P, Q, R, will be found on a line which is constructed as follows: take on each of two intersecting lines three points, L, M, N, and L', M', N', respectively (none being identical with the point common to both lines), and draw the lines through the pairs of points, L and M', M and L', L and N', N and L', M and N', and N and M'; P is then the intersection of the first pair of lines, Q the intersection of the second pair of lines, and R the intersection of the third pair of lines.

David Hilbert, to whom Baker referred in connection with the demonstration of the commutativity of the multiplication of symbols in a geometric space where Pappus' theorem is valid, had used a slightly more specialized version of the latter, which he called Pascal's theorem, and stated thus: 'Let A, B, C and A', B', C' be two triples of points, each lying on one of two intersecting straight lines (the points should not coincide with the point of intersection [of the two lines]); then, if CB' is parallel to BC' and CA' is parallel to AC', also BA' is parallel to AB'' (Hilbert, 1899, Theorem 40). The reason why Hilbert named the theorem after Blaise Pascal was that Pappus' theorem was a special case of a theorem due to Pascal, which applied to the case when the six points L, M, N, L', M', N' were taken on a conic section.

q-numbers, and interpreting the latter in a space of infinitely many dimensions, Dirac was able to visualize a quantum-mechanical relation between the *commuting* variables as a geometrical relation in an adequately chosen flat space. That is, for those special relations he had finally arrived at a definite geometrical picture of what a q-number was like. However, this visualization of q-numbers broke down when relations including noncommuting variables were considered; but even for these cases one could obtain a proper interpretation in terms of relations between geometrical points, provided one admitted peculiar, nonflat spaces of infinitely many dimensions. Although Dirac did not address himself to this point, he was perfectly aware of it, as is evident from several remarks in his first paper of January 1926 on q-numbers (Dirac, 1926a). A short discussion of the mathematical implications may, therefore, help to understand better his ideas and methods.

The properties of nonflat spaces had indeed been discussed by the mathematicians in some detail. David Hilbert, for example, had shown that the validity of Pappus' theorem in flat space could be derived from the assumption that the so-called 'axiom of Archimedes' holds (Hilbert, 1899, Chapter 6). This axiom requires that for two given distances there always exists an integral multiple of the smaller one that exceeds the larger one. It follows, therefore, that in spaces, in which this axiom does not hold, it must be impossible to introduce a suitable measure of distances. As a result one is also unable to define infinitely small angles; hence, in such spaces there will always be infinitely many straight lines that do not intersect a given straight line. Dirac learned about these results from Baker's *Principles of Geometry* (Vol. 1, Chapter II). The consequences for the q-number theory were obvious: the geometrical space associated with noncommuting q-numbers possessed the same non-Euclidean structure as had been found earlier in those spaces in which the validity of the axiom of Archimedes was not postulated. The fact that one cannot measure distances in these spaces is fully equivalent to Dirac's general observation in q-number theory, to which we have already referred earlier, namely, that 'one cannot say that one q-number is greater or less than another' (Dirac, 1926a, p. 562).[210]

[210] Dirac elaborated this statement in more detail in the following paper by saying that 'the dynamical variables cannot be ordinary numbers expressible in the decimal notation (which numbers will be called c-numbers), but may be considered to be numbers of a special kind (which will be called q-numbers), whose nature cannot be exactly specified, but which can be used in the algebraic solution of a dynamical problem in a manner closely analogous to the way the corresponding classical variables are used' (Dirac, 1926b, p. 281).

Although he was unable to specify the exact nature of the q-number space, Dirac knew some of its properties from Baker's *Principles of Geometry*. For example, Baker had drawn attention to the fact that in spaces, in which the axiom of Archimedes does not hold, a notion of order still exists, and he illustrated this point in an example which is of some interest in connection with Dirac's work. In discussing the determination of time with clocks, which do not run steadily, he made the following observations:

> In the case of time we measure intervals by observing the succession of natural phenomena, as for instance by heart-beats, or by a clock. But if all the clocks went twice as fast between what we now call 12 o'clock and 12:30 o'clock, as they now do, so that they indicated 1 o'clock at the time we now call 12:30, and went two-thirds as fast as they now do between the times we now call 12:30 and 2 o'clock, so that they marked 2 o'clock as they now do, we should, if we had agreed to regard as standard time that given by the clocks, reckon the interval between our

These consequences, arising from geometry, were very serious for the q-number calculus because they were equivalent to the following statement: In spite of the fact that for all c-number equations of classical mechanics one could obtain analogous q-number equations, and that one could perform various mathematical operations on these equations in accordance with well-defined prescriptions, it was still not possible to obtain any physical result from the theory *because q-numbers could not be compared with experiment*. It was necessary to add something to this *mathematical theory* in order to transform it into a *physical theory*. Dirac had anticipated this by requiring that: 'In order to be able to get results comparable with experiment from our theory, we must have some way of representing q-numbers by means of c-numbers, so that we can compare these c-numbers with experimental values' (Dirac, 1926a, p. 563). To find the representation in terms of c-numbers the correspondence between the classical and the quantum expressions, which Dirac had obtained earlier (Dirac, 1925d), turned out to be helpful. It led to the definite requirement that: 'The representation must satisfy the condition that one can calculate the c-numbers that represent $x + y$, xy and yx, when one is given the c-numbers that represent x and y' (Dirac, 1926a, p. 563). Evidently, the calculations had to be consistent with the laws defining the sums and products of quantum variables, which Dirac had formulated in fall 1925 (Dirac, 1925d). Still, all these requirements, taken together, did not define completely the representation of the q-numbers describing the quantum variables of a given system in a unique manner, with one exception. There existed one class of systems, the multiply periodic ones, in which one had a unique representation of q-numbers.

In his paper on the fundamental equations of quantum mechanics Dirac had focused, in particular, on the dynamical variables describing multiply periodic quantum systems. On the basis of his earlier discussion, he now concluded:

> If a q-number x is a function of the coordinates and the momenta of a multiply periodic system, and if it is itself multiply periodic, then it will be shown that the aggregate of all its values for all values of the action variables of the system can be represented by a set of harmonic components of the type $x(nm)\exp\{i\omega(nm)t\}$, where $x(nm)$ and $\omega(nm)$ are c-numbers, each associated with two sets of values of the action variables denoted by the labels n and m, and t is the time, also a c-number. (Dirac, 1926a, p. 563)

> present times 12 and 12:30 as being equal to the interval between 12:30 and 2. Though this would lead to modifications in our description of the behaviour of natural bodies, as, for instance, the contractions of the heart, it would not affect our notion of the succession of time, or of the order of events. The questions suggested in this connection, recently also in vogue as arising in discussions of the Principle of Relativity in Physical Science, are inherent in the considerations of the Foundations of Geometry. (Baker, 1922, p. 124)

This passage may not only be used to throw light on the interpretation of q-numbers; it also refers to the origin of Dirac's ideas. Since he had participated in the discussions about the problems of relativity at Baker's tea parties, he was also familiar with those parts of the foundations of geometry that dealt with spaces where the axiom of Archimedes did not hold. Thus, he did not hesitate to apply his geometrical knowledge to discover the laws of quantum mechanics.

Dirac knew that Born, Heisenberg and Jordan had also based their quantum-mechanical scheme, the matrix mechanics, entirely on this representation.[211] Yet he did not want to restrict himself to this representation even for discussing periodic systems like the hydrogen atom. That is, in performing calculations he would not immediately specialize to the harmonic representation of multiply periodic variables, but stick to *q*-number equations as long as possible. A *q*-number, he argued, 'still has a meaning and can be used in the analysis [of a system] when it is not multiply periodic, although there is at present no way of representing it by *c*-numbers' (Dirac, 1926a, p. 563). He demonstrated how his theory worked by treating selected problems, such as the hydrogen atom and the scattering of radiation by free electrons. He always proceeded in the following way: first, the *q*-number variables describing the system under investigation were constructed; the equations of motion for these variables were then integrated; finally, the results were expressed in terms of *c*-numbers and compared with the available empirical information. With this procedure Dirac developed an approach for treating quantum-mechanical problems that was quite different from the one of Born, Heisenberg, Jordan and Pauli. It enabled him to derive certain new results and to confirm those that had been obtained from matrix mechanics.

V.2 The Introduction of Action-Angle Variables

In the early days of quantum mechanics, theoretical physicists gave the dynamical variables describing atomic systems the same names which had been used before in classical theory, although the operations with quantum variables often led to quite different conclusions. They were well aware of these differences, and Paul Dirac, for example, stated that 'the only justification for the names given to dynamical variables lies in the analogy to the classical theory, e.g., if one says that x, y, z are the Cartesian coordinates of an electron, one means only that x, y, z are *q*-numbers which appear in the quantum solution of the problem in an analogous way to the Cartesian coordinates of the electron in the classical solution' (Dirac, 1926b, p. 281). The quantum variables thus occur in the same places in quantum-mechanical equations as the corresponding classical variables, and they also show up similarly in the solutions of the equations. In constructing the quantum variables a difficulty may arise if two or more *q*-numbers replace the same classical variables and thus seem to have the claim to the same name. 'In such a case,' argued Dirac, 'one must decide which of the properties of the classical variables are dynamically the most important, and must choose the *q*-number which has these properties to be the corresponding quantum variable' (Dirac, 1926b, p. 281). He found an example of an ambiguity of this kind when

[211] Heisenberg had informed Dirac about the papers of Born and Jordan (1925b) and Born, Heisenberg and Jordan (1926) already in his letter of 20 November 1925. These two papers appeared in print before Dirac submitted his second paper on quantum mechanics and were quoted by him (see, in particular, Dirac, 1926a, pp. 561, 563).

he discussed the concept of frequency in multiply periodic systems, for both the q-numbers that represent orbital frequencies and the q-numbers that represent transition frequencies correspond to the classical frequencies. In Heisenberg's original proposal, the orbital frequency had been declared to be 'unobservable in principle' (Heisenberg, 1925c, p. 879) and had been excluded from the new quantum theory; in Dirac's approach, however, the situation was different, provided he could come up with the q-number representing the orbital frequency. In classical dynamics, the latter quantity is obtained as the time derivative of the angle variable; hence all one had to do was to introduce angle variables in quantum mechanics. In the matrix theory of Born, Heisenberg and Jordan the introduction of nonperiodic variables was not possible without extending the scheme considerably, but in q-number theory this step presented no difficulty, for, in general, q-numbers allowed one to describe a wider class of variables than those which can be represented by quantum-theoretical trigonometric series. Since Dirac was convinced that he would need action and angle variables for the solution of the equations of motion of realistic atomic systems, he concentrated his efforts on constructing these variables as the first illustration of q-numbers, devoting two papers to this question (Dirac, 1926a, b).

In introducing action and angle variables Dirac followed closely the procedure of classical theory. Therefore, he defined a dynamical system to be multiply periodic in quantum theory when there exists a set of uniformizing variables, J_k and w_k, having the following properties: (i) They are canonical variables, i.e., they satisfy the commutation relations

$$[w_k, J_l] = \delta_{kl}, \qquad [w_k, w_l] = [J_k, J_l] = 0, \tag{142}$$

($k, l = 1, \ldots, f$, where f is the number of degrees of freedom of the system). (ii) The Hamiltonian H is a function of J_k, the action variables, alone, though not necessarily the same function as in the classical theory. (iii) The original position and momentum variables describing the system are multiply periodic functions of w_k, the angle variables, of period 2π; for example, x, the Cartesian coordinate of a particle in the system, can be expressed in one of two ways,

$$x = \sum_\alpha x_\alpha(J) \exp\{i(\alpha w)\} \tag{143a}$$

or

$$x = \sum_\alpha \exp\{i(\alpha w)\} x'_\alpha(J), \tag{143b}$$

with

$$(\alpha w) = \alpha_1 w_1 + \alpha_2 w_2 + \ldots + \alpha_f w_f. \tag{143c}$$

The α-sums on the right-hand side of Eqs. (143a) and (143b) are f-fold sums, in which $\alpha_1, \ldots, \alpha_f$ all assume integral values and the coefficients $x_\alpha(J)$ and $x'_\alpha(J)$ are functions of the action variables only. Because of condition (ii), the equations of motion in the uniformizing variables become identical to the classical equations, that is,

$$\dot{J}_k = [J_k, H] = 0 \tag{144a}$$

and

$$\dot{w}_k = [w_k, H] = \frac{\partial H}{\partial J_k}. \tag{144b}$$

The $\dot{w}_k$'s, given by Eq. (144b) as constants of the motion, correspond to $2\pi\nu$, the orbital frequencies of Bohr's theory. In this context it should be noted that Dirac defined the angle variables such that the variable x, following Eqs. (143), assumes the same value after a given w_k has changed by a period of 2π. Hence the J_k's correspond to the conventional action variables (which are conjugate to angles whose period is 1) divided by the factor 2π.[212]

Having thus introduced the action and angle variables, Dirac had to demonstrate that they allowed a description of all variables of the system consistent with the fundamental equations of quantum mechanics. For instance, he had to show that the time derivative of any x, which is represented by Eq. (143a) or Eq. (143b), reproduces the trigonometric series, the coefficients x_α and x'_α being replaced by $i(\alpha\omega)x_\alpha$ and $i(\alpha\omega)'x'_\alpha$, respectively, where the factors $(\alpha\omega)$ and $(\alpha\omega)'$ become identical to the transition frequencies of the system. To prove this point Dirac made use of the equation

$$\exp\{i(\alpha w)\}F(J) = F\left(J - \alpha\frac{h}{2\pi}\right)\{\exp i(\alpha w)\}, \tag{145}$$

where $F(J)$ is an arbitrary function of the J's $(= J_1, \ldots, J_f)$, and $F(J - \alpha(h/2\pi))$ is the value of this function when each J_k is replaced by $J_k - \alpha_k(h/2\pi)$. He then found that the quantities $(\alpha\omega)$ and $(\alpha\omega)'$ are indeed given by the expressions[213]

$$(\alpha\omega) = \omega\left(J, J - \alpha\frac{h}{2\pi}\right) = \frac{H(J) - H(J - \alpha h/2\pi)}{h/2\pi} \tag{146a}$$

[212] Dirac used that definition in the papers dealing with action variables in quantum mechanics 'in order to save writing' (Dirac, 1926a, p. 567). He also chose h as 'a real constant, equal to $(2\pi)^{-1}$ times the usual Planck's constant' (Dirac, 1926a, p. 561), for the same reason. We do not follow this notation, however. Later on, in his book on *The Principles of Quantum Mechanics*, Dirac denoted $h/2\pi$ by the letter $\hbar$ (Dirac, 1930, p. 96).

[213] Evidently one has

$$\frac{d}{dt}\exp\{i(\alpha w)\} = \frac{2\pi}{ih}(\exp\{i(\alpha w)\}H - H\exp\{i(\alpha w)\}),$$

which, because of Eq. (145), is equal to

$$i\frac{H(J) - H(J - \alpha(h/2\pi))}{h/2\pi}\exp\{i(\alpha w)\} = i(\alpha\omega)\exp\{i(\alpha w)\}.$$

Hence Eq. (146a) is proven. Similarly, one derives Eq. (146b) with the help of the relation

$$F(J)\exp\{i(\alpha w)\} = \exp\{i(\alpha w)\}F(J + \alpha(h/2\pi)).$$

and

$$(\alpha\omega)' = \omega\left(J + \alpha\frac{h}{2\pi}, J\right) = \frac{H(J + \alpha h/2\pi) - H(J)}{h/2\pi}. \tag{146b}$$

The transition frequencies, $\omega(J, J - \alpha h/2\pi)$ and $\omega(J + \alpha h/2\pi, J)$, are q-numbers; that is, they are merely the *same functions* of the action variables as the Bohr frequencies of the classical J's which are c-numbers.

As another confirmation of the application of the general laws of quantum mechanics to a system, described by the action and angle variables satisfying Eqs. (142)–(144), Dirac again calculated by using Eq. (145) the product of two variables, x and y, both represented by Eqs. (143). He obtained the result

$$\sum_{\alpha} x\left(J, J - \alpha\frac{h}{2\pi}\right)\exp\{i(\alpha w)\}\sum_{\beta} y\left(J, J - \beta\frac{h}{2\pi}\right)\exp\{i(\beta w)\}$$

$$= \sum_{\alpha,\beta} x\left(J, J - \alpha\frac{h}{2\pi}\right) y\left(J - \alpha\frac{h}{2\pi}, J - \alpha\frac{h}{2\pi} - \beta\frac{h}{2\pi}\right)\cdot \tag{147}$$

$$\exp\{i(\alpha + \beta)w\},$$

where the coefficients $x(J, J - \alpha h/2\pi)$, $y(J, J - \beta h/2\pi)$ and $y(J - \alpha h/2\pi, J - \alpha h/2\pi - \beta h/2\pi)$ denote respectively the coefficients $x_{\alpha}(J)$, $y_{\beta}(J)$ and $y_{\beta}(J - \alpha h/2\pi)$ used earlier. In Eq. (147) one recognizes Heisenberg's rule of multiplication of quantum variables. Moreover, if one substitutes for each action variable J_k the c-number $n_k h/2\pi$ and denotes the c-numbers replacing $x(J, J - \alpha h/2\pi)$ and $\omega(J, J - \alpha h/2\pi)$ by $x(n, n - \alpha)$ and $\omega(n, n - \alpha)$, respectively, then one can obtain a representation of the q-numbers describing a multiply periodic system by means of c-numbers as follows: One considers 'the aggregate of all c-numbers $x(n, n - \alpha)\exp\{i\omega(n, n - \alpha)t\}$, in which it is sufficient (but not necessary) for the n to take a series of values differing successively by unity, as representing the values of the q-number x for all values of the q-numbers J_k' (Dirac, 1926a, p. 569). This representation is the same as the matrix representation. Dirac noticed that, in general, the c-numbers n_k need not be integers but may differ from integers by a common difference, say η_k, which is smaller than 1. Different values of η_k then distinguish different representations of the q-numbers of the multiply periodic system.[214] Dirac concluded, however, that only one

[214] It is of interest to look at the origin of Dirac's notation. The fact that the quantum-theoretical variables x are 'represented' by the 'entirety' of quantities, $x(nm)\exp\{i\omega(nm)t\}$, he took over from Born, Heisenberg and Jordan. In the three-man paper, §1. *Grundprinzipien*, the authors had stated: 'A quantum-theoretical quantity a—be it a coordinate or momentum or any function of both—will be represented by the entirety of quantities ('*Gesamtheit der Größen*'), $a(nm)\exp\{2\pi i\nu(nm)t\}$. . . .' (Born, Heisenberg and Jordan, 1926, p. 561). Earlier, Born and Jordan had already considered the matrix as the 'representation of a physical quantity, which in classical theory is denoted by a function of time' (Born and Jordan, 1925b, p. 859). On the other hand, the fact that examples of symbolic quantities 'may be formed with aggregates of the numbers of ordinary arithmetic' had been discussed by Henry Baker in *Principles of Geometry* (1922, Vol. I, Section III, p. 66), where quadratic matrices were also treated among the examples.

representation is of physical importance, and it is determined by the property that all $x(nm)$ are zero when any of the n_k or m_k is smaller than a certain value, say n_{k0}, denoting the ground state of the system.

The question now arises of whether the representation of the variable x is unique. Since the coefficients $x_\alpha(J)$ and $x'_\alpha(J)$ in the expansions (143a) and (143b) determine the intensity of the radiation having a frequency $\omega(n, n-\alpha)$ and polarization along the direction of the x-axis, the physical situation requires that the intensity calculated from either expansion should be the same. The relation between the coefficients $x_\alpha(J)$ and $x'_\alpha(J)$ follows from Eq. (145), which leads to the identity

$$x = \sum_\alpha x_\alpha(J)\exp\{i(\alpha w)\} = \sum_\alpha \exp\{i(\alpha w)\} x_\alpha\left(J + \alpha\frac{h}{2\pi}\right). \tag{148}$$

That is, if one denotes the quantity $(J + \alpha h/2\pi)$ by J'_k, and rewrites the $x'_\alpha(J)$ in Eq. (143b) as $x'_\alpha(J')$, then $x'_\alpha(J')$ is the same function of the J'_k as $x_\alpha(J)$ is of the J_k. Hence, if one substitutes the values n_k for the J_k in $x_\alpha(J)$ one would obtain the same c-number as the one obtained by substituting the values $n_k - \alpha_k\, h/2\pi$ for the J_k in $x'_\alpha(J)$. Thus, the frequency $\omega(n, n-\alpha)$ is symmetrically related to the two states given by $J_k = n_k$ and $J'_k = n_k$, and, as a result, the intensity of radiation will be the same with either of the representations, Eq. (143a) and Eq. (143b), for x.

The scheme of action-angle variables is supposed to describe not only the emission, but also the absorption of radiation in a manner analogous to the procedure in classical theory. In the latter, the amplitudes of the emitted and absorbed radiation having the same frequency, say $(\alpha\omega)$, are related to each other by the mathematical process of complex conjugation: that is, the absorbed amplitude, which appears in the description together with the periodic function $\exp\{-i(\alpha\omega)t\}$, is the complex conjugate of the emitted amplitude, which appears together with the periodic function $\exp\{i(\alpha\omega)t\}$. In order to obtain a similar relation in quantum mechanics, Dirac postulated that the dynamical variables are *real* quantities. The reality property had to be defined more carefully than in classical theory because the imaginary unit occurs in one of the fundamental laws, the quantum conditions. Thus, although one can always exchange the imaginary factor i arising from the number $\sqrt{-1}$ by the factor $-i$ in an equation of quantum mechanics (just as in an equation of classical mechanics), this is not so with the factor i in the quantum condition. However, since the factor i in the quantum condition is multiplied with Planck's constant h, Dirac noted that from any quantum-mechanical equation 'one may obtain another equation by writing $-i$ for i wherever it occurs and at the same time $-h$ for h, or by reversing the order of all factors and writing $-h$ for h, or by applying the two operations together, which reduces to reversing the order of all factors and writing $-i$ for i' (Dirac, 1926a, p. 565). Then he defined that this operation, when applied to a q-number x, yields the 'conjugate imaginary,' $\bar{x}$; he called x 'real' if it is identical to $\bar{x}$ (Dirac, 1926a, p. 565).

If x and J are real numbers, as postulated above, it follows from Eq. (143a) that

$$x = \sum_{\alpha} x_{\alpha}(J)\exp\{i(\alpha w)\} = \sum_{\alpha} \exp\{-i(\alpha w)\}\bar{x}_{\alpha}(J), \tag{149}$$

where $\bar{x}_{\alpha}(J)$ denotes the conjugate imaginary of $x_{\alpha}(J)$. By using Eq. (145) to bring $\bar{x}_{\alpha}(J)$ to the left of the exponential function and comparing the coefficients of the same exponentials, say $\exp\{-i(\alpha \text{w})\}$, one obtains the equation

$$\bar{x}_{\alpha}\left(J + \alpha\frac{h}{2\pi}\right) = \bar{x}\left(J + \alpha\frac{h}{2\pi}, J\right) = x_{-\alpha}(J) = x\left(J, J + \alpha\frac{h}{2\pi}\right), \tag{150}$$

which expresses a particular symmetry of the amplitudes with respect to the two variables, J and $J + \alpha(h/2\pi)$. If definite values (c-numbers) of the J's are inserted, Eq. (150) assures that the amplitude of transition from a stationary state with quantum numbers n_k to another stationary state with quantum numbers m_k has the same absolute value as the inverse amplitude, connected with the transition from the state described by m_k to the state described by n_k. Hence, in agreement with the observations, the absorbed and emitted intensities become identical.

Although the action-angle scheme satisfied the formal requirements, which the correspondence to the analogous classical scheme required, the crucial tests were passed only after a successful application to real atomic systems had been made, that is, after one had calculated the results for the energy states and the intensities of the transitions between these states that were confirmed by the spectroscopic data. To start with, Dirac chose the example of the hydrogen atom, where he expected to be able to proceed as in the classical calculations, for already in the old quantum theory the correct energy states had been obtained. The hydrogen atom consists of an electron having the charge $-e$ and mass m_e, both of which he assumed to be c-numbers, orbiting in the (xy)-plane around a heavy nucleus of charge $+e$. Dirac assumed that the nonrelativistic Hamiltonian of this system is given in terms of x and y, the Cartesian coordinates of the electron, and the canonically conjugate momenta, p_x and p_y, by the equation

$$H = \frac{1}{2m_e}\left(p_x^2 + p_y^2\right) - \frac{e^2}{\left(x^2 + y^2\right)^{1/2}}. \tag{151}$$

Then he transformed the Cartesian variables into polar coordinates, r and θ, and their conjugate momenta, p_r and k. He defined the respective q-numbers by the equations

$$r = \left(x^2 + y^2\right)^{1/2}, \qquad \cos\theta = \frac{x}{r}, \qquad \sin\theta = \frac{y}{r} \tag{152a}$$

and

$$p_r = \tfrac{1}{2}(p_x\cos\theta + \cos\theta\, p_x) + \tfrac{1}{2}(p_y\sin\theta + \sin\theta\, p_y), \qquad k = xp_y - yp_x. \tag{152b}$$

That is, he kept the algebraic relations for the position coordinates of classical theory and constructed the quantum-theoretical momenta in such a way that the quantum Poisson brackets of canonical variables were satisfied.[215] In the new variables the Hamiltonian becomes

$$H = \frac{1}{2m_e}\left(p_r^2 + \frac{k_1 k_2}{r^2}\right) - \frac{e^2}{r}, \tag{153}$$

where

$$k_1 = k + \frac{h}{2\pi} \qquad \text{and} \qquad k_2 = k - \frac{1}{2}\frac{h}{2\pi}.$$

The equations of motion for the hydrogen atom are, therefore,

$$\frac{dr}{dt} = [r, H] = \frac{p_r}{m_e}, \tag{154a}$$

$$\frac{dp_r}{dt} = [p_r, H] = \frac{k_1 k_2}{m_e r^3} - \frac{e^2}{r^2}, \tag{154b}$$

$$\frac{d\theta}{dt} = [\theta, H] = \frac{k}{m_2 r^2}, \tag{154c}$$

and

$$\frac{dk}{dt} = [k, H] = 0. \tag{154d}$$

Since H does not depend on the angle θ and, consistent with it, also commutes with the momentum variable k $(= m_e r^2\dot{\theta})$, the latter becomes a constant of motion.

In order to obtain the energy states of the hydrogen atom, Dirac had first to calculate the expressions for the variables r and p_r, and he proceeded in the following way. He assumed that the quantum radial coordinate r possesses a structure which is analogous to the classical solution. The latter is given by the equation

$$\frac{1}{r} = l^{-1} + l^{-1}\epsilon\cos\{\theta - \chi\}, \tag{155}$$

which describes an ellipse with the *latus rectum* l, eccentricity ϵ, and χ as the angle between the major axis of the ellipse and the line $\theta = 0$. Dirac replaced this

[215]Since r and θ are only functions of the coordinates x and y, they commute, hence $[r, \theta] = 0$. Moreover, it is straightforward to prove that also $[r, k] = 0$. In addition, one has the equation

$$r[\exp\{i\theta\}, k] = ir\exp\{i\theta\};$$

hence it follows that the quantum-theoretical Poisson bracket, $[\theta, k]$, assumes the value unity. Similarly, it can be shown that $\exp\{i\theta\}$ and p_r, and k and p_r, commute, respectively, while $[r, p_r] = 1$.

equation in quantum mechanics by the equation

$$\frac{1}{r} = a_0 + a_1 \exp\{i\theta\} + a_2 \exp\{-i\theta\}, \tag{156}$$

and determined the q-numbers a_0, a_1 and a_2 with the help of the equations of motion, Eqs. (154a) to (154d), as

$$a_0 = \frac{m_e e^2}{k_1 k_2}, \tag{156a}$$

$$a_1 = \frac{1}{2} a_0 \frac{k_2}{k}\left(1 - \frac{k_1^2}{P^2}\right)^{1/2} \exp\{-i\chi\}, \tag{156b}$$

$$a_2 = \frac{1}{2} a_0 \frac{k_1}{k}\left(1 - \frac{k_2^2}{P^2}\right)^{1/2} \exp\{+i\chi\}, \tag{156c}$$

where the q-number P is defined by the equation

$$2\, m_e H = -\frac{m_e^2 e^4}{P^2}. \tag{156d}$$

The analogy of the quantum-mechanical solution, Eqs. (156), to the classical solution, Eq. (155), becomes apparent if one identifies the corresponding quantities in both solutions. One notes, in particular, that a_0 is the quantum-mechanical quantity corresponding to l^{-1}, the inverse *latus rectum*, χ corresponds to the classical angle denoted by the same letter and the expressions, $(k_2/k)(1 - k_1^2/P^2)^{1/2}$ and $(k_1/k)(1 - k_2^2/P^2)^{1/2}$, correspond to the classical eccentricity. Hence the quantum radial variables, r and p_r, fully corresponded to the classical variables.

Dirac discovered that this close analogy between quantum and classical variables ensures that the number of energy levels obtained from quantum mechanics coincides with the number deduced from the observed hydrogen spectrum. That is, apart from relativistic corrections, which he had neglected, there arose only *one* sequence of energy states from the Hamiltonian H, Eq. (153). The situation would be different if he had started instead from the slightly different Hamiltonian H',

$$H' = \frac{1}{2m_e}\left(p_r^2 + \frac{k^2}{r^2}\right) - \frac{e^2}{r}, \tag{157}$$

which can also be considered as the quantum reformulation of the classical Hamiltonian in terms of polar variables. From H', however, one derives an orbit of the electron that is an ellipse with a rotating apse line; hence two angle variables are required to describe the energy spectrum of the system and *two* sequences of states would arise in contradiction with experiment.[216]

Apart from giving the correct term spectrum of the free hydrogen atom, Dirac's procedure clarified an important point in quantum mechanics: in this specific case he proved that, in order to describe a quantum system, one can employ any set of dynamical variables. That is, Cartesian variables are as good as polar variables, and one can always transform from one complete set to another complete set. In this sense, the Hamiltonian H', given by Eq. (157), is not correct because it cannot be obtained by a contact transformation from the Hamiltonian in Cartesian coordinates, given by Eq. (151). In order to employ it, one has to assume that the system of polar variables plays a preferred role in quantum mechanics, and there exists no possible justification for such a step on the basis of the principles and fundamental equations of the theory.

In order to calculate the energy states of the free hydrogen atom, Dirac had to eliminate the radial variables completely and *to express the Hamiltonian H as a function of one uniformizing action variable alone*. The latter, evidently, cannot be the angular momentum k defined by the second relation in Eq. (152b). Dirac had rather to search for a new q-number having two special properties: first, it should determine the energy completely; second, the differential quotient of the Hamiltonian with respect to it should yield a constant q-number. He identified this q-number with $\dot{w}$, the time derivative of the uniformizing angle variable w, given by Eq. (144b), that is, with the orbital frequency of the electron in the hydrogen atom. The hydrogen problem is thus solved once $\dot{w}$ is expressed as a function of the uniformizing action variable.

To obtain $\dot{w}$, Dirac made use of the fact that every variable of the system may be expressed in a Fourier series (as in Eq. (143a)), which, in general, is a sum of one constant and several, if not infinitely many, periodic terms. Thus, from a relation between the polar angle θ and the uniformizing angle w, he obtained the equation

$$\frac{d\theta}{dt} = \frac{k}{m_e r^2} = \dot{w} + \dots, \tag{158}$$

where the dots on the right-hand side indicate the periodic terms. He evaluated the constant term (proportional to r^{-2}) in Eq. (158) with the help of Eqs. (156),

[216] If one attempts to solve the equations of motion that follow from the Hamiltonian H', Eq. (157), which deviates from H, Eq. (153), but also leads to the classical Hamiltonian in the limit of high quantum numbers, then the *Ansatz* for r, Eq. (156), has to be replaced by the *Ansatz*

$$\frac{1}{r} = a_0 + a_1 \exp\{i\theta'\} + a_2 \exp\{-i\theta'\}.$$

Since the angle θ is given by the equation

$$\theta' = \theta \frac{k'}{k} \qquad \text{with} \qquad k' = \left(k^2 + \frac{1}{4}\left(\frac{h}{2\pi}\right)^2\right)^{1/2},$$

the orbit of the electron is an ellipse with a rotating apse line.

and obtained the result

$$\dot{w} = \frac{m_e e^4}{P^3} . \tag{159}$$

The right-hand side of Eq. (159) is now identical to the differential quotient dH/dP calculated from Eq. (156d). Referring to Eq. (144b), Dirac concluded that the q-numbers P and w constitute the pair of uniformizing action-angle variables for the free hydrogen atom. Again, these quantum variables correspond completely to the classical action-angle variables. With the help of Eqs. (146b) and (156d), then, the quantum transition frequency may be expressed as

$$\omega\left(P + \alpha \frac{h}{2\pi}, P\right) = \frac{m_e e^4 \pi}{h} \left\{ \frac{1}{P^2} - \frac{1}{(P + \alpha(h/2\pi))^2} \right\}, \tag{160}$$

as a function of the action variable J and the c-number α, which assumes integral values.

In order to complete the calculation of the hydrogen spectrum, Dirac had to know the values which the action variable P assumes. By introducing P in Eq. (156) for the radial variable r, he noticed that the latter disappears when the value $P = 0$ is taken. Consequently, each of the amplitudes, say the amplitudes $x(nm)$ of the Cartesian coordinate x, vanishes in the c-number representation of the coordinates when n or m is zero. This result suggests, Dirac argued, that the state in which the action variable assumes the value $h/2\pi$ is the normal state of the system. Provided it can be proved that $x(nm)$ is zero when m is a negative integer and n a positive integer, 'one would have to put P equal to an integer multiple of h [= Planck's constant divided by 2π] . . . , and one would then obtain the observed frequencies of the hydrogen spectrum,' he concluded (Dirac, 1926a, p. 579).

In January 1926, when Dirac submitted the paper containing what he called 'a preliminary investigation of the hydrogen atom,' he knew that Wolfgang Pauli had already obtained the hydrogen spectrum from matrix mechanics (Dirac, 1926a, p. 570, footnote). Nevertheless, he published his own calculations because he wanted to present a simple but realistic example in which the q-number methods worked properly. The result itself, that is, the derivation of the frequency spectrum, Eq. (160), did not signify anything new to him, as it only confirmed the well-known calculation of the semi-classical Bohr theory. This was hardly surprising, for he had obtained it in closest analogy to the classical result. The main goal, which he had in mind when developing the action-angle scheme in quantum mechanics, was, of course, to be able to discuss with its help atoms with several electrons, which had not been treated successfully in the Bohr–Sommerfeld theory. The example of the hydrogen atom just served the purpose of illustrating his methods, which, he hoped, would be general enough to be useful, say for calculating the energy states of helium. In the following paper, while preparing himself to deal with the problem of many-electron atoms, he closed the gap that had remained in his calculation of the hydrogen atom: he proved that the lowest value of the action variable P, occurring in Eq. (160), is indeed $h/2\pi$, for it is also necessary to yield the correct hydrogen spectrum.

V.3 A Preliminary Investigation of Many-Electron Atoms

Dirac had pursued quantum mechanics because the new theory offered 'the key to the quantum riddle'; that is, it seemed to provide the possibility of obtaining an adequate description of the energy states of all kinds of atoms existing in nature. In the example of the hydrogen atom he had learned that all one had to do for deriving quantum-mechanical results was to follow closely the analogy with the classical solution of the equations of motion, which implied the transformation of the quantum system to action and angle variables. The question now arose whether he could extend this successful method for treating the equations of motion of many-electron atoms. To deal with it, he had first to isolate clearly the essential steps of the quantum-mechanical method and to determine whether they could be suitably generalized.

Dirac noticed that for the calculation of the energy states of the hydrogen atom it had been important that the Hamiltonian H in polar coordinates, following Eq. (153), did not include the angle θ. This allowed one to regard k, the canonically conjugate variable, as a constant of motion and to reduce the mathematical problem of solving four partial differential equations (the Hamiltonian equations of motion) in the variables r, p_r, θ and k, to the simpler problem of solving only two such equations, namely, the equations of motion for r and p_r, the radial variables. Dirac knew well that a similar simplification of the problem could always be effected in classical theory even in the case of the equations of motion describing atoms with several electrons. The procedure was first described by Carl Gustav Jacob Jacobi in connection with the problem of the motion of two planets around the sun: this method was referred to as 'elimination of the nodes' in Whittaker's *Analytical Dynamics*, for the angle which is eliminated from the Hamiltonian describes the node of the fixed (xy)-plane with the plane of the three bodies.[217]

The crucial question for Dirac was whether he could imitate the classical procedure of elimination of the nodes in the quantum-mechanical treatment of

[217]Whittaker covered this point in detail in his *Analytical Dynamics*, §158, under the heading 'Reduction [of the three-body problem] to the 8th order by use of the integrals of angular momentum and elimination of the nodes' (Whittaker, 1904, p. 332). The three-body system in classical mechanics is described by nine equations of motion for the position variables, each of which is of the second order, or, in Hamilton's canonical formalism, by eighteen first-order differential equations. Since the centre of gravity moves in a straight line, and since the integration of this motion yields six constants, the system of eighteen differential equations of motion can be reduced to twelve. Moreover, the total angular momentum of the three-body system is constant throughout the motion, and the resulting three constants (one in each of the perpendicular directions chosen in space) reduce the system of differential equation to one of the ninth order. However, one of the angular variables conjugate to the angular momentum components is 'ignorable,' i.e., it does not occur in the Hamiltonian of the

atoms with two or more electrons. Evidently, in order to arrive at the result that the Hamiltonian of a given atom be independent of one of the angle variables, he had first to introduce, instead of the Cartesian coordinates and their canonically conjugate momenta, new variables similar to the polar variables defined by Eqs. (152) for the hydrogen atom. In his paper on 'The Elimination of the Nodes in Quantum Mechanics,' Dirac again took note of the variables used earlier in classical theory, these being r, the distance of each electron from the centre, with p_r, the radial momentum, as the conjugate variable; and L_z, the component of the total angular momentum of the system in the z-direction, with the azimuth about this direction, ϕ, as the conjugate variable. Then, in the case of an atomic system with a single electron, the variables, in addition to the radial variables, are k, the magnitude of the angular momentum of the electron, with θ, the angle in the orbital plane, as the conjugate variable. However, in the case of an atomic system with two electrons, the variables are k and k', the angular momenta of the two electrons; θ and θ', the conjugate angles between the radius vectors and the line of the nodes; and j, the total angular momentum, with ψ, the azimuth of the line of the nodes about the direction of j, as the conjugate variable. When more than two electrons exist in an atom, one can, for many practical calculations, use the same description as for the two-electron atoms by considering all the electrons except one, the series (valence) electron (in whose properties one is interested), as forming a compound system or core that plays the role of the second electron.[218] If Dirac succceeded in translating these variables into q-

---

system, as was first noticed by Jacobi, who wrote:

> By applying the method, which I have just discussed, to the three-body problem, one reduces the latter to the investigation of the problem of motion of two bodies, which has remarkable properties. From the three equations furnished by the conservation of angular momentum, one sees that: 1. The intersection common to the planes of the orbits of the two bodies constantly remains in a fixed plane: this is the invariant plane of the system; 2. The inclinations of the planes of the two orbits with respect to the fixed plane are rigorously determined by the parameters of these orbits, considered as variable ellipses.
>
> Let us consider as the variables of the problem the inclinations of the two orbits with respect to the invariant plane, the two radial vectors, the angles which [the radial vectors] make with the common intersection of the planes of the two orbits—situated in the invariant plane—and, finally, the angle between this intersection and the fixed normal to this (invariant) plane. One finds *that the latter angle entirely drops out from the system of differential equations and is determined* [*only*] *after the integration by a quadrature*. Thus, in this new form of differential equations, there appears no trace of the nodes. The six differential equations of second order, which express the relative motion of the three bodies, are [now] reduced to five equations of the first order and one of second order. As a result, one has reduced the order of the system of differential equations of the three-body problem. The known integrals [of motion] being only four in number, one can say that one has effected a new reduction of the order of differential equations of the three-body problem. Such a reduction can be applied to the problem of an arbitrary number of bodies. (Jacobi, 1843; *Gesammelte Werke* 4, pp. 298–299)

[218]The approach based upon an atomic core had been very successful in the earlier analysis of many-electron spectra, which Dirac had studied. Dirac's notation coincided, to a large extent, with Heisenberg's (1924e; 1925b). The following table summarizes their respective notations for action and angle variables:

Dirac (1926b)	k	θ	k'	θ'	j	ψ	p	ϕ
Heisenberg (1924e)	k	ϕ_k	r	ϕ_r	j	ψ	m	—

Dirac denoted the z-component of the total angular momentum by p, but we shall not follow this notation here.

numbers, then he could hope to calculate the energy states of two-electron atoms, such as the helium atom, and even the states of atoms having arbitrarily many electrons by employing the core-valence electron approximation.

The way to achieve this goal was to follow the same procedure, which he had successfully applied to the hydrogen atom: namely, to obtain the new quantum variables by postulating that they obey the same algebraic equations involving the Cartesian coordinates of the electrons as do the corresponding classical variables. Dirac further required that the quantum variables be real, that the angle variables, θ, θ', ψ and ϕ, be so chosen that the Cartesian coordinates are multiply periodic in the angles with the period 2π, and finally that the action and angle variables be canonical variables. Under these conditions, the practical solution of a many-electron problem would follow along the lines of the hydrogen calculation. That is, after introducing the radial variables and the canonically conjugate momenta, as well as the above-mentioned action and angle variables, the Hamiltonian of the system under investigation becomes a function of the r's, p_r's and the action variables only; then one completes the evaluation of the Hamiltonian by performing another contact transformation, with the help of which the radial variables disappear. The Hamiltonian would finally depend entirely on the uniformizing action variables, just as in the example of the hydrogen atom.

An important portion of Dirac's work consisted in constructing the proper action and angle variables for a given multiply periodic system. He began by re-interpreting certain classical equations, defining the dynamical variables, as equations between q-numbers. One of these equations was the following: if a Cartesian coordinate, say x, is periodic in an angle variable w of period 2π, then in classical theory the second derivative of x with respect to w is given by $-x$. In quantum mechanics, the second derivative is expressed as a 'double Poisson bracket' with J, the action variable conjugate to w; hence the classical relation can be reformulated as

$$[J, [J, x]] = -x. \tag{161}$$

Equation (161) may be used to select the proper quantum action variable from a set of q-number candidates, each of which approaches the classical action variable in the correspondence limit. For the purpose of constructing the q-number representing the quantum angle variable w, Dirac referred to the classical relation

$$\exp\{i\theta\} = \sqrt{\frac{x+iy}{x-iy}}\,, \tag{162}$$

between the polar angle θ and the Cartesian coordinates, x and y, for the motion in the (xy)-plane (see Eq. 152a)). He generalized this equation in such a way that arbitrary periodic motions could be described by admitting other quantities, say a and b, instead of x and y, and reformulating the resulting equation in quantum mechanics as

$$\exp\{iw\}(a - ib)\exp\{iw\} = a + ib, \tag{163}$$

where w is the angular variable. In Eq. (163) a and b are two real q-numbers, which, together with the action variable J, satisfy the equations

$$[J,a] = b \tag{163a}$$

and

$$[J,b] = -a, \tag{163b}$$

arising from the differential equations between the corresponding classical variables a and b.[219] Then, from Eqs. (161) and (163), there immediately follows the equation

$$[\exp\{iw\},J] = i\exp\{iw\}, \tag{164}$$

which expresses the fact that w can be considered as the angle variable which is canonically conjugate to the action variable J.

Dirac tested the application of Eqs. (161) and (163) for constructing the action and angle variables for certain realistic atomic systems (Dirac, 1926b, §§5, 6, 7). By considering the action and angle variables, k and θ, defined by Eqs. (152a, b), he had already performed the necessary calculations in the case of the one-electron atom. The variable k indeed satisfies Eqs. (161) with either of the Cartesian coordinates x or y; moreover, θ is actually obtained from Eqs. (163) with the variables x and y substituted for the q-numbers a and b. For the two-electron system, say the helium atom, the action variables k, k', j and L_z are defined by the equations

$$k^2 - \frac{1}{4}\left(\frac{h}{2\pi}\right)^2 = l_x^2 + l_y^2 + l_z^2, \tag{165a}$$

$$k'^2 = \frac{1}{4}\left(\frac{h}{2\pi}\right)^2 = l_x'^2 + l_y'^2 + l_z'^2, \tag{165b}$$

$$j^2 - \frac{1}{4}\left(\frac{h}{2\pi}\right)^2 = L_x^2 + L_y^2 + L_z^2, \tag{165c}$$

[219] According to Eq. (152a), the Cartesian coordinates, x and y, satisfy the equations

$$-\frac{\partial x}{\partial \theta} = -\frac{\partial(r\cos\theta)}{\partial\theta} = r\sin\theta = y = [J,x]$$

and

$$-\frac{\partial y}{\partial \theta} = \frac{\partial(r\sin\theta)}{\partial\theta} = -r\cos\theta = -x = [J,y].$$

Hence x and y can serve, in this special case ($w = \theta$), as the variables a and b in Eqs. (163a, b).

and

$$L_z = l_z + l'_z, \tag{165d}$$

where l_x $(= yp_z - zp_y)$, l_y $(= zp_x - xp_z)$, l_z $(= xp_y - yp_x)$, and l'_x, l'_y, l'_z are the components of the angular momentum of the first and the second electron, respectively; and L_x $(= l_x + l'_x)$, L_y $(= l_y + l'_y)$ and L_z are the components of the total angular momentum. The canonically conjugate angle variables, θ, θ', ψ and ϕ, are obtained from the equations,

$$\exp\{i\theta\}(q \cdot l' - i[k, q \cdot l'])\exp\{i\theta\} = q \cdot l' + i[k, q \cdot l'], \tag{166a}$$

$$\exp\{i\theta'\}(q' \cdot l - i[k', q' \cdot l])\exp\{i\theta'\} = q' \cdot l + i[k', q' \cdot l], \tag{166b}$$

$$\exp\{i\psi\}(\mu_z - i[j, \mu_z])\exp\{i\psi\} = \mu_z + i[j, \mu_z], \tag{166c}$$

and

$$\exp\{i\phi\}(L_x - iL_y)\exp\{i\phi\} = L_x + iL_y, \tag{166d}$$

with

$$\begin{aligned} q \cdot l' &= xl'_x + yl'_y + zl'_z, \qquad q' \cdot l = x'l_x + y'l_y + z'l_z, \\ \mu_z &= l_x l'_y - l_y l'_x, \qquad \text{and} \qquad L_y = [L_z, L_x]. \end{aligned} \tag{166e}$$

When a several (or many-) electron atom can be approximated by a quasi-two-electron system—the core consisting of the entire atom except the valence electron—one would formally employ the same variables as in the two-electron system. Thus, the valence electron is described by the radial variables, r and p_r, and the action-angle variables, θ and k, while the core by the variables r', p'_r, θ' and k'. The global action variables of the many-electron atom, j and L_z, are not necessarily composed directly of the parts contributed by the valence electron and the core, respectively. Because of the relative importance of the couplings between the angular momenta of the individual electrons, one had previously—in the semiclassical interpretation of spectroscopic data—considered different ways of obtaining the total angular momentum of the atom. For example, one could first add the angular momenta of the two outer electrons, and then add this sum to the resultant angular momentum of the remaining electrons, in order to arrive at the components, L_x, L_y, L_z, that enter into Eqs. (165c), (165d), (166d) and (166e); or one could follow any of the other methods successfully explored before (see, e.g., Heisenberg, 1925b). Dirac assumed that in quantum mechanics one has to proceed in an analogous manner.

The transformation of the original Cartesian variables of an atom to radial variables (e.g., $r, p_r; r', p'_r$) and action-angle variables (e.g., $\theta, k; \theta', k'; \psi, j; \phi, L_z$), just as in the case of the hydrogen atom, did not yet yield the values of the energy states; the latter would follow only after the final transformation to uniformizing variables (the action and angle variables, P and w, defined by Eqs. (156d) and (158) in the case of the hydrogen atom) was carried out. However, in the earlier approach, based on the correspondence principle, many problems—especially the ones connected with the relative intensities of the multiplets of many-electron atoms—had already been solved with the help of the equations that were available *before* the radial variables were removed; the reason being that in calculating the relative intensities from the position variables, x, y and z, the radial variable drops out, as the ratio $x^2 : y^2 : z^2$ determines the required intensities. The question now arose whether the same method could be translated into quantum mechanics. To answer it, Dirac treated three special problems in atomic physics: first, the problem of the boundary values of the action variables, which leads to the correct assignment of quantum numbers to these variables; second, the energy splitting of complex line multiplets of many-electron atoms in a weak magnetic field; third, the problem of the relative intensities of complex multiplets (Dirac, 1926b, §§8, 9, 10). To get at the results, all he had to do was to obtain the equations for the variables considered in terms of the radial and the action variables (k, k', j, L_z). For each of the action variables he would then substitute the appropriate set of c-numbers, which, to a certain extent, defined the stationary states of the system under investigation.

From general considerations about the action variables in quantum mechanics, which we have mentioned in Section IV.4, Dirac had derived the result that the different c-numbers representing a given action variable form an arithmetical progression with a constant difference equal to $h/2\pi$ (Dirac, 1925d, §6). He had also required that the c-numbers must be bounded, at least in one direction (namely, from below); then the system, of which J is a dynamical variable, will possess a state of lowest energy, representing the *normal* state of the atom. In order that this requirement be satisfied one had to prove that *all* terms in the Fourier expansions of the Cartesian coordinates, which correspond to the transition from a stationary state to one below the normal state, must vanish. Although this looked like a difficult task, Dirac succeeded in carrying out the desired proof, as all equations that transform the Cartesian variables of a many-electron system into action and angle variables could be written down in a particular form. That is, if J and w are a pair of action and angle variables, wherever the function $\exp\{iw\}$ occurs it has immediately in front of it a factor $(J_2 - c)$ with a c-number c, and wherever $\exp\{-iw\}$ occurs it has immediately behind it the factor $(J_2 - c)$ or immediately in front of it the factor $(J_1 - c)$, J_1 and J_2 being identical to the q-numbers $J + \frac{1}{2}h/2\pi$ and $J - \frac{1}{2}h/2\pi$, respectively. Because of this situation, the series of values of the action variable J terminates at the special value $c + \frac{1}{2}h/2\pi$. That is, the transition amplitude denoted by the pair, $(c + \frac{1}{2}h/2\pi, c - \frac{1}{2}h/2\pi)$, vanishes because the factor $(J_2 - c)$ in front of $\exp\{iw\}$ or behind $\exp\{-iw\}$ vanishes if one inserts $J = c + \frac{1}{2}h/2\pi$. In a similar manner Dirac discussed the equations involving the periodic functions $\exp\{\pm 2iw\}$,

$\exp\{\pm 3iw\}$, etc., which determine the transition amplitudes between the stationary states whose J-values differ by the amount $2h/2\pi$, $3h/2\pi$, etc. He concluded that 'all the amplitudes in the Fourier expansions that are related to a value of J greater than c and a value less than c vanish' (Dirac, 1926b, p. 300). For this reason he called the value $J = c$ the 'boundary value' of the action variable.[220]

In the case of two action variables, J and J', the situation was slightly more complicated. Dirac found that the transformation equations could then always be written in such a way that the periodic function $\exp\{iw\}$—with w the angle conjugate to J—is preceded by the factor $F(J,J')$, where F is a function of both variables. (Of course, there also existed the alternate possibility, namely, to write the transformation equations with the term $\exp\{-iw\}F(J_2,J')$.) Evidently, the equation

$$F(J,J') = 0 \tag{167}$$

will then fix the boundary value for the action variable J. That is, if in Eq. (167) one inserts a given value for J', it will yield the boundary value of J. In particular, the *lowest* value of J is obtained by using the lowest value of J' in Eq. (167). On the other hand, by investigating the equations containing the periodic functions, $\exp\{\pm iw'\}$, with w' the angle conjugate to J', Dirac discovered that there always occurred the terms $F(J,J_2')\exp\{iw'\}$ or $\exp\{-iw'\}F(J,J_2')$. He therefore concluded that Eq. (167) could likewise be used to determine the boundary values of the action variable J', provided one inserts the values for J. It is evident how, from a similar discussion of the equations describing periodic functions of more than two degrees of freedom, the boundary values for the appropriate action variables can be found.

To show how the boundary values are obtained in practice, Dirac first chose the example of the one-electron atom, on which a force may be imposed in the z-direction. Unlike the free hydrogen atom, which we have discussed in the previous section, the motion of the electron in this case will not remain restricted to the (xy)-plane. Hence the situation must be described by two action variables: k, the q-number given by Eq. (165a) and related to l_x, l_y, l_z, the components of the angular momentum of the electron, and l_z. Their conjugate angle variables, θ and ϕ, are defined by the equations

$$z + i[k,z] = rk^{-1/2}\left(k_2^2 - l_z^2\right)^{1/2}\exp\{i\theta\}k^{-1/2} \tag{168a}$$

[220]The amplitudes related to the transitions $(c + \frac{1}{2}h/2\pi, c - \frac{3}{2}h/2\pi)$ and $(c + \frac{3}{2}h/2\pi, c - \frac{1}{2}h/2\pi)$ are obtained by inserting the J-values, $c + \frac{1}{2}h/2\pi$ and $c + \frac{3}{2}h/2\pi$, into the factors in front of $\exp\{2iw\}$. Now, in the transition equations the periodic function $\exp\{2iw\}$ arises from squaring the product, $(J_2 - c)\exp\{iw\}$, thus yielding the result

$$(J_2 - c)\exp\{iw\}(J_2 - c)\exp\{iw\} = \left(J - c - \frac{1}{2}\frac{h}{2\pi}\right)\left(J - c - \frac{3}{2}\frac{h}{2\pi}\right)\exp\{2iw\}.$$

This term indeed vanishes if the above-mentioned values for J are taken.

and

$$l_x + il_y = \left(k^2 - l_{z,2}^2\right)^{1/2} \exp\{i\phi\}, \tag{168b}$$

with the Cartesian coordinate z and q-numbers $k_2 = k - \frac{1}{2}h/2\pi$ and $l_{z,2} = l_z - \frac{1}{2}h/2\pi$.[221] From the factors, which multiply the periodic functions, $\exp\{i\theta\}$ and $\exp\{i\phi\}$, on the right-hand side of Eqs. (168a) and (168b), one obtains the equation

$$k \pm l_z = 0, \tag{169}$$

which determines the boundary values k and l_z. As a result, for a given value of the action variable k, l_z assumes $2|k|$ values ranging from $|k| - \frac{1}{2}$ to $-|k| + \frac{1}{2}$.[222] Now, by considering a one-electron system—which is described by the Hamiltonian of a free hydrogen atom, Eq. (151), plus a small perturbing term proportional to r^{-2}—Dirac concluded that $|k|$ assumes half-integral quantum numbers, i.e., the values $\frac{1}{2}h/2\pi$, $\frac{3}{2}h/2\pi$, $\frac{5}{2}h/2\pi$, etc. Moreover, 'there will be thus $1, 3, 5, \ldots$ stationary states for $S, P, D, \ldots$ terms when the system has been made non-degenerate by a magnetic field [in the z-direction], in agreement with observation for singlet spectra' (Dirac, 1926b, p. 300).

Dirac derived two further consequences from his theory. First, the Eqs. (168a) and (168b) state that, in a one-electron system, no transitions can occur from an S-state to another S-state, that is, from a state with $k = \frac{1}{2}h/2\pi$ and $l_z = 0$ to a state with $k = -\frac{1}{2}h/2\pi$ and $l_z = 0$ because the factors $(k_2^2 - l_z^2)$ and $(k^2 - l_{z,2}^2)$ assume, in this case, the value zero. Second, the values of J [$= P$], the uniformizing variable, which should completely determine the energy spectrum of the free hydrogen atom, are given by the equation

$$J = P = |k| + \frac{1}{2}\frac{h}{2\pi}, \tag{170}$$

[221] Equation (168b) is obtained by writing Eq. (163) as $(a + ib) = \{(a + ib)(a - ib)\}^{1/2}\exp\{iw\}$ then inserting $a = l_x$ and $b = l_y$, and taking into account the equation

$$(l_x + il_y)(l_x - il_y) = l_x^2 + l_y^2 - l_z^2 + \frac{h}{2\pi}l_z = k^2 - l_{z,2}^2.$$

Equation (168a) is derived in a similar way.

It should be noted that Eq. (163) can also be written as

$$(a - ib) = \{(a - ib)(a + ib)\}^{1/2}\exp\{-iw\},$$

which, for example, leads to a second-order equation defining ϕ, that is,

$$l_x - il_y = \exp\{-i\phi\}\left(k^2 - l_{z,2}^2\right)^{1/2},$$

which again has the structure that Dirac needed to derive the 'boundary value' for l_z.

[222] The relation between k and l_z can be derived easily from Eq. (165a) and the fact that l_z assumes values between $+l$ and $-l$, whereas the square of the angular momentum, $(l_x^2 + l_y^2 + l_z^2)$, assumes the values $l(l+1)(h/2\pi)^2$ (see, for example, Born, Heisenberg and Jordan, 1926, Chapter IV.1). Now l commutes with r^2 and k^2, and assumes values that are integral multiples of $h/2\pi$. Hence the lowest $|k|$-value must be $\frac{1}{2}h/2\pi$, provided the lowest l-value is zero.

which is just the boundary value equation derived from Eq. (156b) or Eq. (156c). Hence J must assume the values $h/2\pi$, $2h/2\pi$, $3h/2\pi$, etc., in agreement with the observed spectrum.

The extension of these considerations to atoms having two or more electrons led Dirac to results that were well confirmed by experiment. For example, he found that in the case of a two-electron system, whose individual electrons are described by the values of the action variables k and k', the action variables of the total system, j and L_z, assume the following values:

$$j = |k + k'| - \frac{1}{2}\frac{h}{2\pi}, \ldots, |k - k'| + \frac{1}{2}\frac{h}{2\pi} \tag{171a}$$

and

$$L_z = |j| - \frac{1}{2}\frac{h}{2\pi}, \ldots, -|j| + \frac{1}{2}\frac{h}{2\pi}. \tag{171b}$$

In Eqs. (171a, b) the values are denoted by the same letters as the variables.

Besides giving the observed relations between the values of the action variables (or quantum numbers in the old Bohr–Sommerfeld atomic theory), Dirac accounted for an important empirical result concerning the anomalous Zeeman effect of many-electron systems. That is, he derived the well-known formula describing the splitting of complex multiplets in a weak magnetic field. In order to obtain this result, Dirac adopted 'the usual model of the atom, consisting of a normal series [valence] electron and a core in which the ratio of magnetic moment to mechanical angular momentum is double the normal Lorentz value' (Dirac, 1926b, p. 301). Since the additional term, which is contributed to the energy of the atom by the action of a weak magnetic field in the z-direction, is proportional to the sum of l_z (the value of the angular momentum of the series electron in the z-direction) and $2l_z'$ (double the value of the core's momentum), Dirac had to calculate this sum in his theory. He obtained the result

$$l_z + 2l_{z'} = L_z\left\{1 + \frac{1}{2}\,\frac{j_1 j_2 - k_1 k_2 + k_1' k_2'}{j_1 j_2}\right\}, \tag{172}$$

which allowed him to express—apart from a factor containing the field strengths, the parameters of the electron and the velocity of light—the magnetic splitting of a multiplet in terms of the values assumed by the action variables L_z, j, k and k'.[223] The factor within the curly bracket on the right-hand side of Eq. (172) is

[223] Equation (172) follows from first-order perturbation theory, according to which the correction in the energy is obtained by time-averaging ΔH_1, the magnetic perturbation term, expressed in terms of the action and angle variables. That is, in the Fourier expansion of ΔH_1 in the angles θ, θ', ψ and ϕ, only the constant terms have to be taken, giving the energy correction ΔE_{mag}. Now, ΔE_{mag} consists of two contributions, one being (const.)l_z, the other (const.)$2l_z'$. The l_z'-contribution is evaluated with the help of the equation

$$j[j, \mu_z] = (L_x^2 + L_y^2 + L_z^2)l_z' - (L_x l_x' + L_y l_y' + L_z l_z')L_z + \frac{1}{2}\frac{ih}{2\pi}\mu_z.$$

Observing that μ_z (see Eqs. (166)) and the left-hand side of this equation do not have any constant

identical to Alfred Landè's g-factor (Landé, 1923a, p. 193, Eq. (8); see Eq. (252) in Volume 1, p. 483), by means of which the spectroscopic data had been explained so well. Unlike the previous theoretical derivations of this formula based upon correspondence arguments, no detailed model-dependent assumptions, other than the one about the anomalous magnetic moment of the core, entered into Dirac's calculation.

As a final exercise in his preliminary treatment of atoms with the help of quantum action and angle variables, Dirac calculated the relative intensities of the components of a complex line multiplet in a magnetic field. To do so, he expanded the quantities $(x + iy)$ and z, which describe the radiation emitted from the atom with electric field vectors perpendicular and parallel to the acting magnetic field, as Fourier series in the angle variables ψ, ϕ and θ. In the case of a one-electron atom, where the action variables are $j = k$ and l_z, and the conjugate angles $\psi = \theta$ and ϕ, and $r = (x^2 + y^2 + z^2)^{1/2}$, he obtained the results

$$2\left(\frac{x+iy}{r}\right) = -\left[F^-(k, k + l_z)\exp\{i(\theta+\phi)\} + \exp\{-i(\theta-\phi)\}F^+(k, k - l_z)\right], \tag{173a}$$

$$2\left(\frac{x-iy}{r}\right) = \left[\exp\{-i(\theta+\phi)\}F^+(k, k + l_z) + F^-(k, k - l_z)\exp\{i(\theta-\phi)\}\right], \tag{173b}$$

$$2\frac{z}{r} = F_0^-(k, l_z)\exp\{i\theta\} - \exp\{-i\theta\}F_0^+(k, l_z), \tag{173c}$$

with

$$F^\pm(k, k \pm l_z) = \sqrt{\frac{(k \pm l_z \pm \frac{1}{2}h/2\pi)(k \pm l_z \pm \frac{3}{2}h/2\pi)}{k(k - h/2\pi)}} \tag{173d}$$

and

$$F_0^\pm(k, l_z) = \sqrt{\frac{(k + l_z \pm \frac{1}{2}h/2\pi)(k - l_z \pm \frac{1}{2}h/2\pi)}{k(k - h/2\pi)}} \tag{173e}$$

---

terms, only the first and second terms on the right-hand side contribute to l_z'. Thus, L_z' is found to be given by the equation

$$l_z' = \frac{L_x l_x' + L_y l_y' + L_z l_z'}{L_x^2 + L_y^2 + L_z^2} L_z = \frac{\frac{1}{2}(j_1 j_2 - k_1 k_2 + k_1' k_2')}{j_1 j_2} L_z.$$

Hence the sum $l_z + 2l_z'$, or $L_z + l_z'$, becomes identical with the right-hand side of Eq. (172).

The relative intensities of the magnetic components of the multiplet lines are obtained by taking the absolute squares of the right-hand sides of Eqs. (173a, b, c) and inserting the appropriate values for k and l_z, the action variables of the one-electron atom.

In the case of an atom consisting of a core and a radiating series electron, the harmonic expansions of the q-numbers, $((x \pm iy)/r)$ and z/r, whose amplitudes determine the intensities of the emitted magnetic components of multiplet lines, exhibit a similar structure as Eqs. (173). In these harmonic expansions, Dirac found the specific terms $F_{+1}\exp\{-i(\theta+\psi)\}$ and $F'_{+1}\exp\{i(\theta-\psi)\}$, $F'_{-1}\exp\{i(\theta+\psi)\}$ and $F_{-1}\exp\{-i(\theta-\psi)\}$, and $F_0\exp\{-i\theta\}$ and $F'_0\exp\{i\theta\}$, respectively, where F_{+1}, F'_{+1}, F_{-1}, F'_{-1}, F_0 and F'_0 are given functions of the k, k' and j.[224] 'The ratios of the amplitudes obtained in this way,' he concluded, 'are in complete agreement with . . . experiment' (Dirac, 1926b, p. 305). He had succeeded in deriving the same intensity formulas as had been previously obtained by Ralph Kronig (1925a) and others on the basis of correspondence arguments and detailed model assumptions concerning the mechanical structure of many-electron atoms.

With all these results Dirac became convinced that his quantum-mechanical scheme of action-angle variables did account for the properties of atoms. The quantum scheme differed from the classical action-angle variables; k and k' were not equal to the angular momenta of the electrons, but were related to the latter by Eqs. (165a) and (165b), and a similar relation, Eq. (165c), held between the action variable j and the angular momentum of the whole system. This observation not only accounted for the various additional terms, $\pm\frac{1}{2}h/2\pi$, which had to be added to the classical variables in the expressions derived from the Bohr–Sommerfeld theory in order to obtain agreement with the spectroscopic data, but

[224] The notation is Dirac's (1926b, p. 304, Eqs. (58), (59) and (60)). For $F_{+1}, F'_{+1}, F_{-1}, F'_{-1}, F_0$ and F'_0 Dirac found the expressions:

$$F_{+1} = \frac{1}{4}\frac{(k+k'+j+\frac{1}{2}h/2\pi)^{1/2}(k-k'+j+\frac{1}{2}h/2\pi)^{1/2}(k+k'+j+\frac{3}{2}h/2\pi)^{1/2}}{(j+\frac{1}{2}h/2\pi)^{1/2}k^{1/2}(k+h/2\pi)^{1/2}}$$

$$\times(k-k'+j+\tfrac{3}{2}h/2\pi)^{1/2} \quad \text{in the numerator;}$$

$$F'_{+1} = \frac{1}{4}\frac{(k+k'-j-\frac{1}{2}h/2\pi)^{1/2}(-k+k'+j+\frac{1}{2}h/2\pi)^{1/2}(k+k'-j-\frac{3}{2}h/2\pi)^{1/2}}{(j+\frac{1}{2}h/2\pi)^{1/2}k^{1/2}(k-h/2\pi)^{1/2}}$$

$$\times(-k+k'+j+\tfrac{3}{2}h/2\pi)^{1/2} \quad \text{in the numerator;}$$

F_{-1} and F'_{-1} are obtained from F_{+} and F'_{+1} by replacing h by $-h$.

$$F_0 = \frac{1}{4}\frac{j^{1/2}(k+k'+j+\frac{1}{2}h/2\pi)^{1/2}(k+k'-j+\frac{1}{2}h/2\pi)^{1/2}(k-k'+j+\frac{1}{2}h/2\pi)}{j_1^{1/2}j_2^{1/2}k^{1/2}(k+h/2\pi)^{1/2}}.$$

$$\times(-k+k'+j-\tfrac{1}{2}h/2\pi)^{1/2} \quad \text{in the numerator;}$$

F'_0 is obtained from F_0 by replacing h by $-h$.

it also provided the answer to the question whether the Hamiltonian in quantum mechanics is the same function of the quantum action variables as the classical Hamiltonian is of the classical variables. Dirac noticed that for atoms, whose potential energy arises because of a central force, the quantum Hamiltonian is the same function of the angular momentum of an orbiting electron as in classical theory (see Eqs. (153) and (165a) for the hydrogen atom). Now in classical theory the action variable k is identical with the angular momentum $l\left(=\sqrt{l_x^2+l_y^2+l_z^2}\right)$; however, this is not so in quantum mechanics. Hence the Hamiltonian of a quantum system *does not exhibit* the same functional dependence on *the action variables as the corresponding classical Hamiltonian.*[225]

After presenting in his paper on the elimination of nodes those applications that required only minor calculations with the q-number equations describing multiply periodic systems, Dirac was ready to attack more difficult problems. In particular, he thought of the problem of calculating the energy states of the helium atom. To do so, he had first to find the proper expressions for r, p_r, r' and $p_{r'}$, the radial variables of the two-electron system; then he had to eliminate the radial variables entirely from the Hamiltonian by performing the transformation to uniformizing variables. Dirac had executed all these steps in the case of the one-electron system, but for the helium atom this task seemed to involve rather tricky technical calculations. Moreover, it was known that in dealing with many-electron atoms relativity effects would be important. Therefore, it was advisable to investigate whether the q-number methods could be extended to relativistic systems. Still another problem, which the Bohr–Sommerfeld theory had not been able to solve, could now be attacked with the help of Dirac's action-angle scheme: this was the problem of calculating the absolute intensities of the radiation emitted by the free hydrogen atom. But Dirac did not consider treating this problem first; he rather decided to calculate the intensity of the Compton radiation scattered by free electrons. To accomplish this task, he proposed a q-number scheme describing systems with relativistic kinematics; he then applied it to the case of Compton scattering and obtained results in agreement with observation.

V.4 Quantum Time and 'Relativity Quantum Mechanics'

In Dirac's work published until fall 1925 the arguments based upon the principle of relativity had played a fundamental role. They had either helped him in obtaining certain new results or—in case he had used other theoretical reasoning in the derivation—in confirming the results he had already obtained. His

[225] In his first paper on quantum mechanics, Dirac had stated that the energy of an atom is the same function of the action variables as in classical theory (Dirac, 1925d, p. 652). Heisenberg, in his letter of 1 December 1925 to Dirac, objected to this. Dirac corrected his previous statement in the paper on the hydrogen atom by saying that the Hamiltonian function H 'is not necessarily the same function of the J's as on the classical theory with the present definition of the J's' (Dirac, 1926a, p. 567, footnote).

investigations on quantum mechanics in fall 1925 and spring 1926 seemed to depart from this practice: that is, in carrying out the programme of establishing the laws of quantum mechanics and their preliminary application to physical problems, Dirac did not refer to any relativistic argument. This fact appears to be all the more surprising as one of the principal phenomena to be described by the new quantum mechanics was radiation, and radiation was known to be described only by means of relativistic concepts. The reason why it had been possible to avoid—thus far at least—a relativistic treatment of radiation was that only certain very specific aspects of radiation had been considered. First, all the theoreticians involved had discussed only one type of radiation, the radiation emitted by atoms when the latter passed from one stationary state to another. And, even of this particular radiation, just two properties had been considered: its frequency, given by the difference of two energy terms of the atom, and its amplitude, which was assumed to be determined by the appropriate transition amplitude of the Cartesian coordinates of the emitting electron. No effort had been devoted, however, to treating the problems of what happened to the radiation on leaving the atom, its propagation in space and interaction with matter. Indeed, the question concerning the very nature of radiation had been temporarily dismissed, which is not very surprising if one keeps in mind the fact that the new theory sprang from the previous approach based on the correspondence principle. In the latter, difficulties had arisen to account for those phenomena that seemed to be described quite well by means of the light-quantum hypothesis, and it was believed that quantum mechanics would remove the difficulties only gradually in the course of time.

Dirac, however, had never shown any prejudice against the light-quantum hypothesis, nor had he paid much attention earlier to the great concern which Niels Bohr and his circle exhibited in this regard. He had made use of certain properties of the light-quantum in his work in 1924 and 1925: for instance, in the calculation of the frequency shift undergone by radiation when scattered by the free electrons of the solar atmosphere (Dirac, 1925b). Then, after Bohr's lectures at Cambridge in May 1925, he became aware of the serious problems that the light-quantum presented to a unified and consistent theory of all quantum phenomena. Still he kept some interest in Einstein's concept, for he gave a talk entitled 'The Light Quantum Theory of Diffraction' at the 107th meeting of the Kapitza Club on 15 December 1925.[226] On this occasion Dirac discussed the scattering of light from a periodic grating in William Duane's theory, in which the momentum transferred to the grating by the light-quanta was assumed to be quantized (Duane, 1923). Dirac chose to speak on this subject although Duane's theory had appeared two years previously, and further work by Paul Epstein and Paul Ehrenfest had shown that diffraction phenomena, which have to do with the coherence of radiation, could not be explained by this theory (Epstein and

[226]The pros and cons of the light-quantum hypothesis were discussed on several occasions in 1925 at the Kapitza Club. For example, R. W. Ditchburn spoke on 'Geiger and Bothe's Experiment on the Compton Effect' at the 96th Meeting on 12 August 1925, and H. W. B. Skinner discussed Gustav Mie's treatment of the Compton effect (Mie, 1925a) at the 98th Meeting on 6 October 1925.

Ehrenfest, 1924). The immediate reason, of course, was that Dirac had just read Duane's work and thought that it would be worthwhile to call attention to it. As a result he was reminded, amidst his deep involvement in the new mechanics of atomic systems, not to forget those problems of quantum theory in which the light-quantum hypothesis had been successful.[227]

While the final decision about the nature of radiation in quantum mechanics seemed to be far off in the future, Dirac approached a more modest problem concerning a specific property of radiation right away. How, he asked, are the frequencies of the radiation emitted by a moving periodic quantum system related to the frequencies emitted by the same system at rest? In classical theory the answer is contained in the so-called Doppler effect, and Schrödinger had shown that the same shift occurs in the frequencies of the radiation emitted by a Bohr–Sommerfeld atom, if he assumed the radiation—in full agreement with the light-quantum hypothesis—to come out in a definite direction with a definite momentum (Schrödinger, 1922a). Dirac had himself obtained a modified derivation of the same result, in which he had made use of the relativistic wave vector of the radiation in order to avoid an explicit reference to the light-quantum (Dirac, 1924c). However, in fall 1925 the new quantum-mechanical scheme provided a clear prescription for obtaining, at least in principle, the transition frequencies of an atom, and it seemed evident that one should use the same prescription whether the atom is at rest or moving with a uniform velocity. For calculating the relation between the frequencies corresponding to the same transitions, Dirac did not consider complicated atomic systems, but turned to the simplest system available: the quantum harmonic oscillator.[228]

In classical theory E_0, the 'proper' energy (or energy in the rest frame) of a harmonic oscillator vibrating in the direction of the y-axis and moving with a uniform velocity v along the x-axis, is given by the equation

$$E_0 = A + \tfrac{1}{2}m_0\left\{ y^2 + \mu^2\left(\frac{dy}{ds}\right)^2\right\}. \tag{174}$$

[227] In fall 1925 Dirac also discussed publicly the scattering of light from the point of view of the classical radiation theory. At the 100th Meeting of the Kapitza Club on 6 November 1925 he spoke about 'Debye's Note on the Scattering of X-rays by Liquids.' Debye (1925) had accounted for the observed interference effects by calculating the scattering of waves by more or less geometrical patterns of atoms, which he assumed to exist in liquids. Thus, Dirac talked, within a short time, about the description of essentially the same physical situation in terms of two existing, however completely different and in many respects contradictory, theories of radiation.

[228] Dirac's undated, handwritten, two-page manuscript, dealing with the 'Radiation from a Moving Planck Oscillator,' is preserved at Churchill College, Cambridge (AHQP, Microfilm Copy M/f, No. 36, Section 9, in *Bohr Archives*, Copenhagen). These notes—apart from the few scattered equations, attached to the manuscript on the Bose–Einstein statistical mechanics, in which Dirac expressed the correspondence between classical and quantum-theoretical quantities (we have referred to them in Section IV.1)—are the earliest available notes of Dirac on quantum mechanics. From the way the equations are formulated it becomes evident that the manuscript on the moving Planck oscillator was written either shortly before or shortly after the completion of the published paper on the fundamental equations of quantum mechanics (Dirac, 1925d); hence it must be dated around November 1925. The main reason is that in it Dirac did not use the q-number method which he developed between November 1925 and January 1926.

In Eq. (174), A is a constant denoting the energy of the nonoscillating system, and μ is a constant that measures the force which drives the oscillating mass m_0 back to the origin of the y-coordinate. Moreover, the differentiation of the position variable y is taken with respect to the proper time s $(= (1 - v^2/c^2)^{1/2}t)$. Since y possesses a periodic expansion containing the two terms, $C_1 \exp\{i\omega(t - x/c)\}$ and $C_2 \exp\{-i\omega(t - x/c)\}$, where ω is 2π times the frequency of the moving oscillator, the proper energy may also be written as

$$E_0 = A + m_0 C_1 C_2 \left[1 + \mu^2 \omega_0^2\right] + \left[1 - \mu^2 \omega_0^2\right] \times \text{the periodic term.} \tag{175}$$

This expression is a constant of the motion only if the quantity ω_0, given by

$$\omega_0 = \omega \frac{1 - v/c}{\sqrt{1 - v^2/c^2}}, \tag{176}$$

assumes the value μ^{-1}. Evidently, $\omega_0/2\pi$ denotes the frequency of the oscillator at rest, and its relation to $\omega/2\pi$, the frequency of the oscillator moving with velocity v in the x-direction, agrees with the Doppler principle.

Dirac knew that the energy of the quantum-theoretical harmonic oscillator is also given by Eq. (174), if only the classical variable y and its differential quotient are replaced by expansions containing the terms, $C(n, n \pm 1) \exp\{i\omega(n, n \pm 1)(t - x/c)\}$, where the indices n and $n \pm 1$ denote the stationary states of the system. By applying the quantum product rule he obtained the result

$$\begin{aligned} E_0(n) = A + \tfrac{1}{2} m_0 \{ & C(n, n-1) C(n-1, n) \left[1 + \mu^2 \omega_0^2(n, n-1)\right] \\ & + C(n, n+1) C(n+1, n) \left[1 + \mu^2 \omega_0^2(n, n+1)\right]\} \\ & + \text{periodic terms} \end{aligned} \tag{177}$$

for the proper energy of the quantum oscillator in the stationary state n. Since the periodic terms contain the factors, $[1 - \mu^2 \omega_0^2(n, n \pm 1)]$, the quantum energy is a constant of motion provided $|\omega_0(n, n \pm 1)|$, the absolute values of all transition frequencies in the rest frame, multiplied by 2π, are identified with the oscillator constant μ^{-1}.

In the quantum equations relating the frequencies of the oscillator in the rest frame to the corresponding frequencies in the moving frame, that is,

$$\omega_0(n, n \pm 1) = \omega(n, n \pm 1) \cdot \frac{1 - v(n, n \pm 1)/c}{\sqrt{1 - v^2(n, n \pm 1)/c^2}}, \tag{178}$$

there enter the quantities $v(n, n \pm 1)$ which are identified as the differential quotients of the position variable x with respect to time t, associated with the transition from the stationary state n to the stationary states $n \pm 1$. Dirac proposed to determine $v(n, n \pm 1)$ with the help of the frequency condition, $\omega_0(n, n \pm 1) = 2\pi h^{-1}[E_0(n) - E_0(n \pm 1)]$, in the rest frame. Further assuming that the frequencies in the moving system are also given by the energy differences divided by Planck's constant, he concluded that 'the frequency of the emitted radiation is the same as on the ordinary theory' (Dirac, unpublished manuscript, 1925, p. 2, AHQP, M/f No. 36, Section 9).[229]

After this early application of the new quantum-mechanical methods to physical systems whose motion is not purely periodic, Dirac did not proceed along these lines for several months. He concentrated on working out the q-number calculus and constructed with its help the action-angle scheme, deriving numerous consequences which agreed with the experimental and theoretical results previously obtained. Having completed the most general and sophisticated investigation of multiply periodic systems—in his paper on the elimination of nodes (Dirac, 1926b)—he was now ready to extend q-number methods in such a way that they would also apply to atomic problems that could not be treated by action-angle variables alone. These problems included, of course, the relativistically moving harmonic oscillator, which he had investigated in a preliminary fashion in fall 1925. Another question, with which Dirac was very familiar, was the scattering of radiation by free electrons.

There was still a different class of systems that could be handled by means of generalized q-number methods, namely, those described by a Hamiltonian which depends explicitly on the time. Such systems are very common since all processes involving the dispersion of radiation by atoms belong to this class, the radiation being represented by a periodic wave function. In their joint paper, Born, Heisenberg and Jordan had indicated how the dispersion situation had to be treated in matrix mechanics (BHJ, 1926, Chapter 1, Section 5). In particular, they had discussed the interaction of an atom (described by q_k and p_k, the position and momentum matrices, respectively, with k denoting the degrees of freedom)

[229]To understand Dirac's conclusion, one must insert Eq. (178) in the frequency condition for $\omega_0(n, n \pm 1)$. Thus one obtains

$$\frac{h}{2\pi}\,\omega(n, n \pm 1) = \left[1 - \frac{v^2(n, n \pm 1)}{c^2}\right]^{-1/2}\left[E_0(n) - E_0(n \pm 1) + v(n, n \pm 1)\cdot\frac{E_0(n) - E_0(n \pm 1)}{c}\right].$$

Now the right-hand sides of these equations are indeed identical with the energy differences of the moving oscillator, $E(n) - E(n \pm 1)$, if one takes over the classical relativistic equation connecting E and E_0, that is,

$$E = \left(1 - \frac{v^2}{c^2}\right)^{-1/2}(E_0 + vp_x)$$

(where $v = dx/dt$ and p_x is the momentum in the x-direction), interprets the quantities $(1/c)[E_0(n) - E_0(n \pm 1)]$ as the differences of p_x, the momentum variable of the oscillator, before and after the emission of radiation in the x-direction, and finally identifies all $v(n, n \pm 1)$ with the constant velocity v. In case the radiation is not emitted in the x-direction but, say, in the z-direction, then $v(n, n \pm 1)$ will depend on the two states denoted by n and $n \pm 1$. These consequences confirm that the quantum radiation possesses the specific properties attributed to it by the light-quantum hypothesis, though Dirac did not mention it explicitly in his manuscript.

with an incident wave of radiation having a constant amplitude and a varying phase, $2\pi\nu_0 t$, where ν_0 is a constant frequency. They had then proposed to replace the description of the total system, atom plus radiation, by a different one, in which a new degree of freedom was formally introduced by adding the canonical pair of matrices q' and p'. The position coordinate q' was identified with the function $\cos\{2\pi\nu_0 t\}$—a rather special matrix having only diagonal terms, $\cos\{2\pi\nu_0 t\}$—and the conjugate momentum was defined as the partial differential quotient of the Hamiltonian H' with respect to q'. The relation between H' and the original Hamiltonian H is given by the equation

$$\mathsf{H}' = \mathsf{H} + 2\pi\nu_0\sqrt{1 - \mathsf{q}'^2}\,\mathsf{p}'. \tag{179}$$

Dirac knew about this procedure of the Göttingen physicists, and it stimulated him to treat such problems also with his methods; the more so, since the application of the matrix scheme appeared to him to be limited, for Born, Heisenberg and Jordan had neglected to take into account the reaction of the atom on the incident radiation.[230] Dirac hoped to adopt a more general point of view than the matrix scheme, a point of view that had emerged in a natural manner from his previous work on uniformizing variables in quantum mechanics. He had become convinced that he could treat *all* variables, including the nonperiodic ones, which appeared in the quantum-theoretical reformulation of an arbitrary classical system, as q-numbers. Why should he not consider the time t, the most fundamental variable in any dynamical system in the same way? This idea was suggested by a recently published paper of Max Born and Norbert Wiener on the operational calculus in quantum mechanics, where the authors had proposed to express the quantum dynamical variables as functions of a fundamental pair of operators, the time t and the differential operator $D = d/dt$. They had made a successful application of this calculus to the uniform motion of a particle in a straight line, an extreme example of aperiodic motion (Born and Wiener, 1926a, b). It was not difficult for Dirac to conclude from Born and Wiener's work that t and $(ih/2\pi)D$ had to be considered as a pair of conjugate quantum variables.[231] But there existed in his opinion a still more compelling reason to introduce time as a quantum variable. 'The principle of relativity,' he argued, 'demands that the time shall be treated on the same footing as the other variables, and so it must be a q-number' (Dirac, 1926c, p. 407). With this guiding idea in mind Dirac entered into a detailed discussion of the two new classes of dynamical systems in quantum mechanics: those described by a Hamiltonian which involves the time explicitly and the ones whose kinematics is relativistic.

The extension of the Hamilton–Jacobi scheme of classical mechanics to

[230] Dirac quoted the particular section of the three-man paper at the appropriate place in his work on 'Relativity Quantum Mechanics' (Dirac, 1926c, p. 409).

[231] Born and Wiener's paper appeared in March 1926 and Dirac quoted it (Dirac, 1926c, p. 405). Since he was familiar with Heaviside's calculus, he had no difficulty in following Born and Wiener's arguments.

systems whose potential energy contains a term depending on the time t, not just via q_k and p_k ($k = 1, \ldots, f$), the position and momentum variables, had been known since long. Sommerfeld, for instance, had mentioned the recipe in his *Atombau und Spektrallinien* as follows: 'One must then take t as an independent variable and replace W by $-\partial S/\partial t$ in the [Hamilton–Jacobi] equation' (Sommerfeld, 1924d, p. 801). For a system of f degrees of freedom, the Hamilton–Jacobi differential equation would then be written as

$$H(q_1, \ldots, q_f; p_1, \ldots, p_f; t) = W = -\frac{\partial S}{\partial t}, \tag{180}$$

where S is the transformation or principal function by which the integration of the equations of motion is achieved. From this equation, together with the equations of motion for q_k and p_k, one obtains the result

$$\frac{dW}{dt} = \sum_{k=1}^{f}\left(\frac{\partial H}{\partial q_k}\frac{dq_k}{dt} + \frac{\partial H}{\partial p_k}\frac{dp_k}{dt}\right) + \frac{\partial H}{\partial t} = \frac{\partial H}{\partial t}. \tag{181}$$

Hence, $-W$ is the variable conjugate to the time variable t. These were just the equations that Dirac needed and he concluded: 'In the solution of the problem there will be now complete symmetry between the new pair of variables t and $-W$ and the original pairs, except for the fact that when one performs the contact transformation to the uniformizing variables, the coordinate t itself must be one of the new variables' (Dirac, 1926c, p. 407).[232] He then defined the classical Poisson bracket of x and y, the two variables of a time-dependent system, by the expression

$$(x, y) = \sum_{k=1}^{f}\left(\frac{\partial x}{\partial q_k}\frac{\partial y}{\partial p_k} - \frac{\partial x}{\partial p_k}\frac{\partial y}{\partial q_k}\right) - \left(\frac{\partial x}{\partial t}\frac{\partial y}{\partial W} - \frac{\partial x}{\partial W}\frac{\partial y}{\partial t}\right) \tag{182}$$

and formulated the equation of motion for x, an arbitrary variable, as

$$\frac{dx}{dt} = (x, H - W). \tag{183}$$

The essential reason for replacing the Hamiltonian H in the bracket on the right-hand side of Eq. (183) by the difference $H - W$ is that the time differenti-

[232]Sommerfeld discussed time-dependent systems in Appendix 7 of *Atombau und Spektrallinien*, which he entitled '*Hamiltonsche Partielle Differentialgleichung*' (Sommerfeld, 1924d, p. 779). There he also extended the scheme to account for systems on which forces, which are not derived from a potential, act; for example, such forces occur when magnetic fields are applied to atoms. Dirac had worked on the latter problem and quoted Sommerfeld's book in his paper on 'The Adiabatic Hypothesis for Magnetic Fields' (Dirac, 1925c), submitted in November 1925. The fact that Whittaker in *Analytical Dynamics* had not discussed time-dependent systems in detail was the reason why Dirac followed Sommerfeld closely. He described the situation, similar to Sommerfeld, as follows: 'On the classical theory it is known that one may solve the problem by considering the time to be an extra coordinate of the system, with minus the energy (or perhaps a slightly different quantity) W as conjugate momentum' (Dirac, 1926c, p. 407).

ation of the variable t must yield the result 1, which is indeed guaranteed since the equation

$$\frac{dt}{dt} = \frac{\partial(H - W)}{\partial(-W)} = 1 \tag{184}$$

is valid. The replacement does not change the other equations of motion because W is an independent variable and does not depend on q_k, p_k and t.

Finally, Dirac noticed that his equations were in agreement with the procedure followed by Born, Heisenberg and Jordan in their special case of a time-dependent system. Indeed, the Hamiltonian H' of Eq. (179) can be written (in classical theory) as the sum, $H + (dq'/dt)p'$, and, on identifying the new variable q' with t, this sum becomes equal to $H - W$.[233]

Dirac took over these results into quantum mechanics in the manner to which he had become accustomed. He described a time-dependent quantum system of f degrees of freedom by $2f + 2$ quantum variables, q_k, p_k, t and $-W$, which satisfy, besides the usual quantum conditions (i.e., the commutation relations between q_k and p_k), the additional equations

$$tW - Wt = -\frac{ih}{2\pi}, \tag{185a}$$

$$tq_k - q_k t = tp_k - p_k t = 0, \qquad Wq_k - q_k W = Wp_k - p_k W = 0. \tag{185b}$$

He defined $[x, y]$, the quantum Poisson bracket of two quantum variables, x and y, by the equation

$$xy - yx = \frac{ih}{2\pi}[x, y] \tag{186}$$

and found that the quantum equation of motion for the arbitrary variable x will assume the form of Eq. (183), with the quantum Poisson bracket replacing the classical one.

Dirac drew attention to a special point which has to be considered when using the quantum scheme for time-dependent systems. 'The fact,' he noted, 'that a dynamical system is now specified by a Hamiltonian equation $H - W = 0$ instead of by a Hamiltonian function H here leads to a difficulty, since the Hamiltonian equation is not consistent with the quantum conditions' (Dirac, 1926c, p. 408). For example, if one considers a variable x, which is a function of q_k and p_k only, then it has to commute with the variable W but not necessarily with the Hamiltonian H. Such a situation, Dirac noticed, had not occurred before in quantum mechanics because so far it was always assumed that if two

[233] The fact that in the equations of motion for time-dependent systems the Hamiltonian H had to be replaced by the difference $H - W$ was standard in the literature, but Dirac did not take it over because it was neither reported in Sommerfeld's book, nor in Whittaker's, nor even in Born's *Atommechanik* (Born, 1925), the books available to him. Thus, he just tried it out and was confident that it would work, as the example of Born, Heisenberg and Jordan confirmed.

sides of an equation were equal, so were also the Poisson brackets of the two sides with an arbitrary variable. Evidently, that equality did not hold in the case of what Dirac, in agreement with Sommerfeld, called the 'Hamiltonian equation.'[234] This difficulty did not arise in classical theory because there the Hamiltonian equation [i.e., the Hamilton–Jacobi partial differential equation] is not an identity. Thus, one can perform algebraic operations on both sides of Eq. (180), but one must not differentiate them, for if that were an allowed procedure all dynamical variables would be constants of motion. 'There must be a corresponding restriction on the use of the quantum Hamilton equation,' argued Dirac, 'although it cannot easily be specified, as there is no hard-and-fast distinction between algebraic operations and differentiations on the quantum theory' (Dirac, 1926c, p. 408). Yet he claimed that this uncertainity will not present any difficulty if one stays in the discussion of the quantum system so closely to the classical theory 'that it will be immediately obvious whether any quantum operation corresponds to a legitimate classical operation or not' (Dirac, 1926c, p. 408).

With this extra precaution, then, one could establish a Hamiltonian scheme of a time-dependent system of f degrees of freedom by reformulating the classical method of introducing $2f+2$ uniformizing variables, $J_0, \ldots, J_f, w_0, \ldots, w_f$, where w_0 is just the time variable. As in classical theory, the Hamiltonian equation leads to the equation

$$H_0 + J_0 = 0, \tag{187}$$

where H_0 is a function of the action variables $J_1, \ldots, J_f$ only, which is obtained by introducing uniformizing variables into the original Hamiltonian H; and J_0 is equal to $-W$ plus a quantity independent of W.

Dirac derived two consequences from this result. First, he argued that Eq. (187), together with Eqs. (185b), allowed Born, Heisenberg and Jordan to treat the time consistently as a c-number in their approach to time-dependent systems. Second, if in quantum mechanics one considers the general Hamiltonian equation,

$$F(q_1, \ldots, q_f; p_1, \ldots, p_f; W; t) = 0, \tag{188}$$

for time-dependent systems, one must, before any evaluation can be done, bring this equation to the standard form in which the variable W is completely isolated from the rest. One reason is that the expression $H - W$ is needed explicitly in order to calculate correctly with the equations of motion. The other reason is that one may not succeed in expressing $F(q_1, \ldots, q_f; p_1, \ldots, pf; W; t)$ as a function only of the $J_1, \ldots, J_f$ with the help of a canonical transformation, and F may

[234]The reason why Dirac called the equation $H - W = 0$ the 'Hamiltonian equation' was that it corresponded to the classical Hamilton–Jacobi differential equation, Eq. (180), which Sommerfeld had also called '*Hamiltonsche Gleichung*' in his *Atombau und Spektrallinien* (1924d, p. 785).

turn out to be the product of such a function multiplied by a factor depending on both the J's and the w's. Situations of this kind occur in relativistic systems in particular, to which Dirac turned next.[235]

From the very foundation of relativistic kinematics it is evident that the time t must be treated on the same footing as the position variables. For a system moving as a whole the principle of relativity requires that the two variables, x_4 and p_4, defined as

$$x_4 = ict, \qquad p_4 = i\frac{W}{c}, \tag{189}$$

where i is a square root of -1 and c the velocity of light (a c-number), are introduced in addition to the Cartesian coordinates of the centre of gravity, x_1, x_2, x_3, and the conjugate components of the total momentum, p_1, p_2, p_3, and that *complete symmetry* exists between the pair (x_4, p_4) and the pairs (x_1, p_1), (x_2, p_2) and (x_3, p_3). Thus, in classical theory, there results the equation

$$\left(ict, i\frac{W}{c}\right) = 1 = -(t, W), \tag{190}$$

which, because of Eq. (186), leads to the quantum condition (185a) for a relativistic quantum system. In this case, relativity theory also yields the Hamiltonian equation

$$\frac{W^2}{c^2} - p_1^2 - p_2^2 - p_3^2 = m_0 c^2, \tag{191}$$

which is obtained from the elementary kinematic fact that the scalar product of the four-vector (p_1, p_2, p_3, p_4) with itself is a relativistic invariant. This invariant is determined by the quantity m_0, which is usually called the 'rest mass' of the system under investigation. The rest mass, and with it the 'proper energy,' m_0c^2, are functions of the internal variables of the system under investigation; hence both quantities are in general q-numbers in quantum mechanics, except for systems consisting of single particles since there are no internal variables in those cases. From Eq. (191) Dirac also concluded that 'the variables p_1, p_2, p_3, W and x_1, x_2, x_3, t may be taken to be uniformizing variables, as they satisfy all the conditions for this except the multiply periodic conditions for the x's, which they obviously cannot be expected to satisfy,' and that 'the remaining uniformizing variables will be functions of the internal variables only' (Dirac, 1926c, p. 410).

Dirac also extended the theory of relativistic systems to include those on which external fields of forces act. This extension would not give rise to any difficulties if the corresponding classical equations of motion could be put into the Hamiltonian form of Eq. (183). As an example he discussed the action of an

[235] An example of this case is given by the energy equation for the relativistic hydrogen atom, where W occurs in a quadratic expression. (See, e.g., Sommerfeld, 1924d, p. 410, Eq. (9).)

electromagnetic field, described by a four-potential with components A_1, A_2, A_3, A_4 (the first three being the components of the vector potential, and the last being equal to ic times the scalar potential ϕ), on a system of mass m_0 and total charge e, thought to be concentrated at the centre of gravity of the system. (The charge e is a c-number in Dirac's scheme.) In classical theory the Hamiltonian equations of motion for the coordinates x_ν and the conjugate momenta p_ν then assume the form

$$\frac{dx_\nu}{ds} = \frac{\partial F}{\partial p_\nu}, \qquad \frac{dp_\nu}{ds} = \frac{d}{ds}\left(m_0 \frac{dx_\nu}{ds} + \frac{e}{c} A_\nu\right) = -\frac{\partial F}{\partial x_\nu} \qquad (\nu = 1, \ldots, 4). \tag{192}$$

In Eqs. (192), s is the proper time (given by the equation $ds^2 = c^2 \sum_{\mu=1}^{4} dx_\mu dx_\mu$), and F represents the Hamiltonian function (replacing $H - W$ in Eq. (183) for nonrelativistic systems) defined by the equation

$$F = \frac{1}{2m_0} \sum_{\nu=1}^{4} \left(p_\nu - \frac{e}{c} A_\nu\right)\left(p_\nu - \frac{e}{c} A_\nu\right) + \frac{1}{2} m_0 c^2. \tag{193}$$

This Hamiltonian satisfies the Hamiltonian equation, $F = 0$, which has the structure necessary for integrating the equations of motion.[236] The classical Eqs. (192) remain valid in quantum mechanics, provided the classical variables are replaced by the corresponding quantum variables. To simplify F, the Hamiltonian function defined by Eq. (193), one may use the fact that the components of A_ν, the four-potential of the electromagnetic field, commute with p_ν, the momentum components, a fact which Dirac concluded from the zero divergence of the classical potential.[237]

[236] Unlike the nonrelativistic case (where the equation $H - W = 0$ holds), the equation $F = 0$ is not an identity, hence its partial differential quotients with respect to the variables are not zero. For example, Dirac found that the quotient $\partial F / \partial m_0$ becomes equal to the constant c^2 (Dirac, 1926c, p. 411, last equation.)

[237] Dirac rewrote the classical condition as

$$\sum_{\nu=1}^{4} \frac{\partial A_\nu}{\partial x_\nu} = \sum_{\nu=1}^{4} (A_\nu, p_\nu) = 0,$$

where p_ν is the variable conjugate to the position coordinate x_ν. He then identified the p_ν in this equation with the momentum coordinates of the system that enter into Eqs. (192) and (193). Albert Einstein and Paul Ehrenfest, on studying Dirac's paper on 'Relativity Quantum Mechanics,' wondered whether this procedure was allowed. Ehrenfest wrote to Dirac and asked the following question:

> In the equation $\partial \kappa_\mu / \partial x_\mu = 0$ (*) [Dirac had used the letter κ_μ to denote the components of the four-potential and dropped the summation symbol when two identical indices occurred] the letters x_μ mean the coordinates of a *point of the "aether"*. But to get from that equation, your equation $[\kappa_\mu, p_\mu] = 0$, one has, if I make no mistake, first of all the equation (*) to write in the form:
>
> $$\frac{\partial \kappa_\mu}{\partial x_\mu} \frac{\partial p_\mu}{\partial p_\mu} - \frac{\partial \kappa_\mu}{\partial p_\mu} \frac{\partial p_\mu}{\partial x_\mu} = 0 \ (???)$$
>
> but here x_μ mean suddenly the *coordinates of the moving particle*. Why is this change of meaning

In applying the foregoing quantum time scheme to atomic systems, which emit and absorb radiation, the main question to be asked first is what time variable should one use, especially when the radiating electron possesses a velocity comparable to c, the velocity of light. One may get some information in this regard from the classical treatment of the problem. There the intensity of the radiation emitted in a given direction, say along the x_1-axis, is obtained from the Fourier expansion of the total polarization of the atom, having the components, $\frac{1}{2}C_\alpha \exp\{i(\alpha\omega)(t - x_1/c)\}$, $\alpha = \pm 1, \pm 2, \ldots$, where $(\alpha\omega)$ is the frequency of the radiation and x_1 the position where the charge is supposed to be concentrated. If the charge is also moving uniformly, a nonperiodic term, which is a constant times $(t - x_1/c)$, must be added. With this assumption one arrives at the correct description of the physical situation in which the atom interacts with the incident radiation; the nonperiodic term does not contribute at all, while the periodic terms yield the well-known dispersion formulae. I, the intensity of the radiation passing through a fixed point at a distance r from the charge, is finally obtained from the equation

$$I = \frac{e^2(\alpha\omega)^4}{8\pi c^3 r^2}|C_\alpha|^2. \tag{194}$$

Before taking over the classical result into quantum theory, one has to make sure that the variables entering the equations (leading to the result) are uniformizing variables. This refers especially to the time variable, $t - x_1/c$. However, already in classical theory a difficulty arises with respect to the correct definition of $t - x_1/c$, for the significance of the coordinate x_1 is not immediately clear when more than one charge contributes to the total polarization of the atom. While in the case of one electron x_1 is just equal to its x_1-coordinate, in the case of a several-electron atom one has to consider the polarization as an expansion in a sum of Fourier series, each of which describes the contribution from one electron whose x_1-coordinate enters into the time variable. Since, in general, the relative displacement of the electrons in an atom is small, one may approximate the sum of the Fourier expansions by a single one, in which x_1 is identified with the coordinate of the centre of gravity, but this approximation is not always a good one.[238] Keeping this in mind, Dirac went on to derive a general result about the emission of radiation from free atoms, whose motion as a whole is determined by the Hamiltonian of Eq. (191).

When one carries out the canonical transformation of the original transformation of the original variables, $x_1, x_2, x_3, t, p_1, p_2, p_3, -W$, to the uniformizing

permitted? Give, please, an explanation and please not too short!! (Ehrenfest to Dirac, 1 October 1926)

We have not been able to find Dirac's answer to this question, but it could be argued in favour of his replacement that the ether point does not have a separate significance from the position of, say, an electron of the atomic system under investigation. After all, it is the electron on which the potential A_ν acts.

[238] For example, if the velocity of the atom as a whole is of the same order as c, one can use, as Dirac noted, the above approximation for the periodic term but not for the nonperiodic term.

variables, that is,

$$\begin{aligned} x_1' &= x_1, \quad & x_2' &= x_2, \quad & x_3' &= x_3, \quad & t' &= t - \frac{x}{c}, \\ p_1' &= p_1 - \frac{W}{c}, \quad & p_2' &= p_2, \quad & p_3' &= p_3, \quad & -W' &= -W, \end{aligned} \tag{195}$$

the right-hand side of the transformed Hamiltonian equation will still be a function of p_1', p_2', p_3' and $-W$, the momentum variables of the total atom alone. Since one now has the correct time variable in the harmonic expansion of the quantum-theoretical polarization of the atom—provided one interprets the variables as q-numbers—one can calculate the properties of the radiation emitted in x_1-direction. Without specifying the variable x_1, there is, in particular, one result which follows from the fact that any component of the radiation, say $\exp\{iw\}$, depends only on the internal variables and, therefore, commutes with p_1', p_2', p_3'. As a consequence, the c-number values which the uniformizing momenta, p_1', p_2', p_3', assume do not change during the emission of radiation from the system. However, this statement takes into account the fact that p_1' includes a change of p_1, the momentum component of the atom, by a term which is equal to $1/c$ times the change in the energy W (see Eq. (195)). 'Hence,' Dirac concluded, 'according to the present theory, the system experiences a recoil when it emits radiation, in agreement with the light-quantum theory' (Dirac, 1926c, p. 414).

Thus, the systematic q-number treatment of atoms emitting radiation confirmed and extended the simple considerations dealing with the harmonic oscillator, with which Dirac had entered the subject of 'relativity quantum mechanics' several months previously. However, his principal interest in spring 1926 was directed towards another goal, namely, to develop a quantum-mechanical theory of the Compton effect. He achieved this goal by straightforward calculations without further difficulties. He was helped by the fact that no complication arose in this case from the definition of the position variable x_1, because the system under investigation, to a very good approximation, consists of high frequency radiation scattered by one free electron which provides the unique coordinate.

The properties of the incident radiation are fully described by the components of the electromagnetic four-potential, A_ν, $\nu = 1, 2, 3, 4$. If the direction of propagation coincides with the x_1-axis and the electric vector oscillates along the x_2-axis, the only nonzero component, A_2, is given by the equation

$$A_2 = a \cos\left\{ \frac{2\pi\nu_0}{c} (ct - x_1) \right\}, \tag{196}$$

where ν_0/c is the wave number of the radiation, which can be measured and must, therefore, be considered as a c-number in quantum mechanics. The same is true of the amplitude a because it may be obtained from I_0, the observable intensity of the incident radiation, by the formula

$$I_0 = \frac{a^2(2\pi\nu_0)^2}{8\pi c}. \tag{197}$$

The Hamiltonian equation for the electron, with mass m_e and charge $-e$, in the electromagnetic field is found by replacing p_2, the momentum component of the free electron in Eq. (191), by the sum $p_2 + (e/c)A_2$. Since Dirac wanted to calculate the frequency and intensity of the scattered radiation in a given direction, say the one determined by the direction cosines l_1, l_2, l_3, he introduced the uniformizing variables by the equations

$$\begin{aligned}
x'_1 &= ct - x_1, & (1-l_1)p'_1 &= -p_1 + l_1 \frac{W}{c}, \\
x'_2 &= x_2, & (1-l_1)p'_2 &= l_2 p_1 + (1-l_1)p_2 - l_2 \frac{W}{c}, \\
x'_3 &= x_2, & (1-l_1)p'_3 &= l_3 p_1 + (1-l_1)p_3 - l_3 \frac{W}{c}, \\
t' &= t - (l_1 x_1 + l_2 x_2 + l_3 x_3)/c, & (1-l_1)W' &= W - cp_1,
\end{aligned} \tag{198}$$

where the unprimed quantities, $x_1, x_2, x_3, t, p_1, p_2, p_3, -W$, represent the variables of the free electron.[239] With these, the Hamiltonian equation assumes the form

$$H - W' = 0, \tag{199}$$

with

$$H = -\frac{1}{2}c\left(m_e^2 c^2 + B\right)A^{-1}, \tag{199a}$$

where

$$A = (1-l_1)p'_1 + l_2 p'_2 + l_3 p'_3 + l_2 \frac{ea}{c} \cos\left(\frac{2\pi\nu_0 x'_1}{c}\right), \tag{199b}$$

[239] The motivation for these new quantum variables is quite straightforward. First, the uniformizing time variable must be given by $t - (1/c)(l_1 x_1 + l_2 x_2 + l_3 x_3)$ because this expression will occur in the periodic functions describing the harmonic components of the scattered radiation. Then one calculates the variable W' by inserting the *Ansatz*

$$W' = c_4 W + c_1 p_1 + c_2 p_2 + c_3 p_3,$$

where c_1, c_2, c_3, c_4 are constants, into the quantum Poisson equation, that is,

$$-1 = [t', W'] = [t, c_4 W] + \left[\frac{l_1 x}{c}, c_1 p_1\right] + \left[\frac{l_2 x}{c}, c_2 p_2\right] + \left[\frac{l_3 x}{c}, c_3 p_3\right].$$

Since the incident radiation is in the x_1-direction, it will influence only the electron variables W and p_1 but not p_2 and p_3; hence c_2 and c_3 are zero. The remaining constants are related by the equation $c_1 = cc_4$, due to the fact that any change in Δp_1, the momentum of the electron, arising from the interaction with the radiation is accompanied by a change in the energy, $\Delta W = c\,\Delta p$. Thus, one can find c_4 from the equation $-1 = -c_4(1 - l_1)$, in agreement with the equation for W'. In a similar way one constructs the variables conjugate to x'_1, x'_2 and x'_3, taking into account the fact that p'_1, p'_2 and p'_3 have to commute with W' and with the nonconjugate position variables.

and

$$B = p_2'^2 + p_3'^2 + 2\frac{ea}{c}\, p_2' \cos\left(\frac{2\pi\nu_0 x_1'}{c}\right), \tag{199c}$$

when one neglects terms proportional to a^2 as being small.[240]

Equation (199) has the desired form for further evaluation. That means, in particular, that the Hamiltonian function contains all the information about the frequencies of the radiation emitted by the electron, that is, the Compton radiation. In order to calculate the frequencies, Dirac had just to introduce *uniformizing action variables*, namely, the ones which commute with the Hamiltonian H. Two of them are p_2' and p_3', while a third one, J, can be constructed from the variables p_1', p_2', p_3' and x_1' as

$$J = \frac{1}{2\pi\nu_0/c\cdot(1-l_1)}\cdot A\cdot\frac{m_e^2c^2 + p_2'^2 + p_3'^2}{m_e^2c^2 + B}. \tag{200}$$

Dirac showed that p_2' and p_3' commute with J and the canonically conjugate variable w, where

$$w = \frac{2\pi\nu_0}{c}x_1' + \frac{2(ea/c)p_2'\sin\{(2\pi\nu_0/c)x_1'\}}{m_e^2c^2 + p_2'^2 + p_3'^2}. \tag{201}$$

In terms of the uniformizing variables the Hamiltonian becomes

$$H = -\frac{c}{2}\,\frac{m_e^2c^2 + p_2' + p_3'^2}{(2\pi\nu_0/c)\cdot(1-l_1)J}. \tag{202}$$

Equations (200), (201) and (202) immediately lead to the following results. First, there is only one (frequency) component of the scattered radiation in the direction defined by the cosines l_1, l_2, l_3 because only one action variable exists, the one given by Eq. (200). Second, since p_2' and p_3' commute with the expression, $\exp\{iw\}$, their c-numbers remain unchanged during the emission (or scattering) process. On the other hand, from the action-angle scheme of multiply periodic systems it follows that the value of the action variable J will be reduced by the

[240]Since the expression A commutes with the transformed energy W', Dirac concluded that a Hamiltonian H', which differs from H, Eq. (199a), by having the factor A^{-1} in front of—instead of after—the bracket $(m_e^2c^2 + B)$ would also be admitted. This would seem to imply a certain ambiguity in the choice of the Hamiltonian because A^{-1} does not commute with B. However, Dirac showed that both Hamiltonian functions, H *and* H', do lead to the same values for the frequency and the intensity of the scattered radiation; hence the formal ambiguity does not imply any physical consequences. The reason for this is that H' and H are connected by the relation

$$H' = A^{-1}HA,$$

which can be interpreted as meaning that H' is obtained from H by a contact transformation. This transformation, of course, will not change the c-number results of the calculation.

amount $h/2\pi$, and this change implies—according to Eqs. (200) and (199b), with $\Delta p_2' = \Delta p_3' = 0$—a change of the uniformizing variable p_1' by an amount $-h\nu_0/c$. Finally, by identifying the change of the Hamiltonian, calculated from Eq. (202), with $h\nu'$, where ν' is the frequency of the scattered radiation, Dirac obtained for the changes of the original electron variables the results

$$\begin{aligned} c\Delta p_1 &= h\nu - l_1 h\nu', \quad & c\Delta p_2 &= -l_2 h\nu', \\ c\Delta p_3 &= -l_3 h\nu', \quad & \Delta W &= h\nu - h\nu'. \end{aligned} \tag{203}$$

When the correction to the momentum component p_2 due to the electromagnetic field can be neglected (i.e., one can drop the term $(e/c)A_2$ against p_2), Eqs. (203) state simply the conservation of momentum and energy in the sense of the light-quantum hypothesis. 'The present theory,' Dirac concluded, 'thus gives the same values for the frequency of the scattered radiation and the recoil momentum of the electron as the light-quantum theory' (Dirac, 1926c, p. 419).

To obtain the amplitude of the scattered radiation as well, Dirac had to calculate the electric moment of the electron induced by the incident radiation in two perpendicular directions, which are both perpendicular to the direction of emission as given by the three direction cosines l_1, l_2, l_3. He chose the directions given by the triples of cosines,

$$\left(l_3, -\frac{l_2 l_3}{1-l_1}, \frac{l_2^2}{1-l_1} - l_1 \right)$$

and

$$\left(l_2, \frac{l_3^2}{1-l_1} - l_1, \frac{l_2 l_3}{1-l_1} \right).$$

Then X and Y, the two position variables of the electron in these directions, that is,

$$\begin{aligned} X &= l_3 x_1 - \frac{l_2 l_3}{1-l_1} x_2 + \left(\frac{l_2^2}{1-l_1} - l_1 \right) x_3, \\ Y &= l_2 x_1 + \left(\frac{l_3^2}{1-l_1} - l_1 \right) x_2 - \frac{l_2 l_3}{1-l_1} x_3, \end{aligned} \tag{204}$$

exhibit periodic parts, X_{per} and Y_{per}, which arise from the interaction of the electron with the incident radiation. They determine the periodic components of the induced electric momentum and thus the components of the emitted Compton radiation. Dirac evaluated X_{per} and Y_{per} in terms of the uniformizing variables p_2', p_3', J and w—the latter being brought in by replacing the periodic part of the variable x_1' with the help of Eq. (201)—as

$$X_{\text{per}} = -2i\frac{ea}{c}\left[\frac{p_2'p_3'J}{m_e^2c^2 + p_2'^2 + p_3'^2}\exp\{iw\} - \exp\{-iw\}\frac{p_2'p_3'(J - h/2\pi)}{m_e^2c^2 + p_2'^2 + p_3'^2}\right],$$

$$Y_{\text{per}} = i\frac{ea}{c}\left[\frac{(m_e^2c^2 - p_2'^2 + p_3'^2)J}{m_e^2c^2 + p_2'^2 + p_3'^2}\exp\{iw\}\right. \tag{205}$$

$$\left. - \exp\{-iw\}\frac{(m_e^2c^2 - p_2'^2 + p_3'^2)(J - h/2\pi)}{m_e^2c^2 + p_2'^2 + p_3'^2}\right].$$

By summing together the absolute squares of the two amplitudes in Eqs. (205) he obtained the result

$$C^2 = \frac{4(ea/c)^2J(J - h/2\pi)}{\left(m_e^2c^2 + p_2'^2 + p_3'^2\right)^2}\left[1 - \frac{4m_e^2c^2p_2'^2}{\left(m_e^2c^2 + p_2'^2 + p_3'^2\right)^2}\right]. \tag{206}$$

If the electron is initially at rest—which is more or less the case in the actual situation where the Compton effect is observed—C^2 becomes

$$C^2 = \frac{e^2a^2}{m_e^2c^24\pi^2\nu_0^2}\left(1 - l_2^2\right)\frac{\nu_0}{\nu'}, \tag{207}$$

where ν_0/ν' is the ratio of the frequencies of the absorbed and emitted radiation. By substituting this value of C^2 for $|C_\alpha|^2$ in Eq. (194), taking $(\alpha\omega)$ equal to $2\pi\nu'$, and introducing I_0, the intensity of the incident radiation from Eq. (197), Dirac obtained for the intensity of the scattered radiation of frequency ν' the result

$$I = \frac{e^4I_0}{m_e^2c^4r^2}\left(\frac{\nu'}{\nu}\right)^3\left(1 - l_2^2\right). \tag{208}$$

But for the factor $(\nu'/\nu)^3$, this result is identical with the one obtained from the classical theory.

On comparing his results with the available experimental data Dirac was very satisfied. For example, in referring to Compton's original intensity measurements (Compton, 1923b), he noted that 'the experimental values are all less than the values given by the present theory, in roughly the same ratio (75 percent), which shows that the theory gives the correct law of variation of intensity with angle and suggests that in absolute magnitude Compton's values are 25 percent too small' (Dirac, 1926c, p. 423). Moreover, from 'relativity quantum mechanics' the same polarization of the scattered radiation was obtained as from classical theory. In particular, for unpolarized incident radiation he obtained plane polarized scattered radiation at an angle of 90°, while a previously published theory of George Eric MacDonnell Jauncey, based on an extreme light-quantum hypothesis—the scattering of radiation being considered as similar to the scattering of material particles—had predicted a shift from 90° (Jauncey, 1924c). Dirac considered the available data as being 'slightly in favour of the present theory' (Dirac, 1926, p. 423). Altogether he was confident that q-number quantum mechanics would yield correct results for all relativistic problems in the same way as before in the problems of multiply periodic systems. For a given atomic system, one had just to work out the transformation to uniformizing variables by faithfully following the classical calculation and observing at the same time the conditions imposed on quantum variables. The ultimate solution of all problems in atomic physics would then just be a matter of time, but it would not require any other new principle.

V.5 Towards New Horizons

With the paper on 'Relativity Quantum Mechanics,' submitted in late April 1926, Dirac had completed his theoretical scheme of quantum mechanics and applied it to certain practical problems. Although Dirac himself would refer modestly only about the initial successes obtained with the q-number methods—while many more were to come—he had developed in his papers a closed system of physical and mathematical ideas. The new theory of quantum mechanics satisfied Dirac's sense of beauty, for he had been able to formulate his equations and methods in close analogy to the Hamiltonian scheme of classical mechanics, which he regarded as a most beautiful theory. From the mathematical point of view also his methods were as satisfactory as those used in classical theory, for the q-number calculus was governed by laws that were as simple, consistent and complete as the laws of conventional arithmetic. In May 1926 the research student Paul Dirac submitted his work on 'Quantum Mechanics' as a dissertation to Cambridge University and obtained with it the Ph.D. degree (Dirac, 1926d). Thus, with the completion of quantum algebra, Dirac's quantum-mechanical theory, a certain period of his scientific development also came to an end. He was now a mature scientist who would display his full powers at the new frontiers of quantum theory.

There is another reason which allows one to separate Dirac's papers, submitted between November 1925 and July 1926, from his later papers on quantum mechanics. Already in his next work, entitled 'On the Theory of Quantum Mechanics' (Dirac, 1926f), he would make use of a different theory, Schrödinger's wave mechanics, and the methods connected with this theory would soon be recognized to be easier to handle for dealing with realistic atomic systems. Although Dirac discovered the possibility of incorporating in Schrödinger's theory an important feature of many-electron systems that did not easily fit into his own scheme, he would nevertheless not surrender unconditionally to wave mechanics. He would rather combine quantum algebra with Schrödinger's differential operator calculus to develop a unified quantum theory of superior range and power. In any case, from July 1926 onwards the purely q-number methods would not be the only mathematical tool that Dirac would employ in quantum-mechanical problems. Whenever it seemed necessary or practical for the calculations, he would go beyond the range of his own scheme.

In reviewing the achievements and merits of Dirac's quantum-mechanical theory several points may be mentioned. An important question is how his methods and results were related to those which had been obtained independently at about the same time by the Göttingen physicists and their friends, although both sides became increasingly informed about their mutual progress after November 1925. Clearly the two schemes, the Göttingen matrix mechanics and the Cambridge q-number algebra, worked very differently when applied to problems of atomic mechanics, and in many cases the goals and intentions of the people involved were not the same. For example, Heisenberg and his friend Pauli, in particular, considered matrix theory—certainly in the beginning—only as a *possible* theory for describing the strange kinematical and mechanical behaviour of atomic systems; they believed that the matrix methods were subject to trial and error. Hence they tested their validity by applying them to selected cases, such as the hydrogen atom or complex atoms in a magnetic field. They hoped that the results of these applications would either confirm the mathematical scheme or indicate a necessary modification. Dirac, on the other hand, did not share this attitude, which was somehow rooted in the correspondence approach of earlier years. He was principally interested in setting up right away a fundamental, complete and final theory, a theory which would describe all observed phenomena in atomic physics, provided the proper details of the atomic systems were taken into account. Therefore, in choosing examples to confirm his theory he did not dwell on tricky situations, such as the hydrogen atom in crossed electric and magnetic fields; he rather investigated how far the general features, the periodic properties of atoms, for example, were reproduced by means of quantum algebra. This was also the reason why, immediately after he had obtained his initial results, he went on to discuss the relativistic extension of the new theory. Dirac had no doubt that if quantum mechanics was a fundamental theory—and he was only interested in setting up a fundamental theory—then it had to be adapted to the requirements of relativity.[241]

This great generality in the formulation and, therefore, in the application of Dirac's methods was particularly appreciated by Ralph Fowler, who encouraged him to go on with his theory of quantum mechanics as far and as fast as possible. Fowler gave the main reason in a letter to Niels Bohr, where he stated:

> I think really his method of approach . . . is better and more fundamental than Born's. Born and Jordan have to invent formal differentiations of matrices (of course) which are equally of course crucial at preserving the formal canonical equations. But they are rather artificial. I think it is a very strong point of Dirac's

[241] Of the Göttingen theoreticians it was Born who was closest to Dirac in this attitude of attempting right away to establish a dynamical theory of the greatest generality. For example, immediately after introducing matrix methods into quantum mechanics, Born proposed to extend the theory to relativistic systems (Born and Jordan, 1925b, Chapter IV).

> that the only differential coefficients you need in mechanics are really all Poisson brackets, and that the direct redefinition of the Poisson bracket is better than the invention of formal differential coefficients. (Fowler to Bohr, 22 February 1926)

The point which Fowler emphasized was, however, only one of many examples in which Dirac's methods offered advantages over the matrix methods; specific, artificial-looking assumptions that had been made in matrix mechanics followed directly from the fundamental equations of Dirac's quantum mechanics. Hence it is not surprising that Max Born also 'admired' Dirac's work.[242]

The fact that Dirac had developed his quantum-mechanical theory independently of the simultaneous work at Göttingen does not mean that he ignored matrix mechanics. It is true that he did not apply matrix calculus to obtain any of his results; indeed, he did not even mention the word 'matrix' until he had completed the formulation of his mathematical methods.[243] But he certainly studied in detail the papers of Born, Heisenberg, Jordan, Pauli and Wiener, learned from them, and quoted them in his publications. After all, the authors of these papers dealt with many questions of interest, including those which did not arise naturally from Dirac's own approach. He was thus forced to consider these questions within the context of quantum algebra. As a result, in spite of the differences of mathematical methods, the concepts and language of quantum mechanics soon became unified. Besides creating many concepts and their names by adapting their analogues in classical mechanics or geometry, Dirac borrowed others from the Göttingen theoreticians. For example, he spoke about the 'representation' of quantum variables by c-numbers, in particular about the 'matrix representation' of the aggregate of values of a multiply periodic variable x by a set of harmonic components containing the c-numbers $x(nm)$ and $\exp\{i\omega(nm)t\}$ (Dirac, 1926b, p. 288), in agreement with Born and Jordan (1925b, p. 859). However, from the beginning he was aware of the fact that matrix representation was not the only representation that occurred in quantum mechanics because aperiodic variables of q-numbers could not be represented by matrices. This is the main reason why he did not turn to the matrix methods in spring 1926 and preferred to *deduce* the c-number representation of the special q-numbers describing a given quantum system from the specific laws obeyed by them (Dirac, 1926a, see especially p. 563). Since Dirac did not fix the representation of q-numbers already in writing down the fundamental laws, his approach to problems of atomic physics appeared to have little in common with the one taken

[242] Before going to Cambridge in summer 1926 to deliver a talk on 'The Quantum Mechanics of Collisions of Atoms and Electrons' at the Kapitza Club (130th Meeting on 29 July 1926), Born wrote: 'I should like to know if I would have the opportunity of meeting Dr. Dirac, whose papers I admire very much' (Born to Kapitza, 17 June 1926).

[243] Matrices were discussed in Henry Baker's *Principles of Geometry* as particular examples of quantities obeying symbolic laws, and Dirac knew about them. It is interesting to note, however, that Dirac first mentioned the word 'matrix' only in his third paper on quantum mechanics (Dirac, 1926b, p. 288), which he submitted quite some time after he had seen the first paper of Born and Jordan on matrix mechanics. He had quoted the latter already in his earlier paper, but he referred to matrices there as aggregates of ordinary or c-numbers (Dirac, 1926a, p. 563), a name which had been used in this context also by Baker (1922, p. 67).

by Born, Heisenberg and Jordan. And yet, quantum algebra and matrix mechanics were just two different formulations of the same physical theory: i.e., quantum mechanics.

The use of q-numbers gave Dirac's theory a peculiar structure. It enabled him to write down the equations of quantum mechanics in an elegant mathematical way as equations between symbolic numbers. The solutions of these equations, which he obtained by performing simple symbolic operations, could not be compared directly with experimental data. In order to establish a relation, one had to derive c-number equations from q-number equations expressing the solutions; and this was done by substituting all q-numbers in the latter by suitable c-numbers. For example, in the case of multiply periodic systems, most variables may be represented by quantum-theoretical harmonic expansions whose terms are products of c-numbers. Although the introduction of q-numbers thus makes a further step necessary in the evaluation of the equations of quantum mechanics, which does not happen when one works with the c-number representation right away, it also provides an adequate description in those cases where the purely matrix methods fail. For example, it is possible to provide an adequate basis for the action-angle formalism in quantum mechanics. While Heisenberg, Pauli and Gregor Wentzel, in order to deal with the aperiodic angle variables w, had to add extra rules to the matrix scheme—the expansions in w do not follow genuine matrix laws and have to be treated with proper care (see Heisenberg and Jordan, 1926; Wentzel, 1926a)—Dirac encountered no problem in regarding the action-angle variables as specific examples of q-numbers. Therefore, in dealing with the problems of atomic physics, Dirac preferred to stick as long as possible to the symbolic method, and he continued to do so even after the convenient tool of differential operators became available. One of the first persons to admit that Dirac's methods were more logical and adequate than matrix methods was Heisenberg. On seeing Dirac's calculation of the hydrogen atom he wrote to him: 'Your separation of the problem into two parts—calculation with the "q-numbers", on one hand, and physical interpretation, on the other—seems to me to correspond completely to the nature of the mathematical problem' (Heisenberg to Dirac, 9 April 1926).[244]

In order to compare the results obtained by the Cambridge and the Göttingen schemes of quantum mechanics, respectively, one has to look at the entire set of papers published between fall 1925 and summer 1926: Dirac, 1925d, 1926a, b, c, e; Born and Jordan, 1925b; Born, Heisenberg and Jordan, 1926; Pauli, 1926a; Born and Wiener, 1926a, b; Wentzel, 1926a; Heisenberg and Jordan, 1926. One finds that Dirac frequently reproduced results which had been calculated earlier

[244] In the preface to the first edition of *The Principles of Quantum Mechanics*, Dirac described his preference for the symbolic method in the following words: 'The symbolic method, however, seems to go more deeply into the nature of things. It enables one to express the physical laws in a neat and concise way, and will probably be increasingly used in the future as it becomes better understood and its own special mathematics gets developed. For this reason I have chosen the symbolic method, introducing the representatives later merely as an aid to practical calculation' (Dirac, 1930, pp. vi–vii).

by Born, Heisenberg and Jordan or Pauli, though he did so from a more general point of view. The most striking example of this is provided by the theory of the hydrogen atom. Long before he submitted his paper containing the calculation of the energy states of hydrogen, Dirac had been aware that Pauli had succeeded in deriving the Balmer formula for the hydrogen spectrum with the help of matrix methods. However, he did not know any of the details of Pauli's procedure, for the latter had delayed the submission of his paper until mid-January 1926.[245] Thus, Dirac's publication of the quantum-mechanical calculation of the hydrogen atom was entirely independent, as may be easily seen by checking his mathematical and conceptual procedure against Pauli's, and his results confirmed the ones obtained by matrix methods. Still there remained a fundamental difference of approach between the two calculations. Pauli included more details that might be compared to experiment; thus he discussed not only the hydrogen atom, but also the additional effects produced in the spectrum by the action of electric and magnetic fields. On the other hand, Dirac's action-angle methods were far more general; they could be extended to many-electron atoms, where Pauli's trick of using a quantum-theoretical reformulation of the Lenz vector would not work. Again, in his paper on the 'elimination of the nodes' (Dirac, 1926b), Dirac derived several results that were contained in the previously published paper of Born, Heisenberg and Jordan (1926). For example, he rediscovered the relations between the components of the quantum angular momenta in multiply periodic systems.[246] However, he went beyond when he *proved* that no transitions occur in atoms from states, in which the angular momentum of one electron possesses a positive value, to those states where the values are negative. Born, Heisenberg and Jordan had simply excluded these transitions. Altogether Dirac arrived at a generally applicable method for obtaining the boundary values of all action variables in multiply periodic systems, a method which had not existed before.[247]

[245] Heisenberg already had written to Dirac about Pauli's work in November 1925. He had informed him: 'Pauli has also succeeded in developing a theory of the hydrogen atom and the Balmer formula on the basis of quantum mechanics' (Heisenberg to Dirac, 20 November 1925). He also promised to send Dirac a copy of Pauli's paper as soon as it was available.

Dirac's paper containing the preliminary investigation of the hydrogen atom was received on 22 January 1926 (Dirac, 1926a), while Pauli's paper arrived at *Zeitschrift für Physik* on 17 January 1926 (Pauli, 1926a). Hence Dirac was in no position to be able to read Pauli's paper before submitting his work. However, he mentioned the calculation in a footnote: 'The hydrogen atom has been treated on the new mechanics by Pauli in a paper not yet published' (Dirac, 1926a, p. 570).

[246] Dirac noted in his paper: 'The angular momentum relations of this section have been obtained independently by Born, Heisenberg and Jordan (Zeits. f. Phys., vol. 35, p. 557 (1926))' (Dirac, 1926b, p. 284, footnote).

[247] The need for such a method had already been expressed by Heisenberg in his first paper on quantum mechanics in connection with the discussion of the intensities of complex line multiplets (Heisenberg, 1925c, p. 893), but the Göttingen theoreticians had not succeeded in being able to do so within matrix mechanics. They had gone as far as formulating expressions for the transition coefficients of the electrons in an atom, say $x(nm)$, which could be proved to be zero unless the angular momenta of the electrons assumed the special values, $(\pm a_m \pm 1, a_m)$, the a_m being suitable numbers. They had then just stated without proof that the transitions from and to negative values could be excluded without loss of generality, i.e., that the numbers a_m and $a_m \pm 1$ were always positive (Born, Heisenberg and Jordan, 1926, p. 602).

In the theory of multiply periodic systems, especially in the application to realistic problems of atomic physics, the q-number methods were every bit as powerful as the matrix scheme. Dirac had actually begun to formulate the quantum-mechanical action-angle methods several weeks before his competitors on the Continent, and he applied them immediately to calculate the energy states of the hydrogen atom without relativistic corrections. Then he discussed all the formal steps necessary for dealing with atoms having one or more electrons, be they free or subject to external forces. In his paper of March 1926 on the 'elimination of the nodes' (Dirac, 1926b), Dirac had gone at least as far as Wentzel (1926a) at the same time. Dirac equalled Wentzel, if not surpassed him, with respect to the mathematical virtuosity of his methods. Why did they both not proceed to calculate the energy states of the helium atom? Especially Dirac, who was constantly thinking about solving this outstanding problem of quantum theory. He had already given all the necessary formulae to deal with the two-electron system (Dirac, 1926b, Section 6, pp. 294–298). Why did he hesitate to perform the actual quantum-mechanical calculation, which seemed to follow exactly the lines indicated by the analogous classical calculations of Hendrik Kramers (1923a) and Born and Heisenberg (1923b)?

The answer to this question is rather complex, for several facts played a role in the historical process. However, it is certain that the possibility—which had been frequently mentioned earlier in connection with the difficulties to account for the states of helium—that the two-electron system may exhibit certain aperiodic motions, did not cause Dirac any headache, for his q-number methods were designed to describe periodic and aperiodic motions equally well. In fact he proved this point in his work on the Compton effect, which he completed in late April 1926 (Dirac, 1926c). If there existed any doubts regarding the practical validity of the q-number methods, the successful calculation of the intensities of the radiation scattered by free electrons should have removed them. The reason for the delay in approaching the helium calculation has to be seen in two specific points. First, in order to formulate the problem one had to choose a particular model of the two-electron atom and define in some detail the position of the electron orbits and the nature of the interactions. Some of these features had already been discussed in the earlier semi-classical investigations, but in late 1925 a new question arose: namely, whether or not one should include the magnetic moment or spin property of the electron which had been proposed by George Uhlenbeck and Samuel Goudsmit (1925).[248] The second fact which stopped Dirac's quantum-mechanical treatment of the helium atom was a fundamental

[248] It is surprising that Dirac did not mention the electron spin hypothesis in his papers, although the work of Uhlenbeck and Goudsmit was known in Cambridge. Moreover, Llewellyn Hilleth Thomas, one of Dirac's fellow-students in Fowler's group, did important work on the spin hypothesis in Copenhagen, removing a fundamental difficulty with the help of an argument from general relativity (Thomas, 1926). Niels Bohr wrote to Ralph Fowler about the work of Thomas and expressed happiness 'that his insight and interest in general principles have allowed him to contribute to a problem which to that degree is in the foreground in the discussion about atomic problems' (Bohr to Fowler, 20 February 1926).

technical difficulty that prevented him from formulating the problem properly with the help of the available methods.[249]

This difficulty arose from investigating the possibility of representing the existence of two identical particles, say electrons, in a quantum system (see Dirac, 1926f, §3). If one denotes by the pair of indices (nm) the state of the atom in which one electron is in an orbit labelled n while the other is in the mth orbit, the question arises whether the two states, (nm) and (mn), which are physically indistinguishable because they differ only by the interchange of two electrons, have to be counted as two different states or only as one. In the first case, the two states (nm) and (mn) would give rise to two rows and columns in the matrices representing the multiply periodic variables, and one would be able to calculate the two transitions, from (nm) to $(n'm')$ and from (nm) to $(m'n')$—with $(n'm')$ and $(m'n')$ denoting two different final states of the two electrons—separately. Now, due to the fact that the two electrons cannot be distinguished from each other, the two transitions are also physically indistinguishable; hence only the sum of the intensities of the two transitions will be determined experimentally. Dirac argued that if one wishes to stick to Heisenberg's postulate that only observable quantities should be included in the theory, then one must adopt the second alternative in which (nm) and (mn) are counted as one state.

But if that be the case, another difficulty arises in the following way. One finds that $x_1(nm; n'm')$, the transition amplitude of x_1, the position coordinate of the first electron, must be equal to $x_2(mn; m'n')$, the amplitude of x_2, the coordinate of the second electron, where the index pairs (nm) and $(n'm')$ assume all possible numerical values: for if the two sets of indices, (nm) and (mn), define just one single row and column of the matrices, and the two sets, $(n'm')$ and $(m'n')$, another single row and column, it follows that the coordinates of the two electrons, x_1 and x_2, are equal. As this result contradicted the quantum condition for dynamical variables, Dirac concluded that it must be wrong. A way out of the difficulty exists only if all quantities, which can be calculated from the theory, are symmetric functions of the position and momentum coordinates of the two electrons. Only then would the sum of x_1 and x_2 be represented by a matrix and the inconsistent equation, $x_1 = x_2$, can be avoided. However, since one has to consider unsymmetrical functions as well, the matrix and q-number methods will not work for many-electron atoms. Moreover, as Dirac showed, 'there is no set of uniformizing variables for a system containing more than one electron, so that the theory cannot progress very far on these lines' (Dirac, 1926f, p. 661).[250]

[249] Dirac first mentioned the difficulty in his paper 'On the Theory of Quantum Mechanics,' which was received by the *Proceedings of the Royal Society of London* on 26 August 1926 (Dirac, 1926f). Thus, he encountered the difficulty between April and August 1926.

[250] To prove this point Dirac considered three transitions, (nm) to $(n'm')$, $(n'm')$ to $(n''m'')$, and (nm) to $(n''m'')$, and tried to describe them with the help of the action-angle formalism (Dirac, 1926f, p. 668). He assumed that the first two transitions corresponded to the classical terms, $\exp\{i(\alpha w)\}$, and the last to a term $\exp\{2i(\alpha w)\}$, and restricted himself to the case of one quantum number per

Thus, the dream of calculating the energy states of helium with the tools of quantum algebra did not become a reality. However, at the very moment when Dirac met this fundamental difficulty new methods for overcoming it, those of Schrödinger's wave mechanics, had become available. His next task became a study of Schrödinger's theory and to discover in it a different formulation of quantum mechanics. The technical difficulty with the two-electron system and the fact that wave mechanical methods were able to fill the gap which quantum algebra had left persuaded Dirac to go on with Schrödinger's theory. This work of his would open the door to a general, unified theory of quantum mechanics.

There was another problem of atomic theory, which seemed to fall conveniently within the range of quantum algebra, and one which Dirac did not attack immediately: this was the problem of calculating the absolute intensities of the spectral lines. When he worked on the theory of the hydrogen atom in winter 1925–1926, he did not want to lose time and preferred to publish his results fast, for he knew about the competition from Pauli. Still, in spring 1926 neither Pauli nor anyone else had succeeded with the calculation of the intensities of the spectral lines of hydrogen, and Heisenberg wrote to Dirac: 'With your treatment of the hydrogen problem, it seems to me that there is only a tiny step towards the calculation of the transition probabilities, a step, which you have certainly taken meanwhile' (Heisenberg to Dirac, 9 April 1926). However, Dirac had not gone ahead with that problem; instead he devoted himself to calculating with q-number methods the intensity of the Compton radiation scattered from free electrons, which he claimed to be the only result 'obtained from the new mechanics that had not been previously known' (Dirac, 1926c, p. 421).[251] Thus, it remained for Erwin Schrödinger to calculate the intensities of the hydrogen lines for the first time in a paper submitted in the middle of May (Schrödinger, 1926f), about two weeks after Dirac had completed the paper on the Compton effect. Dirac had thus lost the opportunity of using q-numbers to derive another, previously unknown result.

---

electron only. One should then have

$$n'' - n' = n' - n \qquad \text{and} \qquad m'' - m' = m' - m.$$

Since the state $(n'm')$ may also be labelled $(m'n')$, one obtains the relations

$$n'' - m' = m' - n, \qquad m'' - n' = n' - m,$$

which would lead in general to a different state $(n''m'')$ than the above equations. Hence there is no unique state $n''m''$, and the theory of uniformizing variables does not work in this case.

[251] By 'previously known' results Dirac meant those results which had been obtained in 1924 and 1925 with the help of the correspondence principle; for example, the formulae describing the intensities of the components of a line multiplet. Ironically, even the intensity formula for the Compton effect had been derived already earlier by Gregory Breit, and Dirac added a footnote to the proofs of his paper (May 1926): 'This result for unpolarized incident radiation has recently been obtained independently by Breit from correspondence principle arguments' (Dirac, 1926c, p. 421). Breit had presented his derivation first at the meeting of the American Physical Society, Kansas City, in December 1925; he then submitted his paper to *Physical Review* on 26 January 1926, where it was published in the issue of April 1926 (Breit, 1926).

In view of the fact that quantum algebra was a more general formulation of quantum mechanics than was matrix mechanics—in that q-numbers would allow for a matrix representation as well as for a representation in terms of differential operators—one might ask why Dirac himself did not come closer to inventing the scheme of wave mechanics. This question appears to be reasonable, for Dirac shared many scientific interests and preferences with the author of wave mechanics. For example, like Schrödinger, Dirac loved to pursue relativistic considerations, and they had both worked on problems of statistical mechanics. Dirac also knew the works of Louis de Broglie and Einstein, from which Schrödinger started out in 1925. But in developing their respective schemes of quantum mechanics, they took very different paths. Dirac, unlike Schrödinger, pursued a mechanical approach to describe atomic systems, in which the wave properties of matter seemed to have no place at all. This is particularly evident from his treatment of the Compton effect: the problem on which he concentrated there was to prove that the new theory explained the corpuscular properties of radiation, and—like Max Born and Norbert Wiener, in their work on operator mechanics—he did not spend any thought on deriving a wave-like behaviour of, say, electrons.

There was still another fact which prevented Dirac from getting on to the path of the wave equation. He always preferred to work with abstract symbolic quantities, such as q-numbers, to describe the properties of physical systems. The introduction of the differential and the wave functions would have meant the selection of a special representation of the q-numbers, and he saw no real reasons for such a choice.[252] Wave mechanics, therefore, came to Dirac as a surprise, as it did to the other pioneers of quantum mechanics, but he had no difficulty in absorbing Schrödinger's methods into his own theory. Instead of becoming too deeply involved in the questions of whether the wave properties of matter were real or not and whether the wave concepts of Schrödinger explained quantum phenomena or not, he considered the differential operators and wave functions from a mathematical point of view: namely, as providing in certain cases of dynamical systems a particularly suitable representation of q-number variables. The principal consideration for Dirac was to use the wave mechanical theory to calculate previously unknown results in atomic physics.

For the moment—that is, in spring 1926—after Dirac had completed his first series of papers in quantum mechanics, the new developments were still to come. However, he had already made a name for himself as a quantum physicist of the first rank. Ralph Fowler inquired about the possibility whether his former research student might be able to spend a few months in Copenhagen, and Niels Bohr replied: 'He shall certainly be welcomed to work here next term. We are all

[252] Even after seeing Schrödinger's first paper on wave mechanics Dirac raised objections to the assumption concerning the wave properties of matter. This may be inferred from a letter which Heisenberg wrote to him in May 1926: 'I quite agree with your criticism of Schrödinger's paper with regard to a wave-theory of matter. This theory must be inconsistent, just like the wave-theory of light' (Heisenberg to Dirac, 26 May 1926).

full of admiration for his work. Dirac's coming will also be a great pleasure to Heisenberg, who will arrive here on the first of May in order to overtake [*sic*] Kramers' post. Already last autumn when I spoke with him about his taking up this post he suggested a visit of Dirac as most desirable' (Bohr to Fowler, 14 April 1926). Kramers' departure from Copenhagen to take up a professorship in Utrecht, and the coming of Heisenberg and Dirac to Copenhagen, marked the beginning of a new era of quantum theory at its mecca, where both of them would contribute their share to complete the structure of quantum mechanics.

Part 2

The Reception of the New Quantum Mechanics 1925–1926

Introduction

The theory of quantum mechanics was formulated in the papers of Born and Jordan (1925b), Born, Heisenberg and Jordan (1926), and Dirac (1925d, 1926a). The application and extension of the theory, especially after the inclusion of the hypothesis of electron spin, yielded results like the calculation of the energy states of hydrogen and numerous other problems dealing with atomic systems. The new theory seemed to be capable of replacing the earlier qualitative arguments, which had been based upon incorporating the quantum of action into classical mechanics guided by the correspondence principle. The physicists responded to quantum mechanics immediately, and the hope arose as early as the end of 1925 that the new theory would work in detail in all problems of atomic physics. One heard about the new results from the authors themselves or from people, such as Wolfgang Pauli, who had been close to the development of the new theory and had, to some extent, participated in it. Of course, every interested physicist read the papers of Heisenberg, Born, Jordan, and Dirac as soon as they appeared in print. The reactions of this community of physicists, scattered in many different places, came without delay and they were primarily positive. Moreover, the authors of quantum mechanics sought to arouse a certain interest in their new scheme by delivering lectures on it before various audiences: Heisenberg spoke to the mathematicians in Göttingen; Dirac addressed mainly the theoreticians in Cambridge who worked with Fowler; Born gave lectures to physicists at several universities in the United States. Born's lectures created much interest in the new theory, and quite a few American physicists began to work with quantum mechanics. The unusual mathematical methods of the new theory, and the technical complications arising in the actual computations, hardly deterred certain people. Naturally, critical opinions were also expressed about the new theory; some thought that the scheme represented just another of the games which the theoretical followers of Bohr played with the atomic data, a kind of a crazy apotheosis of the earlier correspondence rules. A formal criticism was also expressed by some concerned theoreticians: matrix mechanics was too ugly, too complicated and too arbitrary to be a proper representation of atomic reality. Another, different theoretical scheme, that is, wave mechanics, which was coming up, seemed to offer a better approach for the description of atomic phenomena. However, from all these conflicting opinions one thing became evident: with the work of Heisenberg, Born, and Jordan and Dirac, a definite breakthrough had been achieved in atomic theory, a breakthrough which could be compared to the one that was brought about by special relativity theory in 1905.

Chapter I

A Welcome to the New Theory: Göttingen and Copenhagen

Since quantum mechanics originated in *Göttingen* on the basis of the correspondence principle, which had been advocated in *Copenhagen* over many years, it was natural that the first response would come from these two places. People there were particularly familiar with the specific difficulties of the Bohr–Sommerfeld atomic theory, its conceptual and mathematical inconsistency and its practical insufficiency. In both places serious attempts had been made to develop a better and more powerful theory. The physicists in Copenhagen and Göttingen did not always follow the same approaches. Niels Bohr and his close collaborators in Copenhagen concentrated their efforts on the detailed analysis of physical phenomena in order to identify the most fundamental physical concepts; in Göttingen, where David Hilbert's mathematical spirit also dominated the theoretical physicists, the emphasis was on the attempt to develop an alternate mathematical scheme that would replace classical mechanics in the atomic domain. Such a scheme, even before 1924, had been called 'quantum mechanics.'[1] From the very day when they became convinced of the breakdown of the semiclassical Bohr–Sommerfeld theory, the Göttingen theoreticians had looked for a complete replacement of the equations of classical mechanics by a scheme that would incorporate the fundamental postulates of quantum theory from the very beginning. In connection with this search for a new atomic mechanics Born (1924b) and Heisenberg (1924e) had asserted that the goal was to discover a new formal system of mathematically consistent rules. In Copenhagen, where the physics of dispersion phenomena was being investigated at the time, one followed with sympathy the attempts that were being made at Göttingen. Bohr, in particular, felt: 'Let these Göttingen people work in that

[1] Max Born once claimed that in his paper (Born, 1924b) he introduced 'for the first time I think' ('*wohl zum ersten Male*') the expression 'quantum mechanics' ('*Quantenmechanik*') (Born, Nobel Lecture, 'The Statistical Interpretation of Quantum Mechanics', Stockholm, 11 December 1954 (Born, 1955b; English translation, p. 258)). However, the almost identical expression 'mechanics of quanta' had been used earlier by H. A. Lorentz, who, in his address on 'The Old and New Mechanics,' delivered at the Sorbonne on 10 December 1923, declared: 'All this has great beauty and extreme importance, but unfortunately we do not really understand it. We do not understand Planck's hypothesis concerning oscillators nor the exclusion of nonstationary orbits, and we do not understand, how, after all, according to Bohr's theory, light is produced. There can be no doubt, the *mechanics of quanta*, a mechanics of discontinuities, has still to be made' (Lorentz, 1925b, quoted in Jammer, 1966, p. 155). Even earlier, Karl Herzfeld used the term 'quantum mechanics' in the announcement of his lectures at Munich: '*Quantenmechanik der Atommodelle*' ('Quantum Mechanics of Atomic Models,' *Phys. Zs.* **22**, 1921, p. 591).

direction [of a final, mathematically consistent scheme] and let us see what comes out' (Heisenberg, Conversations, p. 330). But, although they knew that a consistent quantum theory was necessary, Bohr and his collaborators in Copenhagen had refrained from seeking to formulate immediately a complete theoretical scheme. The differences in attitudes and expectations explain the differences in the receptions given to quantum mechanics in the scientific circles of Göttingen and Copenhagen.

Heisenberg's work on the quantum-theoretical reformulation of kinematics and mechanics (Heisenberg, 1925c) attracted a great deal of attention among the physicists and mathematicians in Göttingen. Ralph Kronig, who visited Göttingen in July 1925, reported about the activities of the theoreticians: 'Every day they had a new result, that was very exciting' (Kronig, AHQP Interview, p. 9). While Born and Jordan, later on joined by Heisenberg, worked to formulate matrix mechanics, the mathematicians followed their efforts with interest and understanding. After all, they had long been familiar with the details of the methods that the physicists needed. For example, Richard Courant, the co-author of *Methoden der mathematischen Physik* (Courant and Hilbert, 1924)—the first chapter of which dealt with the results on quadratic forms—told Heisenberg 'that the mathematicians had known that there must be some Hermitian form which had a continuous and discrete spectrum at the same time; some part being discrete the other continuous' (Heisenberg, Conversations, p. 319). Such a Hermitian form would represent the quantum-theoretical Hamiltonian of a hydrogen atom. However, although Hilbert, who had developed the theory of infinite quadratic forms about twenty years earlier, had searched for an explicit example of a form having both discrete and continuous spectra, it was not found until Schrödinger proposed the differential operator in his wave equation for the hydrogen atom (Schrödinger, 1926c). Thus, the Göttingen mathematicians, in spite of the fact that they were greatly interested in the subject, did not actually help in solving the practical problems of quantum theory and left their solution entirely in the hands of the physicists.[2] Hilbert did encourage Born's collabora-

[2]The only mathematician who rendered some help to the physicists was Hermann Weyl in Zurich. After he learned about matrix mechanics from Born in September 1925, he performed a few calculations in the new theory, especially those connected with canonical transformations (see letters: Weyl to Born, 27 September 1925; Weyl to Jordan, 13, 23 and 25 November 1925).

However, although he had worked for his doctoral thesis on the continuous spectra of quadratic forms connected with differential equations (Weyl, 1908), Weyl did not think of replacing the matrix calculations by the problem of solving the equivalent differential equations. As Edward U. Condon recalled later, the physicists received a little hint from Hilbert:

> . . . when [Born and Heisenberg and the Göttingen theoretical physicists] first discovered matrix mechanics they were having, of course, the same kind of trouble that everybody else had in trying to solve problems and to manipulate and to really do things with matrices. So they had gone to Hilbert for help and Hilbert said the only times he had ever had anything to do with matrices was when they came up as a sort of by-product of the eigenvalues of the boundary value problem of a differential equation. So if you look for the differential equation which has these matrices you can probably do more with that. They had thought it was a goofy idea and that Hilbert didn't know what he was talking about. So he was having a lot of fun pointing out to them that they could have discovered Schrödinger's wave mechanics six months earlier if they had paid a little attention to him. (Quoted in Reid, 1970, p. 182)

tors: for example, he invited Heisenberg to deliver several lectures to the mathematicians early in the winter semester of 1925–1926; these lectures, in turn, stimulated Hilbert and his circle to deal with the problems of the mathematical foundations of quantum mechanics.

Apart from the encouragement which Hilbert and the other Göttingen mathematicians provided, they helped the physicists, who struggled with the formulation of quantum mechanics, in determining whether the entire approach was consistent. The particular difficulty which bothered Heisenberg at the very beginning of his work was the fact that the conjugate quantum-mechanical variables did not commute. As this was a completely unusual situation in physics, he did not know whether it led to mathematical contradictions.[3] The mathematicians, on the other hand, had already encountered theories which, like the matrices, contained noncommuting quantities. Moreover, in Göttingen one paid particular attention to the question of the consistency of mathematical schemes, for David Hilbert had turned his attention again in the early 1920s to the axiomatic foundations of mathematics. At that time he tried to cope with the criticism raised by Luitzen Egbertus Jan Brouwer and Hermann Weyl against the methods he had used earlier in arithmetic. The theoretical physicists in Göttingen, among them even those who were not particularly interested in pure mathematics, were acquainted fairly well with the debates between the mathematicians. For example, Heisenberg had attended some of Hilbert's lectures on this subject, and he recalled: 'It was quite obvious in Göttingen that the theoretical physicists also [should attend] the discussions about axiomatics in mathematics. So, when there were discussions, for instance between Brouwer from Holland and Hilbert, everybody of course went because one knew that was something very exciting and very deep' (Heisenberg, Conversations, p. 159). Now, in late 1925 these mathematical discussions contributed importantly to the development of physics, for, as Heisenberg said:

> I remember that in being together with young mathematicians and listening to Hilbert's lectures, I heard about the difficulties of mathematics. There it came up for the first time that one could have axioms for a logic that was different from

---

Although we are certain that Condon did exaggerate—e.g., he claimed elsewhere that Heisenberg had learned things from Emmy Noether, which can be definitely disproved—, Hilbert probably did advise the physicists (see the following Chapter III). On the other hand, Courant, the most active mathematician at the time in Göttingen, did not prove to be very helpful, for as Born remarked: 'I think even at that time Courant didn't care very much' (Born, Conversations, p. 36).

[3] At first, as we have discussed, Heisenberg tried to avoid dealing with noncommuting quantities. Then, through the work of Born and Jordan and Dirac, it became certain that noncommutativity was the essential feature of quantum mechanics. How strange the noncommuting quantities appeared to physicists can be gathered from the following. When, in July 1925, Born mentioned to Pauli, who was well informed about Heisenberg's work, the problem of calculating the difference of the products of pq and qp, with p and q representing the conjugate momentum and position variables, respectively, Pauli answered that this difference 'had to be zero, of course' (Jordan, AHQP Interview, Second Session, p. 20).

> classical logic and still was consistent. That, I think, was just *the* essential step [in physics]. You could think just in this abstract manner of mathematicians, and you could think about a scheme which was different from the other logic, and still you could be convinced that you would always get consistent results. (Heisenberg, Conversations, pp. 328–329)

For many people, especially the physicists, it was an entirely new situation to have the example of a new mechanics, differing from classical mechanics in its fundamental features, which nevertheless represented a consistent scheme. Of course, in physics the situation was not entirely new, for it had already occurred twenty years earlier in connection with the dynamics of fast-moving particles. One had learned then that things happened in a way which contradicted the Newtonian laws of classical mechanics, and the explanation of these phenomena was possible only in terms of altered laws, the laws of special relativity; hence previously paradoxical situations had ultimately been described by mathematically consistent methods. In atomic physics of the early 1920s many phenomena had been discovered—one may just remind oneself of the effects observed by Stern and Gerlach and by Ramsauer—for which there was no explanation in terms of the usual dynamics, whether Newtonian or relativistic. It was in Göttingen that one was most prepared psychologically to think of a new dynamics, a quantum mechanics, on the basis of which these paradoxical effects would be explained naturally. Heisenberg, in particular, was aware of the situation, which he summarized as follows:

> It was a consequence of this whole development—the theory of relativity, the axiomatization of mathematics, etc.—which gave the possibility of distinguishing between paradox and inconsistency. At that time it was rather new and not obvious to everybody. It just came out of all these discussions. I could not say that there was a definite moment at which I realized that one needed a consistent scheme which, however, might be different from the axiomatics of Newtonian physics. It was not as simple as that. Only gradually did the idea develop in the minds of many physicists that we can scarcely describe nature without having something consistent, *but we may be forced to describe nature by means of an axiomatic system which was thoroughly different from the old classical physics and even using a logical system which was different from the old one*. (Heisenberg, Conversations, p. 329)

The paradox concerning the noncommutativity of the products of conjugate quantum dynamical variables presented no conceptual difficulty to minds like Hilbert's. In his work on the foundations of geometry, Hilbert had already referred to axiomatic systems that described mathematical objects which did not commute (Hilbert, 1899; see, especially, Chapter Six, dealing with non-Archimedean systems of numbers and geometries). Since there existed (completely consistent) geometries in which these axioms were satisfied, it appeared to be perfectly reasonable for physicists to accept a mechanical theory in which the variables did not commute. That observation determined, as we have already discussed, the work of Dirac, who followed the analogy to geometry in detail. In Göttingen, on the other hand, the axiomatic results of the mathematicians provided a kind of logical and psychological support for the physical theory under investigation; the backing of the mathematicians helped the theoretical physicists and made them more confident to proceed with quantum mechanics in spite of the unconventional methods involved. They discovered that the physical paradoxes presented by many quantum phenomena were matched exactly by the paradoxes that arose in the new theory when compared to classical mechanics.

While the Göttingen mathematicians received quantum mechanics favourably, the experimental physicists were not so happy. 'Göttingen is divided into two camps,' wrote Heisenberg to Pauli, 'those who, like Hilbert (or also Weyl, in a letter to Jordan), talk about the great success which has been scored by the introduction of matrix calculus into physics; the others, like Franck, who say that one will never be able to understand matrices' (Heisenberg to Pauli, 16 November 1925). The experimentalists, some of whom were personally most favourably disposed towards the Göttingen theoreticians, in the beginning did not see much advantage in the new theory over the correspondence approach of the previous years. Only under the impression of the increasing successes of quantum mechanics would they make peace with the new scheme. Even then, outside the mathematical circle of Göttingen, the feeling remained that matrix mechanics was a typical product of Max Born's formal, theoretical mind. This feeling changed only after wave mechanics confirmed the correctness of the theory.

The situation in Copenhagen was different, for there did not exist any division of the physicists in groups favouring either the more mathematical or the more intuitively physical approach to quantum theory. Though Niels Bohr and his collaborators did not aim so strongly at an immediate consistent mathematical formulation as did Born's group, they welcomed quantum mechanics with affirmative appreciation. Heisenberg's first paper on the theory made a great impression, for it immediately raised hopes that the well-known quantum-theoretical difficulties and paradoxes would soon be understood. Heisenberg, who visited Copenhagen in September 1925, recalled that Hendrik Kramers was most interested in the details of the new theory, while the general feeling was the following: 'A very important new development is going on now, and that is probably leading us towards the solution. It is not yet *the* solution, because its interpretation is still unclear, but it is going towards it' (Heisenberg, Conversations, p. 289). This feeling was first expressed publicly in Bohr's article on 'Atomic Theory and Mechanics,' the extended printed version of his address delivered on 30 August 1925 before the sixth Scandinavian Mathematical Congress in Copenhagen (Bohr, 1925b).[4] Bohr had appended a section entitled 'An Attempt at a Rational Quantum Mechanics,' in which he pointed out that: 'Quite recently Heisenberg . . . has taken a step probably of fundamental importance by

[4]Bohr remarked in a footnote: 'This article represents substantially the contents of an address delivered, on August 30, before the sixth Scandinavian Mathematical Congress in Copenhagen. It should be mentioned, however, that the present elaboration of the text has been notably influenced by the appearance, since the delivery of the address, of an important paper of Heisenberg, to which reference is made below' (Bohr, 1925b, p. 845).

formulating the problems of quantum theory in a novel way by which the difficulties attached to the use of mechanical pictures may, it is hoped, be avoided.' Moreover, this procedure had already led to 'a self-contained theory sufficiently analogous to classical mechanics' (Bohr, 1925b, p. 852). As an important result of the new procedure, Bohr cited Pauli's as yet unpublished calculation of the hydrogen atom. These developments, together with the hypothesis of electron spin, made Bohr very happy. On 28 April 1925 Bohr had written to Rutherford about his worries concerning 'the outstanding difficulties [of quantum theory] which offer such uncomforting indications of essential difficulties in our usual description of natural phenomena.' Nine months later, he wrote to Rutherford again: 'In fact, due to the last work of Heisenberg prospects have with a stroke been realized, which although only vague[ly] grasped have for a long time been the centre of our wishes' (Bohr to Rutherford, 27 January 1926). Not only did Bohr accept the various successes which had been achieved by Heisenberg, Born and Jordan, Pauli, and Dirac, he also sensed a way of proceeding with his own programme of describing properly the paradoxical quantum phenomena, such as the Compton effect or the Stern–Gerlach effect.

Thus, Bohr, though he did not participate personally in performing calculations with matrices, began to work with the new scheme. He left, as he had done for many years, the derivation of detailed results in atomic physics to others and turned to 'the fundamental difficulties involved in the construction of pictures of the interaction between atoms either by means of radiation or by collisions' (Bohr, 1925b, p. 852). With the mathematical description of quantum phenomena seemingly in hand, Bohr wanted to find the correct physical picture, the new 'mechanical model in space and time.'

In Göttingen and Copenhagen quantum mechanics was received with enthusiasm primarily for two reasons. First, in both places the difficulties of the Bohr–Sommerfeld atomic theory were well known and the approach based on the correspondence principle was followed to overcome them. Second, there was no fear of employing unusual mathematical tools; in Göttingen there even existed some familiarity with them.[5] In other places, even where Bohr's views were followed with sympathy and understanding, one encountered greater difficulties with the new theory. When Heisenberg visited Leyden in July 1925 and mentioned his new results in quantum mechanics, Ehrenfest remarked that 'he was like Sir Isaac Newton; not only had he invented a new mechanics, but he also had to invent a new mathematics to go with it' (Goudsmit, AHQP Interview, Second Session, p. 26). However, since everything which Heisenberg did at that time was taken very seriously, the Leyden physicists accepted his theory as well as the mathematical methods connected with it. A little later, Ehrenfest, who

[5]The matrices were completely new to the physicists then in Copenhagen. As David M. Dennison, an American visitor, recalled: 'I did not have the mathematical background to have ever heard of matrices . . . The notion that multiplication might depend on the order was one which I really just didn't have at all' (Dennison, AHQP Interview, Second Session, p. 21). However, in spite of this he went on immediately into matrix calculations in order to obtain the energy states of the rotating molecules.

knew about matrices, even began to work with them and had his student George Uhlenbeck learn them, too.[6] Uhlenbeck did not find matrix mechanics particularly difficult because the mathematics seemed to him to be rather elementary. He recalled: 'We studied it, but it was so clear that you couldn't make [up] any problems with it, even for yourself. Everything became these infinite number of equations that you had then to solve, and so nobody knew exactly how to do it' (Uhlenbeck, AHQP Interview, Second Session, p. 15). Still, it was clear enough to Ehrenfest and his collaborators that quantum mechanics represented a real advantage over the earlier quantum theory, although they started to contribute actively to the new scheme only after the wave mechanical methods became available.

Some physicists were even more puzzled by the mathematical tools of quantum mechanics. For example, Max Planck, who greeted the work of the Göttingen physicists as a positive step towards a final quantum theory, remarked to Alfred Landé: 'I wonder what you will make out of this matrix calculus' (Landé, AHQP Interview, Second Session, p. 3).[7] At that time Landé was occupied in preparing the second edition of his book *Die neuere Entwicklung der Quantentheorie* (Landé, 1926d), in which, as Fritz Reiche, another Berlin physicist and former student of Planck's, noted in a review that he (Landé) had also included 'a brief perspective of the most recent and most significant development of quantum theory: the new quantum mechanics of Heisenberg, Born and Jordan' (Reiche, 1926, p. 794). The wider acceptance of the theory thus became coupled with the question of whether one grasped the mathematical tools or not. On the one hand, the mathematician Constantin Carathéodory, whom Reiche showed Heisenberg's first paper, came to the conclusion that it was 'terribly interesting and so clear' (Landé, AHQP Interview, Second Session, p. 3); on the other, the opinion of a number of physicists was expressed most directly by the experimentalist Bidhu Bhusan Ray, who had previously spent a year and a half in Copenhagen and now wrote to Niels Bohr from Berlin on 6 March 1926: 'I personally shall not be sorry if the whole outburst of Born and Jordan is completely withdrawn, and I fear if atomic physics has to progress on the line of Born and Jordan, you will find very few people left in the atomic physics circle.' However, in spite of Ray's pessimism, matrix mechanics flourished in 1926, for many people were convinced that it represented a more definite approach to atomic theory than had existed before.

[6] Ehrenfest did not particularly like matrices. H. B. G. Casimir told me the following story. Once, long after the discovery of wave mechanics, Born was in Leyden. During a discussion Born mentioned how a certain physical argument could be expressed in the language of matrices. Ehrenfest insisted that it was clearer in wave mechanics. Born said: 'It's a question of habit' ('*Es ist eine Gewohnheitssache*'). Ehrenfest: 'But there are also bad habits' ('*Aber es gibt auch schlechte Gewohnheiten.*') (J. Mehra)

[7] In an address on recent research in physics on 14 February 1926, delivered before the *Akademische Kurse*, Düsseldorf, Planck said about the quantum theory: 'A promising start in this direction [i.e., to formulate the correct theory] is represented by the establishment of the so-called quantum mechanics, which—in the hands of the Göttingen physicists Heisenberg, Born and Jordan—has already yielded beautiful results' (Planck, 1926, p. 260).

Chapter II
Propagation of Quantum Mechanics in Europe

The new ideas in quantum theory became known among the interested physicists with little delay. The initial fundamental papers were published very rapidly after their submission in summer 1925 and winter 1925–1926: Heisenberg's first paper, submitted at the end of July to *Zeitschrift für Physik*, appeared in September; the article of Born and Jordan, in which the matrix formulation was introduced, was submitted at the end of September to *Zeitschrift für Physik* and published in November; Paul Dirac's first paper on the fundamental equations of quantum mechanics, submitted early in November to the *Proceedings of the Royal Society of London*, appeared in December; the memoir of Born, Heisenberg and Jordan, submitted mid-November 1925 to *Zeitschrift für Physik*, was published in February 1926; finally, Born and Wiener's paper on operator mechanics, submitted early in January to *Zeitschrift für Physik*, was published in mid-March.

Apart from the unusually fast publication of the work on quantum mechanics, the main ideas and results were available to many friends and colleagues of the authors even before publication. An exceptional example of this situation arose because of the special relationship of Wolfgang Pauli and Werner Heisenberg. In many letters, written during the period when matrix mechanics was being formulated, Heisenberg informed Pauli about the entire development. Because of this knowledge Pauli was able to enter actively into work on the new theory already in October 1925 with a sophisticated calculation of the energy states of the hydrogen atom. This work, although Pauli delayed its submission to *Zeitschrift für Physik* until January 1926 because he wanted to include the relativistic corrections, constituted the very first response to quantum mechanics, not taking into account Paul Dirac's still earlier reaction to Heisenberg's original ideas, which also came about through information obtained prior to publication. Like Pauli in Hamburg, Hendrik Kramers in Copenhagen also belonged to the privileged circle of people who had access to matrix mechanics before the theory appeared in print. Kramers not only had the opportunity of hearing about the developments from Heisenberg, who stayed in Copenhagen during September and part of October 1925, he also visited Pauli in Hamburg at the end of October and early November when Pauli had just completed his calculation of the hydrogen atom in matrix mechanics. As a result, Kramers' immediate concern with matrix mechanics was directly stimulated by his contacts with Heisenberg and Pauli; his paper dealing with the physical interpretation of the commutation rules, which appeared in the December issue of *Physica*, gives full evidence of that fact (Kramers, 1925c).[8] Similarly, the work of Lucie Mensing on the energy states of the diatomic molecule in matrix mechanics (Mensing, 1926a) originated from her conversations with Pascual Jordan in Göttingen, although she submitted her paper as late as the end of March 1926, after the papers of Born and Jordan (1925b) and Born, Heisenberg and Jordan (1926) had already appeared in print. Finally, one must also consider Gregor Wentzel's treatment of action-angle variables in matrix theory (Wentzel, 1926a) as a consequence of his stay in Hamburg and his discussions there with Pauli.

Although the main part of the early work on matrix mechanics was carried out by those who invented the theory or others whom they stimulated directly, in some cases other physicists, who had less personal contact with the Göttingen group, picked up the subject after seeing the first published papers. Cornelius (Kornel) Lanczos of Frankfurt University was the fastest to respond. His professor, Erwin Madelung, had been together with Max Born in Göttingen a long time ago, but in the mid-1920s he did not have much of an exchange with his predecessor in the physics chair at Frankfurt. Lanczos used to follow the new literature on quantum theory carefully on his own. When he came across the papers of Heisenberg (1925c) and Born and Jordan (1925b)—the latter became available in later November 1925—he discovered the connection of quantum mechanics with field theory, which he had investigated previously in his own work on some aspects of radiation phenomena. The paper on the field-like representation of quantum-mechanical relations, in which Lanczos worked out his connection, was received by *Zeitschrift für Physik* on 22 December 1925 (Lanczos, 1926a). It appeared in the last issue of Volume 35, the same volume which contained the memoir of Born, Heisenberg and Jordan.

In Germany, Austria and Switzerland the papers on matrix mechanics in *Zeitschrift für Physik* were widely read and in many places they received a positive response. This was, of course, the case with Sommerfeld in Munich, who followed the development with interest and sympathy; he did not participate in the work on matrix mechanics, but continued his own programme of analyzing the spectroscopic data.[9] However, contributions to matrix mechanics were made from two other places. One of the authors was Fritz London who, after learning atomic theory in Göttingen and Munich, had gone to Stuttgart to work with Paul Ewald, a former student of Sommerfeld's. London succeeded in reformulating the classical methods of Jacobi's transformation theory of periodic systems in

[8] Also later in 1926, Kramers contributed to quantum mechanics. For example, he assisted Dennison in his work in Copenhagen on the quantum-mechanical rotator and, in Utrecht, he calculated the intensities of Zeeman components with the help of matrix methods. (See W. C. van Geel, 1926, especially p. 878.)

[9] See, for example, the paper of Sommerfeld with his student Albrecht Unsöld on the hydrogen spectrum, which the authors concluded with the remarks: 'The situation, therefore, is the following: the relativistic formula remains valid for hydrogen, X-ray and visible spectra, but its model-dependent foundation seems to contradict the empirically-established quantum theory. We may expect the resolution of this contradiction from the new quantum mechanics, which is in the process of being created, and for which the quantum [– number] organization given above may perhaps serve as a guideline' (Sommerfeld and Unsöld, 1926, p. 275).

quantum mechanics. His paper was submitted in May 1926 to *Zeitschrift für Physik* and appeared in print later that year (London, 1926b).[9a] Otto Halpern of the University of Vienna, on the other hand, worked on the quantization of the rotator in quantum mechanics; his note was received by *Zeitschrift für Physik* on 5 June 1926 (Halpern, 1926b).[10] Halpern, who obtained his doctorate in 1922 from the University of Vienna, had already made contributions to questions of atomic structure and radiation theory.[11] Thus, he was well-acquainted with the Bohr–Sommerfeld semiclassical quantum theory and the correspondence treatment of atomic systems before he saw the papers on the new Göttingen quantum mechanics. He immediately recognized the fact that the problem of calculating the energy states of the rotator had not yet been solved by Heisenberg, Born and Jordan because of the difficulty of considering nonlibration coordinates, e.g., angle variables, as matrices. To avoid that difficulty Halpern proposed to proceed as follows. One expresses the classical Hamiltonian of the rotator in the plane, H $(=(1/2A)p_\phi^2$, where A is the constant moment of inertia and p_ϕ the angular momentum), in terms of the Cartesian coordinates q and p, which are obtained from the pair of action-angle variables p_ϕ and ϕ by the Poincaré transformation,

$$q = \sqrt{2p_\phi}\ \cos\phi, \qquad p = -\sqrt{2p_\phi}\ \sin\phi. \tag{1}$$

Halpern reformulated the transformed classical Hamiltonian, $H(p, q)$, that is,

$$H = (8A)^{-1}(p^2 + q^2)^2, \tag{2}$$

as the Hamiltonian matrix, $\mathsf{H}(\mathsf{p}, \mathsf{q})$, by substituting the matrices p and q for the libration coordinates p and q. He thus obtained the result

$$\mathsf{H} = (8A)^{-1}(\mathsf{p}^4 + \mathsf{p}^2\mathsf{q}^2 + \mathsf{q}^2\mathsf{p}^2 + \mathsf{q}^4), \tag{3}$$

by evaluating the square of the factor $(\mathsf{p}^2 + \mathsf{q}^2)$ and observing the commutation relation for p and q. Halpern then calculated the second time derivative of q, and identified its matrix element denoted by the indices m and n with the product of a factor (which was $4\pi^2$ times the same matrix element of q) and the square of the transition frequency. This yielded the equation for the energy

[9a] Already earlier, London had dealt with a problem of matrix mechanics in a note, entitled '*Energiesatz und Rydbergprinzip in der Quantenmechanik*' ('Energy Principle and Rydberg's Law in Quantum Mechanics'), in which he proved energy conservation independently of the assumption of the validity of the combination principle in atomic theory (London, 1926a).

[10] Halpern announced the paper earlier in a letter of March 1926 to *Naturwissenschaften* (Halpern, 1926a).

[11] Otto Halpern was born in Vienna, Austria, on 25 April 1899. In 1930 he went to the United States and became a professor of physics at New York University (1931–1941), Guest Professor at Columbia University (1947) and Professor of Physics at the University of Southern California (1947–1952).

states W_m,

$$4\pi^2(W_m - W_n)^2 = \frac{1}{A}(W_m + W_n) - \frac{h^2}{16\pi^2 A^2}, \tag{4}$$

having the solution

$$W_m = \frac{(m+\frac{1}{2})^2 h^2}{8\pi^2 A}, \qquad m = 0, 1, 2, \ldots . \tag{5}$$

W_m are the energy states of the rotator. Halpern observed in a footnote added in the proof of his paper that his result deviated from Lucie Mensing's (1926a) by a constant term ($= h^2/32\pi^2 A$) (Halpern, 1926b, p. 11). However, this practically unimportant detail did not affect the impression that the Viennese theoretician had no problem in handling matrix methods for treating quantum systems of practical interest. The same mathematical and theoretical ability was not confined to the theoretical physicists of German-speaking countries; the Russian scientists also successfully took up the new Göttingen mechanics in 1926.

After World War I the Russian physicists had to overcome particularly difficult conditions. They lacked funds for salaries and instrumentation for research as well as for scientific and personal exchange with foreign colleagues.[12] However, from 1922 their contacts with German physicists, in particular, improved considerably, and an increasing number of papers by Russian authors was published in *Zeitschrift für Physik*, *Physikalische Zeitschrift* and other journals.[13] In 1921 the senior physicist Orestes D. Chwolson had submitted an article

[12] In the early 1920s many individual scientists and institutions in Russia sent out calls for help to colleagues in Western countries. For example, on 18 August 1920 S. Boguslawski (from Saratow) wrote to his former professor, Max Born, asking him for a position outside the Soviet Union, claiming that it was impossible to survive on his salary from the local university. Another request was published in the issue of 15 July 1922 of *Physikalische Zeitschrift* (23, p. 296). It read:

> The Physico-Mathematical Society in Kratnodar (Russia) has sent the following appeal for help to the Editor:
>
> "The famine in Northern Caucasia has brought much suffering to the members of the Physico-Mathematical Society in Kratnodar. If they do not receive some help through food supply, the scientific work will cease and the Society will have to be dissolved. In this difficult situation of its members, the Society hopes to receive the sympathy of those colleagues who are in a more fortunate position. Therefore, in agreement with the decision of the assembly [of its members] of 19 March [1922], the Presidium of the Society requests the editor's office [of the *Physikalische Zeitschrift*] to make known the grave situation of its members to the German friends of physics. The food aid can be expedited via the American Relief Administration in Rostow-on-Don to the address: *Physico-Mathematische Gesellschaft*, *Professorenheim*, *Gymnasiumstraße* 47, *Kratnodar*.
>
> "In our Society numerous professors from the Universities of Moscow and Petersburg are working. The President of the Society is Professor B. Rosing, the inventor of the electric telescope, based on the application of cathode rays (see Korn, *Elektrische Telephotographie*).
>
> "Signed: B. Rosing, President; Boris N. Siana, Member of the Society."
>
> We pass on this emergency call for help for the information of our readers and request them to send the gifts intended for our suffering professional colleagues in Russia to the editor's office of this journal [*Physikalische Zeitschrift*]. An initial collection among the Göttingen physicists and mathematicians raised an amount of 1105 Marks.'

[13] One of the reasons for this new situation could have been a political event, the signing of the German–Soviet agreement at Rapallo, Italy, on 16 April 1922, by which full relations were established between the Weimar and Soviet Governments. A little later, in August 1922, the German

on atomic theory to *Zeitschrift für Physik* (Chwolson, 1921), which remained the only Russian contribution until 1923 when a dozen papers of Russian scientists appeared—among them one by G. Krutkow and Vladimir Alexandrovich Fock (Fok) from Leningrad (St. Petersburg) on the Rayleigh pendulum, another by Krutkow on fluctuation theory, and one by Sergei Ivanovich Wawilow (Vavilov) and his student W. L. Lewschin on polarized fluorescence light in colour dissolutions. The number of such papers almost doubled in each of the following years, 1924 and 1925, and many new authors, e.g., Jakow Frenkel, Dimitrii Vladomirovich Skobelzyn (Skobeltsyn) and Igor Tamm, joined the list of contributors to *Zeitschrift für Physik*. In fact, many of the Russian physicists, who later became famous, were introduced to the international scientific community in this way. In return, the development of physics as represented by the *Zeitschrift für Physik*, with its strong emphasis on atomic phenomena and quantum theory, became widely known in Leningrad, Moscow and other places. As a result, the Soviet scientists were soon able to work on the outstanding questions of the day, such as the quantum theory of magnetism or the correspondence approach to the intensities of spectra. At the time when quantum mechanics was formulated, J. Frenkel, one of the internationally well-known younger physicists from Leningrad, spent a year as a Rockefeller Foundation Fellow in Germany at the Universities of Berlin, Hamburg and Göttingen. Though he disliked matrix mechanics and did not make use of it in any calculation, he contributed to the theory by writing a paper on the problem of electron spin and resolving the difficulty of the factor two (Frenkel, 1926).

The first Russian paper on matrix mechanics arrived at the *Zeitschrift für Physik* on 23 April 1926. The author, Igor Evgenevich Tamm of Moscow, dealt in it, independently of the other approaches of the time, with the problem of the rotator in the new theory. He found for the energy states of the rotator a result which deviated from that of Lucie Mensing and also from that of Otto Halpern by an additive constant (Tamm, 1926). Again, one need not pay so much attention to the question of which of these three authors was right—it was proved by Schrödinger (1926d) that the result of Mensing was the correct one—than to the fact that Tamm did not hesitate to go into the unusual and complicated matrix methods and mastered them well. In the same paper he also considered the general problem of canonical transformations in quantum mechanics, which would require an extension of the matrix methods to nonlibration coordinates. Tamm was certainly prepared to work successfully in the new theory because of his familiarity with Bohr's correspondence approach. Indeed, he had completed

---

Physical Society published a request to its members to provide reprints of their publications that had appeared since the outbreak of World War I; these were to be sent to St. Petersburg (Leningrad) for distribution to Russian scientists who had been unable to obtain German journals (*Phys. Zs.* **23**, issue no. 15 of 1 August 1922, p. 312). Another step to re-establish scientific exchange was announced by the Physical Society of Berlin on 8 December 1922. The senior Russian physicist Orestes D. Chwolson was elected honorary member after an address which opened with the words: 'After long years of separation the relations between the science of your nation and ours begin to be re-established' (*Verh. d. Deutsch. Phys. Ges.* (*3*) **3**, issue no. 3 of 31 December 1922, p. 93).

in July 1925 a paper attempting a quantitative formulation of the correspondence principle in order to calculate the intensities of spectral lines (Tamm, 1925).[14] About six months later a young author, Lev Davidovich Landau from Leningrad, submitted his paper on the spectra of diatomic molecules, in which he calculated the energy terms and the intensities of transition lines of band spectra as well as their Zeeman and Stark effects (Landau, 1926).[15] Although these results were published after the methods of wave mechanics had been developed, Landau employed exclusively the matrix methods of Born, Heisenberg and Jordan.

By mid-1926 quantum mechanics had reached all those places in Europe where an interest in atomic theory existed. At many of these the theoreticians began to apply the new methods, but even people who did not work in theory took notice of the results. For example, the experimentalist Fritz Kirchner of Munich University, who investigated the angular distribution of Compton electrons, noted in his lecture at the Düsseldorf meeting of the *Deutsche Naturforscher und Ärzte* (German Scientists and Physicians) of September 1926: 'The experimentally-determined emission distribution of the scattered electrons can be described satisfactorily in relation to the electric [-field] vector [of the incident radiation], if the validity of the momentum conservation for the elementary process is justified. This appears indeed to be achieved within the new quantum mechanics according to the already available attempts of Dirac [1926c]' (Kirchner, 1926, p. 801). In other cases one affirmed the fact that the quantum-mechanical calculations gave results that agreed with experiments—in contrast to

[14] Igor E. Tamm was born on 8 July 1895 in Vladivostock, Russia. He entered the mathematics and physics faculty of Moscow State University in 1914 and was graduated in 1918. During 1919–1920 he served as a physics lecturer at Krim University in Simpheropal, and in 1921–1922 he taught at the Polytechnical University in Odessa. In Odessa he developed a close friendship with L. Mandelstam, which lasted until the latter's death in 1944. In 1922 Tamm went to Moscow; he became a professor at Moscow State University in 1927 and was appointed to the chair of theoretical physics in 1930. When in 1934 the Academy of Sciences of the U.S.S.R. moved from Leningrad to Moscow, Tamm was appointed head of the Lebedev Physical Institute. He remained in this position until his death on 12 April 1971.

Tamm's early scientific work was performed under the guidance of L. Mandelstam; he dealt with the electrodynamics of anisotropic media and with crystal optics in relativity theory. Later, he turned to researches in relativity and quantum theory and, after quantum mechanics was invented, to dispersion theory in solids; he suggested the quantization of elastic waves. Together with Ilya M. Frank he developed in 1937 the theoretical interpretation of the radiation of electrons moving through matter faster than the speed of light (the effect observed by P. Čerenkov in 1934). For this work the three physicists were awarded the Nobel Prize in 1958. Tamm also contributed early to Dirac's theory of the electron and Fermi's theory of weak nuclear forces.

[15] L. D. Landau was born on 22 January 1908 in Baku, Russia. His mathematical talents appeared very early; at the age of fourteen he began his studies at the university of his hometown, and continued them at the University of St. Petersburg or Leningrad, as it was renamed in the mid-twenties. Upon graduation in 1927 Landau joined the Physico-Technical Institute of Leningrad for four years; he then became head of the theoretical department of the Kharkhov Physico-Technical Institute. Earlier, he spent the period 1929–1931 at Bohr's Institute in Copenhagen, the E.T.H. in Zurich, and visited several German universities on a Rockefeller Foundation Fellowship. In 1935 he was appointed head of the theoretical physics department of the Institute for Theoretical Problems of the U.S.S.R. Academy of Sciences, Moscow. Landau was awarded the 1962 Nobel Prize in Physics for his theoretical investigations of condensed matter. After suffering for several years from the effects of a severe automobile accident, Landau died on 1 April 1968 in Moscow.

the earlier results of the Bohr–Sommerfeld semiclassical quantum theory. For example, at Zurich, where Peter Debye renewed his interest in the theory of electric and magnetic susceptibilities and their temperature dependence (Debye, 1926), his student R. Sänger investigated experimentally the dielectric properties of gases consisting of CH_4, CH_3Cl, CH_2Cl_2, $CHCl_3$ and CCl_4 molecules (Sänger, 1926). He confirmed that they obeyed Debye's classical formula and referred to the recent quantum-mechanical calculation of Mensing and Pauli (1926), which, in the case of simple diatomic molecules, yielded a formula identical with the classical one. Though the calculation had not yet been extended to the complex molecules he had studied, Sänger assumed that the outcome would again agree with his data, while the Bohr–Sommerfeld theory would not do so at all. Thus, the experimentalists dealing with the different properties of atoms and molecules took notice of the quantum-mechanical results and confirmed them.

Important as this generally favourable attitude of the community of physicists was towards quantum mechanics, the main propagation of the new theory occurred, of course, through further theoretical work. The leading centres in these efforts continued to be the same places where the theory had been invented and the first pioneering applications made. This was particularly so in Göttingen, where Born, Heisenberg and Jordan mastered the tools of matrix mechanics in a unique manner and tried to extend the methods to describe a wider range of phenomena. In Hamburg, Pauli sought to make use of all his mathematical virtuosity to attack the problem of calculating the intensities of hydrogen lines in the Balmer spectrum. He was also very active and efficient in advertising quantum mechanics and getting other people involved in the new theory. In contrast, Paul Dirac in Cambridge did not stimulate others to work along his lines. However, he himself worked steadily on the generalization of q-number methods. The fourth centre of quantum theory was Copenhagen, though Niels Bohr himself did not directly participate in any calculations. Already since many years Bohr had left the more technical aspects of quantum theory to Hendrik Kramers, his mathematically versatile collaborator. Kramers learned the methods of matrix mechanics very quickly. He was able to follow Pauli's calculation of the hydrogen atom immediately and explain it to Bohr (Bohr to Pauli, 5 December 1925); he connected certain results of the new theory with his dispersion-theoretic work (Kramers, 1925c); and he was in a position to advise those people in Copenhagen and, later on, in Utrecht, who were interested in quantum mechanics. One such case was David M. Dennison, a former student of Oskar Klein at the University of Michigan, then an International Education Board Fellow.[16]

[16] D. M. Dennison was born on 26 April 1900 in Oberlin, Ohio. He studied first at Swarthmore College (1917–1921) and continued at the University of Michigan under O. Klein and W. F. Colby, receiving his Ph.D. in1924. Afterwards he spent three years in Europe, studying at the Universities of Zurich, Munich, Copenhagen, Cambridge and Leyden. In summer 1927 he returned to the University of Michigan, where he served consecutively as instructor (1927–1928), assistant professor (1928–1930), associate professor (1930–1935) and professor (1935–1971). Dennison was the chairman of the physics department from 1955 to 1965, and in 1966 he was appointed the Harrison M. Randall University Professor. He died on 3 April 1976.

Dennison visited Copenhagen twice during his three years stay in Europe, the first time before quantum mechanics was discovered and again shortly after the theory was established. Thus, he personally came in contact with its development. He also met Heisenberg and recalled much later that 'during the first part of the time when I was there, Heisenberg was more concerned with other, in a way, less important problems . . . they were spectroscopic problems of some sort or other, and he was groping for ways of dealing with them. . . . He was not in Copenhagen at the time when I saw . . . his article on matrices, and I studied it pretty much by myself' (Dennison, AHQP Interview, Second Session, p. 20). Dennison applied his new knowledge immediately to the motion of a particle in space at a fixed distance from a centre, that is, the quantum-mechanical rotator in space and the rotations of a molecule possessing an axis of symmetry. He found that a complication arises in these systems because of the constraints, which restrict the original degrees of freedom of freely moving particles. It had therefore to be decided as to how to formulate the quantum conditions correctly. 'Kramers was the one who was initially interested in this notion of systems which had constraints in them,' Dennison recalled. 'He talked to me about it and I think it was through him that I got interested in [writing the] article on the rotation of molecules, which used the method of constraints' (Dennison, AHQP Interview, Second Session, p. 20).

The method followed by Dennison in his paper on the rotation of molecules (Dennison, 1926) emerged from Kramers' and Dirac's earlier discovery of the connection between the quantum conditions in quantum mechanics and the classical Poisson brackets. What Kramers suggested was to calculate (for a system described by t Cartesian coordinates, $q_1, \ldots, q_t$, of which $t - s$ can be expressed as functions of the remaining s coordinates—the system thus having s degrees of freedom and $t - s$ constraint equations) first the Poisson brackets, and then to replace them by quantum-mechanical expressions. In the case of the genuine s degrees of freedom the quantum conditions were, of course, the well-known commutation relations of Born, Heisenberg and Jordan. But, on the advice of Kramers, Dennison considered a further quantum condition which included all the degrees of freedom of the system, that is,

$$\frac{sh}{2\pi i}\mathbf{1} = \sum_{\sigma=1}^{t} m_\sigma(\dot{\mathsf{q}}_\sigma \mathsf{q}_\sigma - \mathsf{q}_\sigma \dot{\mathsf{q}}_\sigma), \tag{6}$$

where m_σ is the mass of the σth particle, q_σ its position matrix and $\dot{\mathsf{q}}_\sigma$ the time derivative. Evidently, Eq. (6) may be interpreted as a generalization of the Thomas–Kuhn sum-rule about which Kramers had thought quite a lot. Dennison now applied the scheme to the simple rotator in two dimensions (that is, a particle rotating in a plane) and obtained for the energy states W_m the result

$$W_m = \left(m^2 + m + \frac{1}{2}\right)\frac{h}{8\pi^2 A}, \tag{7}$$

where A is the moment of inertia given by the product of the rotating mass and the square of its distance from the axis of rotation. The quantum number m may assume either integral or half-integral values. In the first case, the lowest value is given by $m = 0$ or -1; hence there exist two ground states of equal energy. There arises one term from each, as Dennison noted, and he proposed to associate one with the right-hand rotations and the other with the left-hand rotations in such a way that both systems do not combine. In the second case, with m assuming half-integral values starting from $m = -\frac{1}{2}$, there is only one ground state and one term sequence.[17] Thus, although Eq. (7) is the same as in Heisenberg's original treatment of the plane rotator (Heisenberg, 1925c, p. 891, Eq. (29)), the energy values are the same only in the second case.

For the three-dimensional rotator, on the other hand, Dennison obtained the following energy states:

$$W_m = (m^2 + m + 1)\frac{h^2}{8\pi^2 A}. \tag{8}$$

However, the quantum number m in Eq. (8) must always assume integral values starting from $m = 0$ *and* -1, because for half-integral values certain transition amplitudes turn out to be infinite.

Finally, Dennison considered the rotation of a molecule possessing an axis of symmetry and two different moments of inertia, A and C (two of the moments being equal to A). In this case, he calculated a more complex expression for the energy states depending on two quantum numbers m and n, that is,

$$W_{m,n} = \left\{ \frac{1}{A}(m^2 + m + 1) + \left(\frac{1}{C} - \frac{1}{A}\right)n^2 + \frac{1}{2C} - \frac{C}{4}\left(\frac{1}{C} - \frac{1}{A}\right)^2 \right\} \frac{h^2}{8\pi^2}. \tag{9}$$

Again, the values of m and n might be either integral or half-integral numbers, the lowest states assuming $m = n = 0$, and $m = \frac{1}{2}$, $n = \pm\frac{1}{2}$, respectively. Since he did not find a theoretical justification for deciding which term systems to use in practical cases, Dennison proposed to take the information in each case from experiment.

In performing the matrix calculations Dennison frequently used—as had Heisenberg, Born and Jordan earlier—results based on the correspondence approach, especially those which gave the relative transition amplitudes of atomic systems. But in all these situations he assured himself that the fundamental matrix equations were satisfied. Through this procedure he not only avoided complicated matrix calculations, he also remained close to the spirit of Copenhagen physics as represented by Bohr and Kramers. In spring 1926, however, there occurred a substantial change; Kramers, who had been Bohr's intimate collaborator since 1916, left Copenhagen to take up the professorship of theoretical physics at the University of Utrecht. Heisenberg replaced him at Bohr's

[17] Dennison found that any situation might arise with two ground states m_1 and m_2, $-1 \leqslant m_1, m_2 \leqslant 0$, such that $m_1 + m_2 = -1$. The above selected cases have the symmetry property that for each positive value of the quantum number m in Eq. (7) there exists a negative value for which the energy is the same.

Institute, and he brought to Copenhagen some of the methods of approach of the Göttingen theoreticians. He took over the responsibilities which Kramers had had previously: he had to give the courses of lectures and to advise junior physicists who visited the Institute. Copenhagen thus became the place where the most active research in quantum mechanics, especially with matrix methods, was performed.

Among the first to receive Heisenberg's help were a young American lady, Miss Jane M. Dewey, and J. Stuart Foster, a Canadian; both worked on the same problem, the Stark effect of helium lines. Miss Dewey had arrived in Copenhagen on an international fellowship, which was awarded to her by the undergraduates of Barnard College at Columbia University, New York. She was new to the problem while Foster had already had a long affiliation with it; he had carried out experimental research on the pattern and intensities of electrically split lines both at Yale University, New Haven, and McGill University, Montreal.[18] Now, in Copenhagen, he became interested in comparing his data (Foster, 1927a) with the results obtained from the new quantum theory, and he wrote a long paper containing detailed calculations (Foster, 1927c). At the same time, in spring 1927, Miss Dewey, who had earlier measured at Bohr's Institute the intensities of the combination lines of helium in Stark effect (Dewey, 1926), found that her results agreed with quantum mechanics (Dewey, 1927).[19] Although the methods of wave mechanics were available and Erwin Schrödinger had already applied them successfully to calculate the electric splitting of the hydrogen lines (Schrödinger, 1926f), Foster and Dewey did not employ them,

[18] J. S. Foster, born on 28 May 1890 at Clarence, Nova Scotia (Canada), obtained his B.S. degree from Acadia University, Nova Scotia, in 1911. He then served as a schoolteacher until 1915 and did military service from 1917 to 1919. Afterwards, Foster continued his studies as Loomis Fellow at Yale University, receiving the Ph.D. degree in 1921. He became an instructor at Yale in 1920 and held a National Research Fellowship. In 1924 he joined McGill University, Montreal, first as an assistant professor (1924–1930); he then served as an associate professor (1930–1931), and since 1931 as full professor. He spent the 1926–1927 academic year as an International Educational Board Fellow in Copenhagen; in 1930 he visited Yale University, and in 1931 he was a visiting professor at Ohio State University. During World War II, Foster served as Scientific Liaison Officer at M.I.T.'s Radiation Laboratory. In connection with this work he was awarded the U.S. Medal of Freedom in 1947. After the war, in 1946, he became director of McGill University's Radiation Laboratory, which obtained a cyclotron in 1949. In 1955 he was named Rutherford Research Professor at McGill, and in 1960 he retired from university service. Foster then lived in Berkeley, California, where he died on 9 September 1964.

[19] Jane M. Dewey submitted her first paper from Copenhagen on 28 August 1926. This paper already contained a theoretical explanation of the observed intensities on the basis of the Kramers–Heisenberg dispersion formulae (Dewey, 1926). In applying the theory, she received help from Kramers and Heisenberg, which she acknowledged (Dewey, 1926, p. 1124). She also noted (Dewey, 1926, footnote 14, p. 1111) that the dispersion formulae could be derived from matrix mechanics, i.e., from the perturbation theory in the Born–Heisenberg–Jordan paper. Foster, who went to Copenhagen in fall 1926, submitted a paper in late October summarizing the observations of the Stark effect of helium lines (Foster, 1927a). He also hinted at a detailed explanation of the data on the basis of quantum mechanics (Foster, 1927a, p. 65). He discussed this theory in a contribution to the Washington Meeting of the American Physical Society, 22–23 April 1927 (Foster, 1927b). His paper, containing the detailed treatment of the problem, was received by *Proceedings of the Royal Society of London* on 8 August 1927 (Foster, 1927c).

partly because the problem could be easily and adequately handled by matrices, and partly because of Heisenberg's influence. In summer and fall 1926 Heisenberg was averse to using the Schrödinger theory, and he imparted this aversion to his associates.

The problem of calculating the effect of an electric field on the energy states of helium was admittedly a complicated one, and it could not be performed without hard work by any method. However, Foster and Dewey had the advantage of taking over Heisenberg's earlier treatment of the free helium atom (Heisenberg, 1926c). Therefore they knew how to treat the helium atomic system, especially by studying the ortho-helium and parhelium cases separately. Each of the spectra of the free helium atom, so Heisenberg had shown, could be considered as being emitted from a system consisting of a single electron moving in a central field of force, which is created by the doubly-charged nucleus and the interaction of the two electrons and thus differs somewhat from the Coulomb electric field. In order to evaluate the perturbative terms to the first approximation only, all the information one needs to know about this internal field is the difference between the empirically observed helium states and the corresponding states of the hydrogen atom. Altogether, the Hamiltonian matrix of the helium atom in an external electric field may be taken as the sum of three terms: the first, W_0, represents the diagonalized energy matrix of the hydrogen atom; the second, H_1', describes the deviation (from the case of hydrogen) of the central field acting on the series electron in a helium atom; the third, H_1'', finally accounts for the effect of the applied electric field.[20] Now W_0, the matrix of the unperturbed Hamiltonian, is diagonal, and the states denoted by the principal quantum number n are twofold degenerate: there is the degeneracy with respect to the azimuthal quantum number k and the degeneracy with respect to the spatial quantum number m. Both degeneracies are removed by the perturbation terms, H_1' and H_1'', the first taking away the azimuthal degeneracy and the second the spatial degeneracy.

In order to proceed with the practical calculation, Foster made use of the perturbation theory for degenerate systems (as developed in Born, Heisenberg and Jordan, 1926, Chapter 2, Section 2). In particular, he started by diagonalizing, for given quantum numbers n and m, a quadratic $(n-m)\times(n-m)$ submatrix of the perturbation matrix $\overline{\mathsf{H}}_1$, which is the time average of the sum $\mathsf{H}_1' + \mathsf{H}_1''$ over the unperturbed motion. The elements of this submatrix (i.e., $\overline{\mathsf{H}}_1' + \overline{\mathsf{H}}_1''$), were well known to him: namely, the first contribution from $\overline{\mathsf{H}}_1'$ is diagonal because the internal field of the helium atom is centrally symmetric and the diagonal elements coincide with the observed values of the free atom; the interaction with the external electric field, on the other hand, gives rise to the elements of the matrix $\overline{\mathsf{H}}_1''$, which are nonzero on both sides adjacent to the diagonal and could be copied from Pauli's treatment of the Stark effect of

[20]The term H_1'' is, of course, given by $-eF\mathbf{z}$, where $-e$ is the charge of the electron, F the strength of the electric field along the z-direction, and $\mathbf{z}$ the matrix describing the z-component of the vector drawn from the radiating electron to the electrical centre of the helium atom.

hydrogen (Pauli, 1926a, Section 5). Hence Foster had just to determine the roots of the eigenvalue equations of degree $(n - m)$ in order to arrive at the electric splitting of the helium lines. For example, in the case of the parhelium lines, which are emitted from initial states with the quantum numbers $n = 4$ and $m = 1$, he solved a third-order equation and obtained the theoretical displacements (from the D-line) of the Stark components associated with the transitions $2P$–$4D$, $2P$–$4F$, and $2P$–$4P$ in agreement with the experimental data. In fact, he succeeded in reproducing all available data concerning the frequencies.

To calculate the intensities of the Stark components of helium, Foster and Dewey had to determine, at least to a first approximation, the elements of the unitary matrix by which the Hamiltonian of the system is diagonalized. They simplified their task by adopting in the case of the atom without the external field the relative intensities which followed from Ralph Kronig's correspondence formulae (Kronig, 1925a), and only studied the transformation necessary to diagonalize the perturbation term H_1'' due to the electric field. In this way they obtained results which fitted the existing empirical data very well. In particular, the theory explained the vanishing of the parallel component (with respect to the applied electric field) of a line with final state P, when the line is so displaced (in the electric field) as to coincide with the undisplaced P-line; that is, a Stark component, $2P - nQ$, must always have zero intensity if it assumes the frequency of the P–P combination line at zero field.

The successful calculation of the Stark effect was of great interest to people at Bohr's Institute in Copenhagen. Kramers, for instance, already had obtained the intensities of the Stark components of hydrogen in 1920 on the basis of correspondence arguments (Kramers, 1920b). Heisenberg had thought about the problem for many years, at least since his discussion with Niels Bohr in Göttingen in June 1922. There he had raised serious objections to Kramers' treatment, objections which faded away only gradually in the course of sharpening the correspondence principle. Now the whole correspondence approach had been replaced by quantum mechanics. That the enormously complex problem of the Stark effect of helium lines could be solved by means of matrix methods was regarded by Heisenberg as a personal triumph. But, late though it was in coming, it represented at the same time a triumph of the entire development which had started out in Copenhagen. This was acknowledged by none other than Niels Bohr himself, who was considered by most experts as *the* authority in matters of atomic theory. The work on quantum mechanics performed in Copenhagen since spring 1926, together with Bohr's approval, helped to establish the validity of the new theory throughout the scientific world.

Chapter III
Early Reviews and Lectures on Quantum Mechanics

The new quantum theory, which sprang up in Göttingen and Cambridge, immediately received a wide response. On the one hand, the knowledge of its successes quickly spread among the experts; by fall 1925 one already knew about Heisenberg's initial fundamental work in Cambridge, Leyden, Berlin and Vienna. On the other hand, review articles were written soon after the first publications appeared, and these helped the physicists who were not experts on quantum theory to understand what was going on. For example, Alfred Landé, who was known for his contributions to atomic theory in connection with the explanation of complex spectra, acted as an interpreter of the ideas of Born, Heisenberg and Jordan and their unusual mathematical methods. In an article entitled 'New Paths in Quantum Theory' ('*Neue Wege der Quantentheorie*,' Landé 1926c) he presented a short and simplified, but very careful, introduction to the physical principles and mathematical formulation of quantum mechanics. He discussed, in particular, the example of the anharmonic oscillator and outlined the essential steps of its treatment by matrix methods. He showed the connection between the semiclassical theory of Bohr and Sommerfeld and quantum mechanics by interpreting the new algebraic equations in terms of the virtual classical pictures that had been used in the radiation theory of Bohr, Kramers and Slater. Landé's aim was to provide a guide to quantum mechanics for those physicists who had earlier followed Bohr's correspondence approach in their work, and this was a majority of the theoreticians who analyzed spectroscopic data. He rendered a similar service to the readers of the second edition of his book on the recent development of quantum theory, which came out in spring 1926 (Landé, 1926d). After explaining the rules of matrix mechanics, he concluded the last section with the words:

> The formulation of the theory can be achieved only by the extensive use of *matrix calculus*, which had been developed long ago by the mathematicians and which, in fact, appears to be adequate for quantum theory. We experience here again the situation that a new field of physics can first be opened from the ground up by mathematical methods especially adapted for it, just as Riemann's covariant tensor analysis serves as the foundation of general relativity and Newton's fluxion calculus as the foundation of mechanics (Landé, 1926d, p. 167).

Landé's early plea for quantum mechanics and the matrix methods helped to propagate the new theory among scientists who were not specialized in atomic theory; these were physicists who followed the theoretical development without actively participating in it, who just wanted to get an idea of what was going on at the frontiers of the field, or those who were engaged in analyzing the experimental data. Landé was particularly suited for his task as an intermediary because he possessed an excellent reputation as someone who had devoted himself for a long time to research on problems on the borderline between experiment and theory in atomic physics. In addition, Landé knew personally most of the people who had contributed to the formulation of quantum mechanics; he had collaborated with Born, Heisenberg and Pauli, and he was able to see through complicated calculations and to present their essence in simple language to experimentalists.

With the exception of Landé, the main role in propagating quantum mechanics was played by Born, Heisenberg and Dirac. Pascual Jordan, the other coauthor in fundamental papers, was hindered by his speech impediment from giving talks to a wider public.[21] Born and Heisenberg, in particular, took the opportunity to speak at many places and on many occasions on the recent results of the theory. While Born did so in the United States from November 1925 to March 1926, Heisenberg took over most of the obligations in Germany. For example, he reported on the application of quantum mechanics to the anomalous Zeeman effect, a work which he had performed together with Jordan, to Lower Saxony's Section (*Gauverein Niedersachsen*) of the German Physical Society in Braunschweig on 14 February 1926. In April Heisenberg was invited to give a talk at the Physics Colloquium of Berlin University. Finally, he was booked to deliver the main lecture on 'Quantum Mechanics' at the 89th Assembly of German Scientists and Physicians (*Versammlung Deutscher Naturforscher und Ärzte*) in Düsseldorf in September 1926. In contrast to Heisenberg, Dirac, who completed his doctorate only in May 1926, did not travel yet. However, he gave several talks on quantum mechanics at the Kapitza Club: the first on 19 January 1926, at the 109th Meeting, was entitled 'The Fundamental Equations of Quantum Mechanics' like his first paper on the theory; he spoke again at the 119th Meeting on 27 April 1926, this time on 'Kramers' Derivation of the Quantum Conditions,' i.e., the content of Kramers' paper dealing with the physical interpretation of the commutation relations in matrix mechanics (Kramers, 1925c); and finally, on 14 June 1926, at the 125th Meeting, he reported on 'Compton Scattering,' presenting the results which he had obtained in a paper completed two months earlier (Dirac, 1926c).

Born, Heisenberg and Dirac also gave the first series of lectures on quantum mechanics. Dirac gave his course during the Easter term, 1926, to a small group of people at the Cavendish Laboratory, Cambridge. Bertha Swirles, later Lady Jeffreys, recalled that this group included D. R. Hartree, J. M. Whittaker, A. H.

[21] Jordan recalled later: 'In my case, of course, the speech impediment was a hindrance, which limited me in communicating our ideas—apart from the publication of incomprehensible papers—entirely to private conversation and discussion within the intimate circle of friends' (Jordan to Born, 3 July 1948).

Wilson, J. A. Gaunt, N. F. Mott and J. R. Oppenheimer.[22] At the same time as Dirac, Ralph Fowler lectured on 'Quantum Theory.' As Lady Jeffreys recalled later:

> The Easter term lectures were complementary to each other. Fowler had a remarkable gift for rapidly digesting new work of others and imparting his own enthusiasm for it. Dirac gave us what he himself had recently done, some of it already published, some, I think, not. We did not, it is true, form a very sociable group, but for anyone who was there it was impossible to forget the sense of excitement at the new work. I stood in some awe of Dirac, but if I did pluck up courage to ask him a question I always got a direct and helpful answer, with no beating about the bush if I was getting things wrong. (Lady Jeffreys in Eden and Polkinghorne, 1972, p. 2)

Heisenberg's lectures in fall 1925 were addressed to a different group; they originated through David Hilbert's interest in the work on quantum mechanics. 'Hilbert had asked me to give talks to the mathematical section of the faculty on these papers [on quantum mechanics],' Heisenberg recalled. 'So I gave a number of talks and then I wrote the paper for the mathematicians' (Heisenberg, Conversations, p. 292). At that time the mathematicians in Göttingen were very excited about the development in physics, and Heisenberg tried very hard to present the topics of quantum mechanics in a way that appealed to them. 'I remember,' said Heisenberg, 'that after the first lecture Hilbert was very satisfied and said very nice things about the lecture. After the second lecture he said, "Well, the second lecture wasn't good. That kind of lecture I could [give] myself." Apparently I had not succeeded so well in explaining things in the second lecture. I was quite ashamed and tried to improve my lectures' (Heisenberg, Conversations, p. 293).[23] At Hilbert's request, Heisenberg summarized the content of the lectures in a paper entitled '*Über quantentheoretische Kinematik und Mechanik*' ('On Quantum-Theoretical Kinematics and Mechanics'), which was received by the *Mathematische Annalen* on 21 December 1925 (Heisenberg, 1926a).

In his lectures to the mathematicians—and in the subsequent paper—Heisenberg tried to show how the empirical facts observed in atomic physics led to the new laws of quantum mechanics, which had to be expressed in terms of mathematical tools and had hitherto been rather unusual in physics. He took the attitude that the physical phenomena give rise to a certain set of fundamental

[22]Of these, Hartree was the most senior theoretician. John Macnaghten Whittaker, son of Edmund Taylor Whittaker, may have been visiting Cambridge from Edinburgh, Scotland, where he was studying. Alan Harris Wilson was a student of Emmanuel College, Cambridge, from 1923 to 1926. J. A. Gaunt and Neville Francis Mott were students of Ralph Fowler; both joined the Kapitza Club in fall 1926. J. Robert Oppenheimer spent the academic year 1925–1926 with J. J. Thomson at the Cavendish.

[23]In later life Heisenberg still remembered some incidents from these lectures to the mathematicians. For example, as he wrote on the blackboard the chalk made a loud screeching noise which greatly annoyed Hilbert. Hilbert remarked: 'The first qualification of a good lecturer is not to make noise with chalk on the blackboard' (Heisenberg, private communication).

postulates, and that these postulates are satisfied as special algebraic relations. In particular, he demonstrated that the so-called combination principle of Rydberg and Ritz, which governs the spectra of atoms, requires the representation of the dynamical variables of atomic mechanics in terms of infinite quadratic forms or matrices; moreover, according to Bohr's correspondence principle as a second postulate, the laws of quantum mechanics may be obtained from those of classical mechanics by formally replacing the dynamical variables in the latter by matrices. With special infinite matrix equations the quantum conditions, which are the most crucial assumptions in atomic physics, could also be satisfied. After establishing the mathematical structure of quantum mechanics, Heisenberg explained the most important features of the theory: he gave the equations of motion of quantum systems; he derived the frequency conditions and the conservation of energy; he gave the method for calculating the eigenvalues of a given Hamiltonian with the help of transformation theory, i.e., Hilbert's well-known method of transforming infinite Hermitian forms to principal axes; and, finally, he sketched the perturbation theory of matrix mechanics and discussed the question of conservation of momentum and angular momentum in atomic systems.

It might appear somewhat surprising that Heisenberg, who had complained shortly before to his friend Pauli about the fact that quantum mechanics was becoming too mathematical because of the efforts of Born and Jordan (Heisenberg to Pauli, 16 November 1925), took such great care in presenting the theory in a rather abstract and mathematical, even axiomatic, style. However, there were several reasons for this. First, he admired Hilbert very much and he certainly felt honoured to be invited to talk to the mathematicians. He also did not entirely dislike the mathematicians' axiomatic approach, for it helped to understand in some sense the mathematical consistency of quantum mechanics. Third, as Dirac had shown, it was exactly the strict mathematical representation which allowed one to formulate the laws of quantum mechanics in a manner that was not bound to matrices alone; that is, by expressing the equations of quantum mechanics as general linear equations, one could, so hoped Heisenberg, get away from the cumbersome and restrictive techniques of matrix theory.[24] This attitude agreed perfectly with the mathematicians who followed Heisenberg's presentation quite closely. Heisenberg recalled: 'I think Courant was interested ... Courant was completely interested later on when he saw the Schrödinger picture come in. Then, of course, there was the problem of the Hilbert space which was exciting for the mathematicians. ... The mathematicians had the feeling that something [special] had been accomplished by the physicists, and that they should be interested' (Heisenberg, Conversations, pp. 293–294).

[24]In his letter to Pauli (16 November 1925), Heisenberg expressed his dislike about calling quantum mechanics 'matrix physics.' He even suggested removing the word 'matrix' from the theory entirely, replacing it by 'quantum-theoretical magnitude' ('*quantentheoretische Größe*'). He followed this suggestion partly in his paper for the *Mathematische Annalen* (Heisenberg, 1926a), using the word 'matrix' as little as possible.

Remarkably enough, Courant did not become engaged in work on quantum mechanics. He surely could have pointed out to the physicists the connection between infinite matrix methods and the methods of differential equations, and he could have advised them on how to find better and easier ways to solve practical atomic problems. But he remained inactive. For Hilbert, who had shown the equivalence of the matrix and the differential equation methods many years earlier, these matters were less important as they concerned the details of calculation rather than matters of principle. On the other hand, he definitely had a strong interest in the mathematical formulation of quantum mechanics; since, however, he suffered from pernicious anemia at that time, he could not work as hard as he was used to. Still he actively followed the papers on matrix mechanics, and in spring 1926 he announced his first course of lectures on 'Quantum Mechanics.' With these, he started his last great effort in questions of theoretical physics: namely, the formulation in a mathematically complete and consistent manner of what Heisenberg had presented in his lectures in fall 1925. Out of these lectures would then grow the joint paper of David Hilbert, Lothar Nordheim and John von Neumann (1928), which, in turn, would stimulate von Neumann's fundamental work on nonbounded operators, extending Hilbert's mathematical theory in order to satisfy the needs of quantum mechanics.

While Heisenberg's lectures to the mathematicians had slightly delayed consequences for the future development of quantum mechanics, the effect of Born's lectures in the United States was felt immediately. This was mainly due to the fact that his public consisted of physicists who were eager to understand the new theory and to apply it to all kinds of physical problems. Another, perhaps not less important, reason was that Born was an experienced lecturer who presented the results of his team in a very skillful manner. This was Born's second visit to the United States; he had lectured before in summer 1912 on relativity at the University of Chicago. He had been invited to return to America, and in spring 1924 he had contemplated the possibility of going to Cambridge, Massachusetts, for the winter semester (Born to Bohr, 16 April 1924), but the visit had to be postponed for a year. At that time the physics department of the Massachusetts Institute of Technology, which had invited him, was making efforts to go into research in atomic physics.[25] Born was supposed to deliver a series of lectures on problems of atomic theory and the structure of matter. Since he was an expert on the theory of crystal lattices and expected that there would be some interest in it

[25] In the 1925–1926 academic year the M.I.T. physics faculty consisted of (professors) Charles L. Norton, Harry M. Goodwin, William S. Franklin, William J. Drisko and Edward P. Warner, (associate professors) Maurice de K. Thompson, Newall C. Page, Charles P. Burgess and Gordon B. Wilkes, and (assistant professors) Arthur C. Hardy, Paul A. Heymans, William G. Brown, William R. Barss, Max Knobel and John T. Norton. The members of some other departments were also interested in atomic physics, such as Vannevar Bush of the engineering department, and M. Sandoval Vallarta and Norbert Wiener of the mathematics department.

The first European visiting physicist at M.I.T. was Peter Debye, who spent some time there early in 1925. After Born's visit, William Lawrence Bragg went there in 1927.

at M.I.T., he gave ten lectures about the recent progress in that field.[26] However, he devoted twice as many lectures to discussing the current status of the theory of atomic structure. After all, Born had worked strenuously on the formulation of matrix mechanics during the month before he left Göttingen on 28 October 1925: he had completed with Jordan the first paper on the theory and had nearly finished the big memoir with Heisenberg and Jordan. Shortly after his arrival in Cambridge in the beginning of November he was informed by Pauli about the successful calculation of the energy states of hydrogen. Born incorporated all these developments into his lectures on 'The Structure of the Atom.' Thus, Born's students and auditors at M.I.T. were exposed to the first systematic presentation of the new theory long before the scientific public was able to see the papers in *Zeitschrift für Physik*.

Max Born attracted a large audience. Not only did the faculty members and students of the physics department of M.I.T. come to his lectures, but also people from other departments of the Institute attended regularly, among them the mathematician Norbert Wiener whom Born had met earlier in summer 1925 at Göttingen. And then other people, such as Edwin Kemble and John Slater, came over from neighbouring Harvard. Slater was originally interested in what Born had to say about the dynamical theory of crystals because he had worked on the compressibility of alkali halides for his thesis (Slater, 1924b). He hoped that Born perhaps knew an explanation for the repulsive forces which seemed to exist between the ions—in addition to the attractive Coulomb forces—and which were not accounted for by the old quantum theory. Although Born mentioned numerous new results in his lectures on lattice theory and, in his eighth lecture, cited Friedrich Hund's recent discussion of the energy of the crystal lattice (Hund, 1925f, g), he could not answer Slater's question. Evidently the development of quantum mechanics was, in late 1925, the more exciting topic and Slater was also drawn into it. He was happy to meet Born and said later: 'Naturally I became acquainted with him, and had a chance to work on some aspects of the quantum mechanical theory' (Slater, 1975, p. 21). As an experienced professor and researcher, Born was used to advising a large number of students. At M.I.T., he not only collaborated with Wiener on operator mechanics, but also helped other people in their efforts.

Born did more for propagating quantum mechanics than just making it the central part of his lectures. 'I wrote my lectures down,' he recalled later, 'with the intention to publish them and thus earn some money, as the salary I received from M.I.T. was not large, and was insufficient to cover our expenses' (Born, 1978, p. 226).[27] Samuel Wesley Stratton, President of the Institute, told him that

[26] Born's ten lectures on the lattice theory of solids constituted, according to his own testimony (Born, 1926d, preface), a review of certain parts of his encyclopedia article on the same subject (Born, 1923b). However, he also covered the results of work which had appeared since 1923.

[27] Born was accompanied on the trip to the United States by his wife Hedwig. As she got some food poisoning she did not join Born on the long journey from Boston to the West Coast, but returned to Germany to recover in a nursing home in Frankfurt.

they would be published as the first book of a new M.I.T. series. Although this honour was not connected with any fee, Born was still glad to accept. He knew that he would also earn some money from the publication of his lectures in German translation (Born, 1926e). Thus, he sent his manuscript to Ferdinand Springer in Berlin, who paid him well. The twenty lectures on atomic structure therefore not only became the first course of lectures that included matrix mechanics, but were also made available—both in English and German—to a large public. 'It is a happy circumstance for American physicists,' Edwin Kemble noted in his review of Born's M.I.T. book, 'that Professor Born was engaged to deliver these lectures on atomic dynamics just as the first accounts of the new matrix mechanics were appearing in Germany. The prompt publication of the text of the lectures with their summary of the first results obtained by this method should be of great service in helping us to keep up with the stream of thought in a field in which we have been too prone to lag behind' (Kemble, 1926b, p. 424).

Since Born believed that he was in possession of a scheme which promised to be the future, final quantum theory, he was also able to summarize in his lectures the lasting results of the old semiclassical quantum theory. He did so in the first nine lectures.[28] He began by introducing quantum theory through Planck's law describing the frequency distribution of radiation emitted from a black body, and followed it by Einstein's derivation of that law on the basis of the statistical method (Lecture 1). Then he turned to the fundamental equations of Hamiltonian dynamics, the canonical equations of motion and the canonical transformations (Lecture 2); he presented the action and angle variables and the role of quantum conditions for atomic systems (Lecture 3); he discussed the so-called adiabatic invariance and Bohr's correspondence principle (Lecture 4), and the theory of degenerate systems and their perturbations (Lecture 5). In the following four lectures Born summarized the status of atomic theory up to June 1925: the theory of the hydrogen atom including relativistic corrections and the influence

[28]There was almost no difference in the contents of the English and German editions of the *Problems of Atomic Dynamics* (Born, 1926d, e). In both the subject of 'Atomic Structure' was divided into twenty lectures. While the first seven lectures were completely identical, Born split the treatment of the semiempirical approach to the spectra of many-electron atoms into two parts in the English edition: Lecture 8 was devoted to *Aufbauprinzip* (the Building-Up Principle) and the discussion of X-ray spectra; Lecture 9 dealt with the formal theory of multiplet structure as understood in the spring and summer of 1925. The treatment of matrix mechanics started with Lecture 10. In the German edition the content of Lectures 8 and 9 was condensed into one lecture. This was justified in the light of the development which took place after Born had signed the foreword to the English edition on 22 January 1926. In the foreword to the German edition, signed in Göttingen on 21 April 1926, Born especially thanked Pascual Jordan for his help in rearranging the material. Jordan had just completed with Heisenberg a paper on the theory of multiplet structure and the anomalous Zeeman effects on the basis of the concept of electron spin (Heisenberg and Jordan, 1926), replacing the old semiempirical results which Born had discussed in his Lecture 9 at M.I.T. The new Heisenberg–Jordan theory could not be discussed in Lecture 9 of the German edition; hence Born shifted it to its proper place, Lecture 18, after Pauli's treatment of the hydrogen atom and before the exposition of the theory of Hermitian forms. Thus, matrix mechanics began in the *Probleme der Atomdynamik* (Born, 1926e) with Lecture 9.

of magnetic and electric fields (Lecture 6); the approach to the helium atom and many electron atoms (Lecture 7); Bohr's *Aufbauprinzip* (Building-Up Principle) and the theoretical difficulties arising from the empirical multiplet structure of spectral lines (Lectures 8 and 9).

In the nine introductory lectures Born emphasized the use of classical mechanics in the erstwhile atomic theory and he pointed out its various shortcomings.[29] 'With this,' he said in closing the ninth lecture on the interpretation of complex spectra, 'we have come to the limit which can be attained by the development of Bohr's fundamental ideas' (Born, 1926d, p. 66). Since the spectroscopic data had outgrown the theory, there was the need to lay the foundation of a real dynamics of atoms. 'Heisenberg,' Born continued, 'found a short time ago the key to the gate closed for such a long time, which kept us from the realm of the atomic laws. In his brief paper, the leading ideas are clearly stated, but only exemplified because of the lack of appropriate mathematical equipment. The required machinery Jordan and I have discovered in the matrix calculus. Shortly afterwards, as I learned later, Dirac also found an algorithm which is equivalent to ours, but without noticing its identity with the usual mathematical theory of matrices' (Born, 1926d, p. 67).

In the following ten lectures Born went on to discuss the principal ideas and results of matrix mechanics. He started by outlining its physical and mathematical foundations, discussing the observable quantities of atomic systems and their representation in terms of infinite matrices, as well as the most elementary operations with the latter (Lecture 10). Then he set up the commutation rules and the rules of matrix differentiation (Lecture 11), followed by the canonical equations of motion and other equations of the Hamiltonian scheme in matrix mechanics (Lecture 12). In Lecture 13 Born treated the examples of the harmonic and anharmonic oscillators and discussed perturbation theory; then he turned to dispersion theory (Lecture 14) and extended the matrix scheme to several degrees of freedom (Lecture 15). After introducing the angular momentum he proceeded to explain the observed intensities of Zeeman components in atomic spectra (Lectures 16 and 17). Born's course reached its high point in the following two lectures. In November 1925 Pauli had sent him a letter to the United States, informing him about the calculation of the energy states of hydrogen, and Born made public this great triumph of the new theory (Lecture 18) long before it appeared in print in *Zeitschrift für Physik*.[30] He devoted Lecture 19 to his

[29] In the first nine lectures on atomic structure Born largely followed the outline of his book on atomic mechanics (Born, 1925), the selection of the material being influenced by the insights that had emerged since the creation of matrix mechanics. Although he was convinced about the superiority of quantum mechanics, Born did not start his lectures by presenting it. 'To do this would not only be to deny to Bohr's great achievement its due need of credit, but even more to deprive the reader of the natural and marvelous development of an idea' (Born, 1926d, pp. ix–x).

[30] Born mentioned Pauli's letter in a letter to Niels Bohr. He wrote: 'Pauli has sent me his calculation of the hydrogen atom [i.e., its states]; I shall present it also in my lectures.' This letter to Bohr, though undated, must have been written in early December 1925, shortly after Born received Bohr's letter of 25 November requesting Heisenberg's services as Kramers' successor in Copenhagen. Borns' Lecture 18 was delivered still either in December or early in January 1926.

hobbyhorse: the general mathematical formulation of quantum mechanics by using Hilbert's theory of infinite quadratic forms and their transformation to principal axes. This formulation also indicated the first step towards describing aperiodic motions in the new theory. In the last lecture, Lecture 20, Born discussed the results of his recent collaboration with Norbert Wiener, in which they substituted matrix calculus by the general operational calculus and were able to treat the problem of uniform motion of a particle in a straight line in quantum mechanics. He thus concluded his course by discussing in depth the latest progress in the field.

Born had organized the material of his lectures very well. He had already written books on both the principal themes of his lectures, and he made a judicious selection of topics for his treatment. But, most important, he was enthusiastic about the recent developments in quantum mechanics and was proud of the fact that the new theory had originated in his Göttingen group. He was perhaps a bit bothered by the fact that Paul Dirac had come up independently with another formulation. However, Born knew well that Dirac's theory was equivalent to matrix mechanics and it had grown out of the same physical ideas. In the eyes of many of those who listened to his lectures at M.I.T., he was *the* physicist who now possessed the key to the most consistent and complete atomic theory. Unlike many other European professors, Max Born was also an approachable and sociable person. Many years later John Slater still remembered with pleasure the party which the Borns gave for their friends before they left Cambridge, Massachusetts. 'Everybody,' he said, 'enjoyed themselves and decided that the Borns were very nice people' (Slater, 1975, p. 329).[31]

After performing his duties at M.I.T., Born set out on a lecture tour of other places in the United States. He visited the University of Wisconsin at Madison, where Sommerfeld had lectured as Carl Schurz Professor three years previously; he went to the California Institute of Technology in Pasadena, the home of R. A. Millikan, R. C. Tolman and Paul Epstein; the University of California at Berkeley, where G. N. Lewis and R. T. Birge taught; and Washington, D.C., where he met the spectroscopists of the National Bureau of Standards and some physicists of the neighbouring Johns Hopkins University, Baltimore. His lectures

[31] Both Slater and Wiener remembered some details of Born's farewell party. After dining in a restaurant around the corner, the guests, including Edwin Kemble, Vannevar Bush and Manuel Vallarta, returned to the Borns' apartment to play with the Christmas toys which the Borns had bought for their children. Slater recalled:

> The main thing was an electric train, which was intended for their small boy. Professor Born had not been able to make it go, and we couldn't either, until finally Norbert, whom we expected to be the most impractical one of the group, suggested that the trouble might be that it worked with a transformer, and it might be that we were in the part of Boston that still had direct current in the lighting circuit. We telephoned the electric light company, and found that was the trouble.
>
> That wasn't enough to stop M.I.T. Bush and Wiener knew that the electric train would work on direct current of the proper low voltage, which could be produced with a storage battery. Only of course no one had a battery. But being resourceful people, they telephoned around to neighboring garages until they located an automobile battery that could be borrowed for the occasion, and went out and got it. And sure enough the train went, and we had Born, his wife, physicists and mathematicians and engineers, all down on the floor playing with the train. (Slater, 1975, pp. 328–329)

at all these places drew large audiences. John H. Van Vleck went from Minneapolis to Madison to hear Born speak. Carl Eckart recalled that he was originally not very attracted by the papers of Born, Heisenberg and Jordan, 'but in the late winter of 1925 or early 1926 Born came to Pasadena and his lucid lectures raised my interest' (Eckart, AHQP Interview, p. 11). As a result, Eckart, like Van Vleck in Minneapolis and Slater at Harvard, began to work on the new theory. Born's visit to America thus turned out to be a great success, both for quantum mechanics and for him personally.[32] The new theory won adherents in America, and many American physicists were attracted to go to Göttingen to obtain a thorough training in the new methods. Without exaggeration one may say that a new era in atomic physics and physical science had started.

[32]Born really cherished the memory of his visit to the United States and the fact that 'the new quantum mechanics drew big audiences, particularly in Pasadena and Berkeley' (Born, 1978, p. 227).

Chapter IV
Enthusiastic Response in the United States

In Europe quantum mechanics was quickly accepted as a major progress in atomic theory, especially by those scientists who had followed Bohr's correspondence approach with sympathy and endurance. In general, everybody in the field knew what papers in which journals were important.[33] Everything which Bohr, Born, Heisenberg, Pauli and Dirac wrote was taken very seriously, even if it was difficult to follow. This attitude existed both in Europe as well as in the United States, where the advances in atomic theory commanded great interest. However, the exchange of scientific information and ideas between, say, Copenhagen, Göttingen and Cambridge, on the one hand, and any place in America, on the other, in general took more time because of the large distance between the two continents. After his return from Copenhagen to Columbia University, New York, Ralph Kronig wrote that 'conditions are most unfavourable for theoretical work' because '. . . European magazines arrive with 5 to 6 weeks delay, while in Europe one often saw the advance proofs' (Kronig to Bohr, 7 October 1926). This delay in receiving information led to certain consequences. For example, Carl Eckart from Caltech, who sent his paper containing the proof of the equivalence of matrix and wave mechanics to *Physical Review* on 7 June 1926 (Eckart, 1926b), had not yet seen Schrödinger's paper containing the same result, although it had come out in *Annalen der Physik* a full month earlier (Schrödinger, 1926e). He had therefore to add a note in proof on 2 September 1926, stating that: 'In an article, dated March 18, 1926, but which did not reach this institute until the above was in course of publication, Schrödinger has published all the essential results obtained in the above paper' (Eckart, 1926b, p. 726). Such delays in receiving published work on subjects, on which the theoretical physicists in Europe were making fast progress at that time, meant a great disadvan-

[33] It is not that in the 1920s there were no outsiders working on atomic theory. For example, Arthur Korn from Berlin, in many papers in scientific journals, in talks at the meetings of the German Physical Society and in a book (Korn, 1926), sought to describe the constitution of atoms as arising from an interplay of attractive and repulsive gravitational and electric forces. As a rule, the new authors, with ideas different from those of the mainstream, had difficulties in becoming accepted. Thus, in the beginning of 1926, Paul Ehrenfest wrote to Einstein: 'A man named Lanczos has been publishing much recently in the *Zeitschrift der Physik* [sic]. Do you understand what he is after? If yes, can you tell me briefly if it is *childish* or contains something [worthwhile]' (Ehrenfest to Einstein, 7 April 1926). To which Einstein replied: 'Lanczos is a competent man, but I have not yet seen anything of his which makes upon me the impression of having been guessed correctly. He certainly possesses formal talent' (Einstein to Ehrenfest, 12 April 1926). Thus even Lanczos, whose original contribution to quantum mechanics we have discussed in Volume 3, Chapter IV.2, was looked upon with suspicion because of his unconventional approach.

tage on the other side of the ocean. Fortunately, this hindrance was at least partly compensated by the fact that Max Born already brought quantum mechanics to America in winter 1925–1926, before most of the papers on it appeared in print. Many of the physicists, who listened to his lectures at M.I.T. and other places, were surprisingly well prepared to understand the new theory and were willing to accept it. Indeed, they were soon able to learn the necessary mathematical tools and contribute valuable results. Quantum mechanics received more than a hearty welcome in the United States; it started a great period of interest in the fundamental problems of the structure of matter, from atomic to nuclear and elementary particle physics—an interest which soon brought science in America to maturity and, ultimately, to a position of leadership in the world.

Although atomic physics and quantum theory had not yet played a role at American universities comparable to that of many European institutions, favourable opportunities existed in the United States for scientists who wished to work in these fields. The country had emerged with considerable economic strength from World War I; the universities and other scientific institutions had not suffered any war damage, and the young generation of students and researchers was happy to push into new fields. Thus, research on atomic and quantum theory was also pursued here and there; these places included notably Harvard University, Johns Hopkins University, University of Chicago, University of Michigan, Princeton University, University of California at Berkeley, California Institute of Technology, George Washington University at St. Louis, Yale University and a few industrial laboratories. However, only a few persons at these places were responsible for research in quantum physics: thus, for example, the chemist Irving Langmuir of the research laboratory of General Electric Company, Schenectady, New York, made important and widely acknowledged contributions to the understanding of atomic structure around 1920, while the reputation of George Washington University depended in particular on the experimental work of Arthur Holly Compton and George Eric MacDonnell Jauncey on the scattering of X-rays in matter. Sometimes the physicists from neighbouring institutions joined efforts to understand atomic phenomena; an outstanding example of this was provided by the collaboration of the physicists from Johns Hopkins University, Baltimore, and the U.S. National Bureau of Standards in Washington, D.C., where spectroscopic investigations were taken up with great vigour.

The increasing interest in the United States to deal with experimental and theoretical problems in atomic physics after World War I arose from several reasons. One was that this field had then definitely become a frontier of research, and there was much excitement to participate in it. The example and influence of some senior destinguished scientists, whose reputations were built on outstanding achievements, now guided the younger generation: they included men like Albert Abraham Michelson of the University of Chicago and Robert Williams Wood of Johns Hopkins University. Even more interested in atomic theory were Robert Andrews Millikan, who in 1921 left the University of Chicago to become director of the Norman Bridge Laboratory of Physics at Caltech, and Arthur Holly

Compton, who moved to Chicago in 1923. Apart from these eminent experimentalists, the American universities in the early 1920s also had some theoretical physicists who had contributed to work on atomic and quantum theory, and continued to do so afterwards. The most prominent among them were Richard Chace Tolman and Edwin Crawford Kemble.

Tolman, professor of physical chemistry and mathematical physics at Caltech since 1922, was interested in modern physical theories—quantum theory and relativity—from the beginning of his scientific career.[34] For example, he already had published in 1914 a paper dealing with the quantum theory of the specific heats of solids, in which he employed his principle of similitude to derive Debye's formula from Planck's law (Tolman, 1914b).[35] In another paper, which appeared in 1918, he used a generalized theorem of energy partition in quantum theory of radiation (Tolman, 1918). During his early years at Caltech, Tolman again worked on problems of quantum theory. For example, he developed an elementary theory of space quantization involving half quantum numbers to explain in particular the observed specific heat of hydrogen at arbitrary temperatures (Tolman, 1923). In a following paper he calculated the time which the molecules spend in higher quantum states from the intensities of their absorption lines, invoking Einstein's 1916 theory of the emission and absorption processes (Tolman, 1924). He compared the results of this investigation a year later with the ones obtained from the correspondence approach based on the concept of virtual oscillators in atomic systems, finding reasonable agreement (Tolman, 1925b). Tolman's main interest was directed towards the general principles of quantum theory. In 1924 he investigated with Paul Ehrenfest the method of so-called weak quantization, that is, the assumption that atomic systems might not strictly obey the quantization rules of multiply periodic systems, but have only a preferred tendency to do so (Ehrenfest and Tolman, 1924). In 1925 he proposed the principle of microscopic reversibility in atomic processes (Tolman, 1925a), which was equivalent to the principle of detailed balancing of Fowler and Milne (1925). However, when quantum mechanics came along, Tolman did not get into it. He took notice of it though when continuing his earlier calculations of the absolute

[34] R. C. Tolman was born on 4 March 1881 at West Newton, Massachusetts. After studying at M.I.T. from 1899 to 1903, and graduating with the S.B. degree, he went to the Technical University of Berlin and to Crefeld for a year (1903–1904). He returned to M.I.T. as an assistant in 1904; he became a fellow in 1905 and an instructor in theoretical chemistry two years later. He completed his Sc.D. degree in 1910 under A. A. Noyes and then joined the University of Michigan as an instructor in physical chemistry. The following year Tolman became an assistant professor at the University of Cincinnati; he served also as an assistant professor at the University of California at Berkeley from 1912 to 1916, and from 1916 to 1918 as a professor of chemistry at the University of Illinois. In 1918 he joined the Chemistry Warfare Service as Chief of the Dispersoid Section, and from 1919 to 1922 he was associated with the Fixed Nitrogen Research Laboratory, Nitrate Division, Ordnance Department of the U.S. Army, having been made its director in 1920. From this position he went to Caltech in 1922. In his later years, Tolman served twice outside Caltech: in 1940 as Vice-Chairman of the National Defense Research Commission and in 1946 as Scientific Advisor to the U.S. Mission at the U.N. Atomic Energy Commission. He died on 5 December 1948 at Pasadena, California.

[35] Tolman's principle of similitude states that the fundamental entities out of which the physical universe is constructed are of such a nature that from them a miniature universe could be constructed exactly similar in every respect to the present universe (Tolman, 1914a, p. 244).

intensities of spectral lines, which he carried out together with his student Richard M. Badger in winter 1925–1926. After presenting the correspondence approach the authors concluded: 'It should also be noted that the new quantum mechanics of Born and Heisenberg is specially adapted for the calculation of intensities of lines, but we have not yet carried through a calculation along these lines' (Tolman & Badger, 1926, p. 395). It seems that the latest quantum-mechanical development offered little attraction to Tolman; his strength lay in general statistical and thermodynamic rather than detailed mechanical arguments.

Edwin Kemble of Harvard University, the other senior American contributor to quantum theory, had quite different interests.[36] As early as 1916, as a student, he had started to work on the theory of diatomic molecules and wrote papers on this subject, in which he made use of quantum theory (Kemble, 1916a, b). He continued to publish in this field later on (Kemble, 1920). Although Kemble did not work on many different problems of atomic theory, but concentrated on molecular structure (see, e.g., Kemble and Van Vleck, 1923), he became very influential in the field of quantum theory through his teaching. As John Slater, a Harvard graduate of those days, recalled: 'The training in quantum theory at Harvard was given by Edwin C. Kemble, a very able young man a few years older than I, one of the earliest Americans to establish really good instruction in the quantum theory of the day. . . . [It] was very definitely up to date as of the time [1923], as I discovered later when I visited European laboratories and found I was as well trained as the students there' (Slater, 1975, p. 5). In particular, Kemble made his students, among them John H. Van Vleck, aware of what was going on in quantum theory and what were its specific problems and difficulties at the time.

Apart from the fact that after World War I the number of people working in quantum physics at universities and other places had grown considerably, the American institutions made use of the opportunity of inviting outstanding European scientists to give courses of lectures. To some extent, this procedure had already become part of a tradition: H. A. Lorentz had given a series of lectures on the theory of the electron at Columbia University, New York, in 1906, and Max Planck himself spoke there on quantum theory when he delivered a course of eight lectures on theoretical physics in 1909 (Lorentz, 1909b; Planck, 1910b, especially Lecture 7). In 1921 Albert Einstein was invited for the first time to the California Institute of Technology. Charles Galton Darwin visited Caltech a year later. Arnold Sommerfeld spent the period from fall 1922 to spring 1923 as

[36] E. C. Kemble was born on 28 January 1889 in Delaware, Ohio, and received his education at Ohio Wesleyan College (1906–1907) and Case School of Applied Science in Cleveland (1907–1911). He was graduated from the latter with a B.A., and went on to become an instructor at Carnegie Institute of Technology (1911–1913) and Harvard University (1913–1917). He completed his Ph.D. under Percy W. Bridgman at Harvard in 1917. Kemble then worked as an engineering physicist at Curtiss Motor Corporation (1917–1918) and as a physics instructor at Williams College (1918–1919) before joining Harvard University again. He became successively instructor (1919–1924), assistant professor (1924–1927), associate professor (1927–1930) and professor (1930–1957).

Carl Schurz Professor at the University of Wisconsin, Madison, and also delivered lectures at Caltech. In 1923 Niels Bohr went on a lecture tour of America; he visited many places, among them the research laboratory of General Electric Company at Schenectady, Princeton University, the U.S. National Bureau of Standards in Washington, D.C., and the University of Chicago. Paul Ehrenfest from Leyden went to Caltech in 1924. The following year Peter Debye visited M.I.T. for some time, and Max Born went there in fall 1925. They all presented the most recent status of their research and propagated the best available European knowledge about atomic and quantum theory in the United States.

An important role, similar to these visitors, was played by those scientists who went to assume positions at American universities after World War I. One of the first of these was Paul S. Epstein, who was appointed to the faculty of Caltech in 1921.[37] He had already made pioneering contributions to the quantum theory of the atom, especially through his explanation of the Stark effect on the basis of the Bohr–Sommerfeld theory of multiply periodic systems; at Pasadena he continued to treat related problems of atomic dynamics. For example, in 1922 he published a major work on perturbation theory (Epstein, 1922a–d); similar work was performed in Europe at about the same time by Max Born and his collaborators in Göttingen and Hendrik Kramers in Copenhagen. Epstein also taught quantum theory at Caltech and advised students and collaborators in this field. He also attracted new people to work on quantum theory; for example, Carl Eckart went to Pasadena from Princeton in 1925 because Epstein was there. Unlike Epstein, who had left Europe permanently, the Swede Oskar Klein, formerly assistant to Niels Bohr in Copenhagen, spent two years in the United States at the University of Michigan—1923–1924 as instructor and 1924–1925 as assistant professor. But even in this short period he rendered a valuable service to American students of modern physics, for he was an authentic representative of Bohr's views on atomic theory. At a time when Sommerfeld's *Atombau und Spektrallinien* was the bible for everybody doing research on atomic theory and the method of phase-integral quantization was considered as the best approach, Klein emphasized the more critical Copenhagen attitude that relied upon correspondence arguments. David Dennison, then a student at the University of Michigan, recalled later: 'I don't think that there were many people that I came in contact with who understood the power of the correspondence principle or really what it was, with the exception of Klein, who was of course preaching it. I picked it up then from him, and began slowly to use it with greater certainty' (Dennison, AHQP Interview, Third Session, p. 3).[38] Klein was particularly suited for sowing the seeds of doubt on the uncritical application of the semiclassical quantum theory of multiply periodic systems. During his stay at Michigan he wrote the paper on the influence of crossed electric and magnetic fields on electron orbits in the hydrogen atom, which showed that the usually accepted

[37] For a biographical sketch of P. S. Epstein, see Volume 1, footnote 351.

[38] It was also because of the influence of O. Klein that Dennison went to Copenhagen, where he learned more about the correspondence principle during his first visit in spring 1925.

methods—that is, the quantization of the phase integral and the adiabatic hypothesis—led to a serious difficulty of the existing theory (O. Klein, 1924a, b).

Although Otto Laporte, a former student of Sommerfeld, did not teach at a university, but worked on the analysis of spectroscopic data at the U.S. National Bureau of Standards, he helped to propagate the latest developments of quantum theory on arriving in the United States. 'Shortly after he came,' reported William F. Meggers, the senior spectroscopist, to Sommerfeld, 'he organized a colloquium which has been maintained mainly by his energy and enthusiasm' (Meggers to Sommerfeld, 8 July 1926). About fifteen scientists from the Bureau and the neighbouring Johns Hopkins University met together regularly for the colloquium. Otto Laporte and Harold Urey later recalled some of the participants in these meetings: the spectroscopists Meggers, Paul D. Foote, Fred Loomis Mohler and Arthur E. Ruark, and the low-temperature physicist Ferdinand Graft Brickwedde from the Bureau of Standards; the X-ray specialists Ralph Walter Graystone Wyckoff and Samuel King Allison, and the theoretician Gregory Breit from the Carnegie Institution in Washington; and F. Russell Bichowski, Joseph Kaplan, Merle Anthony Tuve and Harold Urey from Johns Hopkins University, Baltimore. The colloquium took place once a week at the Geographical Laboratory or the Laboratory of Terrestrial Magnetism, or in some other department of the Bureau of Standards. 'This was a more important Seminar to us than what was going on at Baltimore,' thought Urey (Urey, AHQP Interview, First Session, p. 18).[39] As Meggers noted, Laporte did the principal work: 'There has been such an avalanche of theoretical developments from Pauli, Heisenberg, Hund, Born and Jordan, Schrödinger and others, that Dr. Laporte has been more than busy keeping himself and others informed. . . . All of us derived great benefit from these meetings' (Meggers to Sommerfeld, 8 July 1926). Indeed, Meggers wanted to offer Laporte a permanent position at the Bureau of Standards at this time, for he regarded the presence of an expert on quantum theory as vital for the future development of the experimental work at the Bureau.[40]

With the help of those experienced scientists who were capable of introducing quantum physics at their institutions, an increasing number of students became

[39] As Urey recalled, the physics department at Johns Hopkins was quite old-fashioned. '[Joseph] Ames was there, and he was a sort of imperial person. No one could say anything that Ames didn't like, and Ames was very much a classical physicist. And [Robert Williams] Wood, of course, knew nothing about modern physics. And [A. H.] Pfund was the other man. Again he didn't know anything about modern physics. It was pretty much of an old fuddy-duddy department in a certain way' (Urey, AHQP Interview, First Session, p. 18).

[40] This plan did not come through. Otto Laporte (born on 23 July 1902 at Mainz, Germany), after obtaining his doctorate from Munich with Sommerfeld in 1924, held a fellowship of the International Education Board at the Bureau of Standards, Washington, D.C. He left Washington in summer 1926 to go to the University of Michigan as an instructor. There he remained, becoming assistant professor in 1927, associate professor in 1935 and professor in 1945. His research work until 1943 was devoted to atomic spectroscopy; in 1944 he shifted to the study of fluid mechanics, in particular subsonic and supersonic flow and the theory of shock waves. Laporte served as Scientific Attaché at the U.S. Embassy in Tokyo (1954–1955), and helped to secure the agreement between the United States and Japan on the uses of atomic energy. He died on 28 March 1971 at Ann Arbor, Michigan.

familiar with this field. John C. Slater and John H. Van Vleck, for example, were taught by Edwin C. Kemble at Harvard. Also at Harvard, J. Robert Oppenheimer was a student from 1922 to 1925, and Gregory Breit spent the academic year 1922–1923 as a National Research Fellow. David Dennison was advised by Oskar Klein at Michigan, and from 1922 Linus Pauling studied at Caltech, receiving his doctorate in 1925. Some of the young Americans had the opportunity of deepening their acquaintance with atomic theory at European research centres like Cambridge, Copenhagen, Göttingen or Munich. They were helped by one or another of the numerous fellowships that were available for that purpose. For example, John Slater, after obtaining his doctorate in 1923, went to Cambridge and Copenhagen as a Sheldon Traveling Fellow; Frank C. Hoyt spent the year 1922–1923 at Copenhagen as a National Research Fellow and the year 1927–1928 at Berlin as a John Simon Guggenheim Fellow; R. de L. Kronig stayed from 1924 to 1926 in Cambridge, Leyden, Tübingen, Copenhagen and Rome, supported by a Behr–Cutting Fellowship; D. Dennison held a fellowship of the International Education Board from 1924 to 1927 at the Universities of Zurich, Munich, Copenhagen, Cambridge and Leyden; and Harold Urey's visit to Copenhagen was made possible by the American–Scandinavian Foundation. The American scholars absorbed the current status of quantum theory in Europe; they worked actively to solve problems at the frontiers of research; and they returned to the United States equipped with new knowledge and a deepened and lasting interest in quantum physics. Thus, they increased the number of privileged scientists who were in a position to appreciate the progress achieved with the help of quantum mechanics.

While the contributions of certain American experimental physicists, such as the discovery of the Compton effect, had been extremely important for the development of the new quantum theory, the contributions of American theoreticians were not as crucial. However, some of the theoretical papers submitted from the United States did add considerably to the understanding of the pressing problems in atomic physics. For example, there were the calculations of E. C. Kemble (1921) and J. H. Van Vleck (1922a, b) that indicated the failure of the quantum theory of multiply periodic systems in the case of the helium atom, even before the final work (Kramers, 1923a; Born and Heisenberg, 1923b) was published in Europe. An even more influential role was played by the idea of the virtual radiation field of atoms, which John Slater proposed while he was in Cambridge, England (Slater, 1924a). This idea was centrally employed in the radiation theory of Bohr, Kramers and Slater (1924), which started the subsequent fruitful development of dispersion theory. In spite of the fact that Slater was not quite happy with the manner in which his original concept was transformed by Bohr and Kramers—especially with the assumption that energy and momentum were only statistically conserved in individual atomic processes—on his return to America he wrote substantial papers treating the properties of atomic systems with the help of methods similar to the ones that Kramers, Born and Heisenberg were using at the time. His former colleague, John H. Van Vleck, at the University of Minnesota since 1923, also worked on the same approach.[41] Stimulated in part by Kramers' dispersion theory, he proposed an extension of the correspondence principle (Van Vleck, 1924a), and applied it to explain the properties of the absorption lines in atomic systems (Van Vleck, 1924b). He then worked on the quantum theory of the polarization of resonance radiation in magnetic fields (Van Vleck, 1925). Finally, he wrote a book on the quantum theory of line spectra (Van Vleck, 1926c), 'the most complete and elegant exposition of the old quantum theory ever produced' (Anderson, 1968, p. 24). Slater, at that time, also worked along lines similar to Van Vleck's: he was interested in setting up a consistent and detailed theory of optical phenomena in atoms based on the classical picture of absorption—that is, by assuming that absorption arises from an interference of the external radiation with spherical wavelets that are excited in the atom (Slater, 1925a). Beyond any doubt, both Slater and Van Vleck proved themselves to be as good experts and pioneers in the correspondence approach as any in Europe; they went as far with it as the physicists in Copenhagen and Göttingen before Heisenberg's discovery of the noncommutativity of the products of quantum variables. For example, Slater definitely felt the necessity of arriving at a new kind of mechanics by replacing the derivatives of the classical theory with the help of discrete differences. But he also confessed that: 'I had not got any of Heisenberg's stuff when [he] came out with it, but I was very much interested, and I thought it was just along [these] lines that one should pursue' (Slater, AHQP Interview, Second Session, p. 6).

The familiarity with the same type of quantum-theoretical considerations as were developed and applied in 1924 and 1925 at Copenhagen and Göttingen enabled the American physicists to follow the results achieved in Europe very closely. Men like Breit, Hoyt, Kemble, Slater and Van Vleck understood the complicated and 'mysterious' papers that were published in *Nature* and *Zeitschrift für Physik* at the time; they knew the spirit of the calculations and the interpretation of the results, and they were aware of the difficulties that were not yet solved. In addition, the Americans worked on the same crucial problems and had the same goal as the physicists around Bohr and Born. This unity with respect to the guiding ideas of atomic theory as well as the principal line of approach, which existed on both sides of the ocean, explains also the quick acceptance without reservation of quantum mechanics in the United States. As Urey recalled, he and the others immediately regarded the papers of Heisenberg, Born and Jordan (and later of Schrödinger) as the final answer to the problems of atomic theory. 'So far as I know,' he said, 'I never talked to any competent physicist who did not almost immediately recognize the outstanding importance . . . I think people realized that we were now getting to the sort of mechanics that describes what we observe' (Urey, AHQP Interview, First Session, p. 15). Of course, Urey, like almost any other physicist, had trouble in reading these papers because of the unusual algebraic methods applied in them. As Laporte recalled: 'This matrix stuff was very, very abstract . . . People kept

[41] For a biographical sketch of J. H. Van Vleck, see Volume 1, footnote 677.

asking me all the time, "All right, what's a matrix then?" . . . I explained it as a tensor' (Laporte, AHQP Interview, Second Session, pp. 23–24). But still, all those who really wanted to understand the theory were not prevented from doing so by the unconventional mathematics.

The American response could not set in, of course, before the papers on quantum mechanics were published, or their content somehow became known. Since there was nobody at that time in the United States who received letters from Heisenberg and Dirac, or the proof-sheets of their papers, one depended on reading the published papers or getting the information from some other source. As we have mentioned earlier, the most important source was Max Born's lectures at M.I.T. and other places in America. Thus, the first to hear about the new theory were those who sat in Born's lectures in Cambridge, Massachusetts: Norbert Wiener, John Slater, Edwin Kemble and the others. John H. Van Vleck heard Born in Madison, Wisconsin, and Carl Eckart in Pasadena, California. On his return from California to Europe, Born stopped in Washington, D.C., and the scientists at the Bureau of Standards and their guests—including Otto Laporte and the members of his Seminar—had the opportunity of learning first-hand about quantum mechanics. Incidentally, Laporte had seen the papers of Heisenberg (1925c) and Born and Jordan (1925b) before Born's visit to Washington. On his way to attend the American Physical Society Meeting in Montreal late in February 1926, Laporte stopped in New York.[42] He recalled later:

> And there I met two young fellows: one had the name [Ralph] Kronig, an instructor at Columbia, and with him he had a student named [Isidor Isaac] Rabi. We said that all this new stuff had come out and I had some new manuscripts which were [also to be] talked over. Kronig, Rabi and I and a few other instructors or graduates from Columbia, whom I didn't know or don't remember, met in some classroom . . . and for the first time, I think, that I even saw this [formula for] $pq - qp$. (Laporte, AHQP Interview, Second Session, p. 24)

From this occasion onwards, Laporte took a particular interest in quantum mechanics, which extended beyond the specific formulation with matrices, and he discussed the new developments in seminars at the National Bureau of Standards.

Soon after they learned the details about the new methods of quantum mechanics, many American theoreticians started to get into it. For example, John Slater, who had listened to Born at M.I.T., recalled: 'Naturally I became acquainted with him, and had a chance to work on some aspects of the quantum mechanical theory. I made a little progress beyond what had already been published, but before I could get my work written up, a paper by a hitherto unknown genius appeared: P. A. M. Dirac's first paper on quantum mechanics. It included not only the small points I had worked out, but much more besides'

[42]This meeting took place on 26–27 February 1926 in Montreal, and Laporte gave a talk there on the interpretation of complex spectra, in which he discussed the application of the recent theoretical work of Pauli, Heisenberg and Hund to the spectra of chromium and iron (Laporte, 1926).

(Slater, 1975, p. 21). Slater was also on the track of making use of Poisson brackets. However, 'Nothing that I did came out, but I was just as active as anybody' (Slater, AHQP Interview, Second Session, p. 6). Frank C. Hoyt, then a National Research Fellow at the University of Chicago, was a little more successful with respect to publishing his work.[43] He had been interested earlier in the correspondence approach for the calculation of the intensities of spectral lines (Hoyt, 1925; 1926a); now he saw the opportunity of using Born and Jordan's matrix evaluation of the harmonic oscillator amplitudes to determine the relative intensities of the mixed oscillation–rotation lines of molecular spectra, especially the so-called (01) and (02) bands of hydrogen chloride. In his talk at the Washington Meeting of the American Physical Society in April 1926, Hoyt reported an agreement between the available data and the matrix calculation (Hoyt, 1926b). However, Hoyt did not publish the details of his work in an extended paper, probably because Lucie Mensing's paper on the same topic became known; it appeared already in a May issue of *Zeitschrift für Physik* (Mensing, 1926a).

The speed with which the European physicists published their results on quantum mechanics could not be matched in the United States. It also affected some of the priority of John H. Van Vleck's paper dealing with the magnetic susceptibilities, which he submitted to *Nature* before he left Minnesota for a summer trip to Europe (Van Vleck, 1926a). In this work Van Vleck arrived at two consequences: first, he found the general result that the average value, which the square of the component of the angular momentum parallel to an applied magnetic field assumes, is identical to one-third of the value of the total angular momentum squared. Hence space quantization in quantum mechanics has no effect on the magnetic or electric susceptibilities of atoms or molecules because the calculation yields the same result as is obtained by (classical) averaging over random orientations of the magnetic and electric moments. From this, Van Vleck concluded that neither the magnetic susceptibility nor the dielectric constant of substances are changed substantially by external fields—in agreement with recent observations on diamagnetic gases (Lehrer, 1926). Second, Van Vleck turned to a diatomic molecule, e.g., hydrogen chloride, and took the matrices for the amplitudes of the rotating dipole—as derived by Lucie Mensing (1926a) or David Dennison (1926)—to compute the dielectric constant using the transformation theory of Born, Heisenberg and Jordan. He ended with the result that only molecules in the state of lowest rotational energy contribute to the polariza-

[43] F. C. Hoyt, born in 1898 in Chicago, Illinois, studied from 1915 to 1920 at M.I.T. (1918–1919, assistant to D. L. Webster; 1919–1920, instructor) and from 1920 onwards at Stanford, where he received his Ph.D. in 1921. After one year as an instructor at the University of Wisconsin, he spent the 1922–1923 academic year as National Research Foundation Fellow with Niels Bohr in Copenhagen. In 1923 Hoyt joined the Ryerson Physical Laboratory at the University of Chicago (National Research Fellow, 1923–1926; research associate, 1926–1928; associate professor, 1928–1947, becoming professor in 1947). He spent the year 1927–1928 on a J. S. Guggenheim Memorial Fellowship at the University of Berlin. When Heisenberg delivered his lectures on the physical principles of quantum theory at the University of Chicago in 1929, their notes were prepared for publication by Frank Hoyt and Carl Eckart (Heisenberg, 1930).

tion; hence Debye's classical formula is obtained from quantum mechanics for that part of the electric polarization of a gas which arises from the permanent dipole moment of the constituent molecules. These results were contained in a note submitted to *Nature*; but on arrival in Copenhagen, Van Vleck received a letter from the editor requesting him to shorten his contribution. 'The resulting delay meant,' he complained later, 'that my article appeared about a fortnight after one with a similar calculation by Mensing and Pauli [(1926)], and practically simultaneously with corresponding ones by Kronig [(1926d)] and Manneback [(1926)]. The delay of a month or so in publication time seems trivial to me now, but was rather distressing to me as a young man' (Van Vleck, 1968, p. 1235).[44]

On one of the problems treated in Van Vleck's note to *Nature* (i.e., on the calculation of the dielectric constant of diatomic molecules in quantum mechanics) two other Americans worked independently at the same time. One of them, as mentioned by Van Vleck, was Ralph Kronig, who had witnessed the emergence of the new quantum theory before he returned from Europe to the United States. He kept himself informed about the enormous progress in these matters by following the literature carefully. As for himself, he continued to work on the dispersion of radiation by atoms. At Columbia he treated, in particular, the dispersion of X-rays on the basis of the Kramers–Heisenberg theory, which he knew to be in accord—at least in first approximation—with quantum mechanics (Kronig, 1926b, c). Then he turned to the new matrix methods and, with their help, worked on the theory of the dielectric constant of diatomic dipolar gases. On the one hand, this problem was close to his interests, for it was connected with optical phenomena in the limit of infinite wavelength radiation; on the other hand, Kronig felt its treatment by Pauli (1921a) on the basis of the old semiclassical theory of multiply periodic systems, or more recently by Linus Pauling (1926a, b), led to unacceptable physical consequences. Therefore, in a note submitted to *Proceedings of the National Academy of Sciences (U.S.A.)* in early July 1926, he calculated the electric polarizability of the rigid quantum-mechanical rotator in a constant electric field (Kronig, 1926d). Starting from Kramers' dispersion formula in the limit of zero frequency—which he knew to be correct also in quantum mechanics in the lowest approximation—he substituted

[44]Van Vleck's article in *Nature* appeared in Volume 118, issue no. 2968, dated 14 August 1926. Mensing and Pauli's article was received by *Physikalische Zeitschrift* on 25 June 1926 and published in its issue no. 15 of 1 August 1926. If one assumes that Van Vleck's letter was sent in the beginning of June, he indeed had his paper ready first. C. Manneback's calculation on the basis of wave mechanics was submitted to *Physikalische Zeitschrift* on 17 July 1926 and published in its issue no. 17 of 1 September 1926. Finally, Kronig's paper was communicated on 3 July to the U.S. National Academy of Sciences and published in the August issue of its *Proceedings* (dated 15 August 1926).

Van Vleck recalled that his original letter to *Nature* was not excessively long and that he complained about the procedure to Niels Bohr in Copenhagen, who replied, 'You should have had me endorse the original version—then it would probably have gone through alright' (Van Vleck, 1968, p. 1235).

At about the same time Van Vleck also submitted a paper applying the new mechanics to solve the problem of the specific heat of molecular hydrogen (Van Vleck, 1926b). This goal was, however, reached only a year later by David Dennison (1927).

Lucie Mensing's formula for the energy states of the field-free system, and obtained the vector matrix of the electric moment in first order of the perturbing field E. The result for $P_z(j, m; j, m)$, the diagonal element of the matrix for the moment in the direction of the external field, turned out to be

$$P_z(j, m; j, m) = \frac{8\pi^2 A E \mu^2}{h^2} \cdot \frac{3m^2 - j^2 - j}{j(j+1)(2j-1)(2j+3)}, \tag{10}$$

where j and m are the quantum numbers of the total angular momentum and its z-component, respectively, A the moment of inertia of the rotator, and μ the electric dipole moment of the atomic system. Kronig noticed that this expression, in the limit of large quantum numbers, j and m, went over into Pauli's semiclassical formula (Pauli, 1921a, p. 324, Eq. (17)). He also noticed that only the states with zero angular momentum contribute to the dielectric constant ϵ of molecules represented by quantum-mechanical rigid rotators. Thus, he obtained the equation

$$\frac{3}{4\pi}\,\frac{\epsilon - 1}{\epsilon - 2} = \frac{N\mu^2}{3kT}\left(1 - \frac{h^2}{24\pi^2 AkT}\right), \tag{11}$$

where k is Boltzmann's constant and N is the number of molecules in a cubic centimetre; Eq. (11) deviates only by the second term—which is too small for usual temperatures and moments of inertia—from Debye's classical formula (Debye, 1912a). Kronig's result agreed, of course, with the one of Mensing and Pauli (1926, p. 512, Eqs. (15) and (16)), which was obtained independently and published at about the same time.

The particular interest of American quantum physicists in the problem of the dielectric constant of matter was stimulated by two facts. On the one hand, there existed in the United States a general interest in the properties of molecules, common to both physicists and chemists; on the other hand, there were available some fine measurements, which Karl T. Compton's student C. T. Zahn had carried out on hydrogen halides (Zahn, 1924). Since these results did not clearly decide between Debye's formula and the previous quantum-theoretical formula of Pauli, Linus Pauling in Pasadena reconsidered Pauli's calculation by introducing half-integral quantum numbers for the angular momentum of the molecules (Pauling, 1926a, b).[45] The resulting formula for the dielectric constant ϵ then

[45] Linus Carl Pauling was born on 28 February 1901 in Portland, Oregon. He studied from 1917 to 1922 at the Oregon State Agricultural College and received his B.S. in chemical engineering in 1922. Then he continued his studies at California Institute of Technology (1922–1925, teaching fellow; 1925, Ph.D.; 1925–1926, National Research Fellow). During 1926–1927 he visited the Universities of Munich (with Sommerfeld), Zurich (with Schrödinger), and Copenhagen (with Bohr) on a Guggenheim Fellowship. Then he returned to Caltech, where he was appointed an assistant professor in 1927, associate professor in 1929, and professor of chemistry in 1931. Pauling became Chairman of the Division of Chemistry and Chemical Engineering and Director of the Gates and Crellin Laboratories of Chemistry at Caltech (1936–1958). He was research professor at the Center for the Study of Democratic Institutions, Santa Barbara, California (1963–1967), and in 1967 became professor of chemistry at the University of California at San Diego. Besides these occupations he held from 1929 to 1933 a lectureship at the University of California at Berkeley, in 1932 a visiting

deviated even more strongly from Debye's classical one, but again the empirical information from Zahn's experiments and from the observations of intensities of molecular spectra did not rule it out. In summer 1926 Pauling went to Europe on a fellowship from the John Simon Guggenheim Memorial Foundation. On a train from Switzerland to Paris he accidentally met John H. Van Vleck and discussed with him the new developments in quantum mechanics, which also threw light on the question of the dielectric constant of polar molecules.[46] This conversation led to an immediate consequence. From Munich, where Pauling went next, he submitted a new paper on 10 September 1926 to *Physical Review*, in which he discussed in detail the influence of external electric and magnetic fields on the polarizability of molecules with a permanent dipole moment, and this time he compared the results of the treatments based on the old and the new quantum theory (Pauling, 1927). In particular, he noted that in the old theory the presence of a strong magnetic field, in a direction at an angle ψ with the electric field, would change the polarization by a factor of $(\frac{3}{2}\cos^2\psi - \frac{1}{2})$; hence, if the angle is 90°, the polarization would be negative and half as large in absolute value as in the case without the magnetic field, a rather unlikely result to hold in reality. However, the explicit matrix evaluation, using the transition amplitudes of the rotators from Mensing (1926a) and Dennison (1926), yielded the conclusion that 'a magnetic field should not influence the dielectric constant of a gas such as hydrogen chloride' (Pauling, 1927, p. 153).[47] Pauling's result was already stated in Van Vleck's earlier note to *Nature* (Van Vleck, 1926a)—more implicitly it was also contained in Kronig's note (Kronig, 1926d); yet it was Pauling's actual calculation, made after his meeting with Van Vleck, that explicitly demonstrated the conclusion.[48]

Thus, the American physicists such as Dennison, Slater, Hoyt, Van Vleck, Kronig and Pauling—who had either been in Europe or had had close connec-

---

professorship at M.I.T., from 1937 to 1938 a lectureship at Cornell University, and in 1948 the George Eastman Professorship at Oxford. Pauling was awarded the 1954 Nobel Prize in Chemistry in recognition of his outstanding theoretical contributions. He was honoured in 1948 with the Presidential Medal of Merit and in 1962 with the Nobel Peace Prize.

[46] Van Vleck later recalled the situation vividly as follows: 'After the scientifically strenuous and rather trying sojourns in Denmark and England I went to Switzerland to recuperate, and make some new trips such as walking over the Joch Pass. Science was involved only in a "border incident." On leaving Switzerland at Delle my eye somehow fell on the passport which was being inspected of someone sitting next to me in the compartment. The name on it was "Linus Pauling," and his wife Ava Helen thought I looked most impertinent to stare at it so hard. Linus and I had previously only corresponded, but on the train a friendship started of over forty years standing. I fear, however, that the trip to Paris was rather boring to Ava Helen while Linus and I talked of the new developments in theoretical physics' (Van Vleck, 1968, p. 1236).

[47] The contributions to the polarization matrix from states with the quantum numbers $j \neq 0$ are in fact not equal to $(\frac{3}{2}\cos^2\psi - \frac{1}{2})$ times the terms in the field-free case. However, all these terms add up to zero ultimately. Thus only the term with $j = 0$ and $m = 0$ remains, and this is identical with the field-free case. The result holds, of course, only to first order in the magnetic field strength.

[48] In Europe, Pauling also contributed to another problem of atomic theory, namely, the question of whether the hypothesis of electron spin would allow one to explain the observed X-ray doublets. In contrast to Gregor Wentzel (1926b), he concluded on the basis of wave mechanical calculations that the answer was in the affirmative (Pauling, 1926c).

tion with the developments there—readily accepted quantum mechanics and contributed to it. It was different in the case of Carl Eckart who, in 1925, arrived at Caltech from Princeton.[49] He became acquainted with the new theory through Born's lectures at Caltech early in 1926 and was particularly interested in the methods of Born and Wiener. 'The result was,' he said later, 'that I spent the spring of 1926 working rather intensively with this operator formulation, and was completely familiar with what is now known as Schrödinger operator [the energy operator] before Schrödinger's papers appeared in Pasadena' (Eckart, AHQP Interview, p. 18). As a result of this involvement in the operator methods, he followed the work of Cornelius Lanczos (1926a) and Erwin Schrödinger (1926c), which had become available at that time. In a paper, communicated to *Proceedings of the National Academy of Sciences* (*U.S.A.*) on 31 May and published in the July issue, he made use of Lanczos' field-theoretical reformulation of matrix mechanics, together with the operator approach of Born and Wiener, to solve the Schrödinger equation for the one-dimensional harmonic oscillator (Eckart, 1926a). Schrödinger had not treated this problem in his first paper on wave mechanics, but he did so in his second communication (Schrödinger, 1926d), which was not available in California before Eckart submitted his work. Eckart recalled how he had proceeded at the time:

> Having gone over the solution to the simple oscillator problem in the Heisenberg version many times, I immediately tried to do it in the Schrödinger version and was blocked by the Hermite equation, which is deceptively simple. It's simple enough after you have learned about differential equations, which I hadn't at that time. In any case, it happened to be the colloquium day, and at the tea preceding the colloquium I found Fritz Zwicky and wrote the equation on the board, asking him how one could solve it. He was equally at a loss, but Paul Epstein came up behind us and rather characteristically remarked: "What have you got there?" I explained, whereupon he turned without a word, went into his office, came back and handed me a copy of his dissertation, told me that he wanted it back, it was his only copy. I opened it and found the whole theory of the Hermite polynomials neatly worked out, and quite promptly applied it to the simple oscillator problem. (Eckart, AHQP Interview, p. 13)

Since the two methods, the quantum mechanics of Born, Heisenberg and Jordan and the wave mechanics of Schrödinger, yielded the same energy opera-

[49]Carl Henry Eckart was born on 4 May 1902 in St. Louis, Missouri. He studied at George Washington University (1919–1923) and Princeton University (1923–1925), where he received his Ph.D. under Karl T. Compton in 1925. After spending two years at Caltech as a National Research Fellow, he went to Europe on a J. S. Guggenheim Memorial Fellowship and visited the Universities of Munich, Berlin and Leipzig. In 1928 he joined the University of Chicago, first as an assistant professor, then becoming an associate professor (1931–1946). In 1942 he moved to San Diego to work on the development of sonar systems with the University of California's Division of War Research. After the end of World War II, Eckart remained in San Diego as Director of the Marine Physical Laboratory. He also served the University of California at San Diego as professor of geophysics and, from 1967, as Vice-Chancellor. Eckart died on 24 October 1973 at San Diego.

tor and the same energy eigenvalues, Eckart concluded that there must be a close connection between them. In a second paper, he gave—again using operator methods—a proof of the equivalence of the two different schemes of quantum mechanics (Eckart, 1926b); and, in a third communication sent to *Physical Review* on 17 July 1926, he calculated the transition amplitudes of the hydrogen lines with the help of the Schrödinger equation, incorporating the spin of the electron (Eckart, 1926c). Although all of these results had already been obtained and published by Schrödinger prior to Eckart, the latter's work constituted an independent and outstanding contribution to quantum mechanics. Indeed, it showed how well the American physicists were at home with the new theory, ready to make their own original contributions in the future.

With the proof of Eckart—and similar proofs by Schrödinger (1926e) and Pauli (in a letter to Jordan, 12 April 1926)—of the equivalence of the matrix- and wave-theoretical formulations of quantum mechanics, the theory gained considerable power as far as applications were concerned, for one could now use both mathematical methods to calculate the properties of atomic systems. In particular, one could repeat many of the previous semiclassical calculations with the help of wave mechanics, as Paul Epstein (1926) did in the case of the Stark effect of hydrogen lines. Later, Epstein took the energy states of the spinning electron from the Schrödinger equation, and confirmed the term formulae anticipated more than a year earlier by Goudsmit and Uhlenbeck (Epstein, 1927). An application of the wave mechanical methods was undertaken by Ralph Kronig and his student Isidor Isaac Rabi.[50] In their paper, submitted to *Physical Review* on 4 November 1926, they calculated the energy states and the transition amplitudes of the symmetrical top and settled, in particular, the question of the values of the rational quantum numbers; that is, they proved that the values of m and n in Dennison's formula, Eq. (9), must always be integral (Kronig and Rabi, 1927).

By fall 1926 the desire to use quantum mechanics to treat atomic systems of considerable complexity—'computomania' as Kronig called it in a letter to Bohr, dated 22 November 1926—had seized the theoreticians both in Europe and America, and most of them began to realize the practical limitations of the matrix methods. Still, not everybody turned to wave mechanics. John H. Van Vleck, for example, continued to use the matrix calculus in his pioneering contributions to the theory of electric and magnetic susceptibilities of matter (Van Vleck, 1927a, b, c; 1928); he also did so in a joint paper with his student

[50] I. I. Rabi was born on 29 July 1898 at Raymanov, Austria. He studied at Cornell University (1916–1919; 1921–1923), where he obtained his B.A. in 1919, and at Columbia University (1923–1927), where he received his Ph.D. in physics under A. P. Wills in 1927. He spent the following two years in Europe, the first as Barnard Fellow at the Universities of Zurich and Munich, the second as International Educational Board Fellow in Copenhagen, Hamburg (with Otto Stern), Leipzig (with Heisenberg), and again Zurich (with Pauli). He returned to Columbia University in 1929, where he became lecturer (1929–1930), assistant professor (1930–1935), associate professor (1935–1937), and professor (1937–1967). In 1944 Rabi was awarded the Nobel Prize in Physics for his experimental work on the application of the resonance principle, together with Otto Stern's molecular beam method, to determine the mechanical and magnetic moments of nuclei.

Edward Lee Hill on the rotational distortion of multiplets in molecular spectra (Hill and Van Vleck, 1928). Van Vleck was also the only person who, other than Dirac himself, employed Dirac's q-number techniques for actual calculations. He recalled this involvement many years later:

> Just before sailing to Europe in 1926 I did see an early paper by Dirac [(1926a)] in *Proceedings of the Royal Society*. It looked important and so I took with me either a reprint or notes I made from the library copy (I forget which). While on the ship I realized that the 'q-number' technique of this article (essentially the quantum-mechanical version of angle and action variables) furnished a method of calculating the mean values of $1/r^2$ and $1/r^3$ needed for determining the relativity and spin corrections to the energy levels of the hydrogen atom, which had been calculated without these corrections by Pauli. I was delighted to find that, as surmised by Uhlenbeck and Goudsmit and by Slater, the combined relativity and spin corrections made the energy levels come out identical (at least to the first approximation in $1/c^2$) with those which Sommerfeld calculated relativistically in the old quantum theory without spin. On my portable typewriter I typed up what would be a paper for publication and went with it to Bohr's Institute in Copenhagen only to be told by him that the same calculation had just been published by Heisenberg and Jordan [(1926)]. (Their method was slightly different from mine in that it used ordinary matrices rather than Dirac's q-numbers.) So a day or two later I again appeared at Bohr's laboratory with a calculation of $\langle 1/r^4 \rangle_{\mathrm{AV}}$ which was needed to determine the quantum defect Δ in the Rydberg formula $-chRZ^2/(n-\Delta)^2$ arising from polarization of inner shells by a valence electron that does not penetrate them. This time I was told that the calculation of $\langle 1/r^4 \rangle_{\mathrm{AV}}$ had already been made by Waller with the (to me) mysterious Schrödinger wave mechanics and was in process of publication [(Waller, 1926)]. A few years later I mentioned to Heisenberg that Dirac's q-numbers could be used to calculate the mean values of negative powers of r (except for $1/r$), and he remarked to me that I should publish this method as the computation of high negative powers of r was quite difficult by wave mechanics. So in 1933 Dirac, by then a Fellow of the Royal Society, communicated to its Proceedings a short article [(Van Vleck, 1934)] in which I finally used his techniques to calculate $\langle 1/r^5 \rangle_{\mathrm{AV}}$ and $\langle 1/r^6 \rangle_{\mathrm{AV}}$, as well of course the lower powers of $1/r$. (Van Vleck, 1968, p. 1235)

In his review of Max Born's lectures on *Problems of Atomic Dynamics*, which appeared in the August 1926 issue of *Physical Review*, Edwin C. Kemble had remarked: 'The prompt publication of the text of these lectures with their summary of the first results obtained by this method [of quantum mechanics] should be of great service in helping us to keep up with the stream of thought in a field in which we have been too prone to lag behind' (Kemble, 1926b, p. 424). By the end of the year, however, the American theoreticians had made up much of the distance to the very frontiers of the new quantum theory.

Chapter V
The Changing Horizon

With the formulation of quantum mechanics during the second half of 1925 and early 1926, the previous period of search for a consistent quantum theory had come to an end. Before that, numerous ideas and methods had been tried to remedy the failures of the Bohr–Sommerfeld semiclassical quantum theory, the more helpful ones being connected with either the use of the correspondence principle or some weakening of the quantum rules. The period starting with the three-man paper or Dirac's paper on the fundamental equations of quantum mechanics marked a basic change in the attitude of those who were interested in atomic physics. Previously, they knew that things were not quite right, that the right clues were being searched for, and that most of the makeshift approaches would have to be abandoned finally. Now, for the first time in many years the quantum physicists felt that in a sense things were all right again. To use a comparison with which Heisenberg liked to describe the situation: one was no longer completely lost in the fog while climbing rocky mountains; one could suddenly see the goal at which one was aiming. 'In the very moment that you have seen that, the whole picture changes completely, because although you still don't know exactly the way how you will make to the rock, nevertheless for a moment you say, "Oh, that's the thing and now I know where I am; I have to go closer to that and I will certainly see that. When I have seen it, then I will find the way to go further"' (Heisenberg, Conversations, p. 321). This was the general feeling of the experts; Heisenberg, Born, Jordan and Dirac were the first to have it, the first to notice this sudden change of affairs in atomic theory. But soon others, who studied the scheme developed by these pioneers, agreed about the progress that had been achieved.

If one compares the development connected with the birth of quantum mechanics with the advances made during the preceding years, one notices a fast acceleration of progress. At least until summer 1925 the physicists in Göttingen and Copenhagen had been convinced that the goal of a consistent quantum dynamics would be reached only by a long sequence of steps, that is, by remodelling the semiclassical theory in successive approximations with the help of the correspondence principle. Niels Bohr, who was not a mathematically minded physicist, was particularly convinced that one would not arrive at a complete and perfect theoretical scheme in one coup; he himself sought to formulate the purely physical concepts continually in a better way in order to achieve a deeper understanding of the problems. In doing so, he was prepared to use successive approximations even when every step was not completely consistent, as long as it led a bit further towards the goal which he had grasped intuitively. At times Bohr believed that the inconsistencies of the theoretical description of quantum phenomena were connected with the progress in their understanding; and, ultimately, he was even willing to accept a sort of 'higher' inconsistency as an integral part of the final theory. However, in the pursuit of successively improved approximations he had arrived, certainly before summer 1925, at the conclusion that this plan would not work, for the entire approach had landed into great difficulties and insurmountable contradictions. He had become convinced that a major advance was necessary and that it would consist in formulating a mathematically consistent scheme, one in which the conflict between classical physics and quantum theory would be removed in one single step. In the new quantum mechanics Bohr recognized such a scheme, although the 'higher' inconsistencies still remained in it. For example, one did not see in the beginning how to incorporate the quantum (corpuscular) features of the radiation, as exhibited in the Compton effect, into the new theory. This situation remained paradoxical even after Dirac had treated the problem in a mathematically consistent way (Dirac, 1926c). Since the end of 1925 there had come about a clear recognition of the difference between a paradox implied in the description of quantum phenomena and an inconsistency of the mathematical scheme employed, and this recognition marked a very important stage of the development. Niels Bohr and the quantum physicists learned that an inconsistency was much worse than a paradox. A paradox might be very disagreeable, but still one could set up a theory from which the results would follow in a logical way, so that two results would never contradict each other. An inconsistency, however, could never lead to a useful theory; if a scheme contained it, one could prove almost anything from it. This fact was, of course, familiar to the mathematicians in Göttingen, where Hilbert was so much preoccupied with the problem of the consistency of mathematical axioms.

The fundamental change that occurred in quantum theory in fall 1925 did not imply that the solution of every problem and difficulty, which had bothered one before, had been found. Especially in the beginning, when only selected problems could be handled in matrix mechanics, all one had was a general glimpse of the new theory. In other words, the physicists then knew how to see the connections between the different quantum phenomena in general. As Heisenberg explained later: 'From my own impression, it makes an enormous difference whether I only see details or whether I see the whole picture. So long as I see only details, as one does in any part of mountaineering, then, of course, I can say, "All right, I can go ahead for the next fifteen yards,"—or hundred yards or perhaps even one kilometre—but still, I don't know whether this is wrong or may be completely off the track. When I have once the connection with the whole, then I know, "Well, that's the right direction"' (Heisenberg, Conversations, p. 322). In fall 1925 the Göttingen theoreticians and Dirac at Cambridge were certain that they were moving in the right direction, and that the actual solution of the problems of atomic theory was only a matter of time. Thus, one expected, for instance, that the energy states of the hydrogen and helium atoms would somehow follow from the new dynamics. When Pauli derived the Balmer spectrum from matrix mechanics, the physicists in Göttingen, Copenhagen and Cambridge were less surprised about the success than the speed with which the result had been obtained.

Although quantum mechanics had been discovered, not everything had yet been found to solve the real problems. There were still many surprises on the way towards a complete description of atomic phenomena. The first surprise, which neither Heisenberg, nor Born and Jordan, nor Dirac, had foreseen, was connected with the electron spin. Heisenberg and Pauli, the foremost theoretical experts on spectra, had expected that the matrix scheme would have to be extended in order to include some kind of 'nonmechanical stress.' By spring 1926, however, the calculations showed that spin was an integral part of atomic dynamics. The second, and even greater, surprise was presented by the advent of a completely different theory of treating atomic phenomena: Schrödinger's wave mechanics. However, as soon as the Göttingen theoreticians saw that its results coincided with those of matrix mechanics—e.g., the dynamical variables in both schemes satisfied the same commutation relations—they accepted Schrödinger's formalism immediately as an alternative manifestation of the theory of quantum mechanics. Pauli, who was among those who proved the equivalence of the two theories, went ahead with the problem of calculating the intensities of the hydrogen lines, Heisenberg evaluated the helium spectrum, and Born discussed the question of atomic scattering with the help of wave mechanical methods. In general, after the introduction of the Schrödinger equation the physicists felt even more confident that they were on firm ground with quantum mechanics, for they were now able to handle more easily than before the realistic models of atoms and molecules and their interaction with radiation.

In his review talk on quantum mechanics, given on 23 September 1926 at the 89th Assembly of German Scientists and Physicians (*Versammlung Deutscher Naturforscher und Ärzte*) at Düsseldorf, Heisenberg presented an impressive list of problems which had already been solved (Heisenberg, 1926d). At the same time he emphasized the remaining open questions. One of the latter concerned the understanding of the paradoxical properties of radiation and matter; that is, the apparently corpuscular nature of radiation as exhibited by the Compton effect, on the one hand, and the possible wave properties of electrons and matter, in general, on the other hand. A further question was how the statistical theory of Bose and Einstein was related to quantum mechanics. And finally, there was the difficult problem of suitably connecting the space-time description of electrons observed in a cloud chamber with the matrix description of electrons in atoms, the latter description rejecting the former as being impossible in principle. In all these questions Albert Einstein became the principal discussion partner of the theoretical quantum physicists. For example, when Satyendra Nath Bose arrived in Berlin in late 1925 he found Einstein very excited about the new quantum mechanics. 'He wanted me to try to see,' Bose recalled later, 'what the statistics of light-quanta and the transition probabilities of radiation would look like in the new theory' (see Mehra, 1975b, pp. 141–142).[51]

Einstein had learned very early about the progress achieved in quantum theory in Göttingen. 'Heisenberg has laid a big quantum egg ("*Heisenberg hat ein grosses Quantenei gelegt*"),' he reported to Ehrenfest already on 20 September 1925. He followed the formulation of matrix mechanics with an open but critical mind. Thus, he wrote to his old friend Michele Besso on 25 December 1925: 'The most interesting [thing], which the theory has yielded recently, is the Heisenberg–Born–Jordan theory of the quantum states. A genuine witches' multiplication table ("*Hexeneinmaleins*"), in which infinite determinants (matrices) take the place of Cartesian coordinates. It [i.e., the Heisenberg–Born–Jordan theory] is highly intellectual and sufficiently protected against the proof of falsification by being very complicated.' In spite of the mathematical complications of matrix theory, in which he did not feel at home, Einstein tried to familiarize himself with it and test its validity. In the beginning of February 1926 he came to the following conclusion:

> I have continued to concern myself very much with the Heisenberg–Born scheme [i.e., matrix mechanics]. More and more I tend to the opinion that the idea, in spite of all the admiration [I have] for it, is wrong. A zero-point energy of cavity radiation should not exist. I believe that Heisenberg, Born and Jordan's argument in favour of it (fluctuations) is feeble, already since the probability for large fluctuations (e.g., the probability for the total energy to be in partial volume V of [the total volume] V_0) certainly does not come out correctly [according to their argument]. Further the fact that the momentum does not vanish is inconsistent with the quantum rules for systems, ($\mathsf{p}_\mu \mathsf{q}_\nu - \mathsf{q}_\nu \mathsf{p}_\mu = 0$). Finally, it seems to me that the theory—even for one degree of freedom—is not covariant with respect to coordinate transformations. (Einstein to Ehrenfest, 12 February 1926)

While he tried, together with his collaborator Jakob Grommer, to construct a one-dimensional system whose energy matrix was not identical with the one obtained after a canonical transformation, he discussed his objections against Jordan's treatment of cavity radiation with the author himself.

[51] S. N. Bose arrived in Berlin in early October 1925 after staying in Paris for about a year. On 8 October 1925 he wrote a letter to Einstein requesting for an appointment to see him. Their meeting took place a few weeks later when Einstein returned to Berlin from his annual visit to Leyden. For Bose 'the meeting was most interesting. He [Einstein] wanted to know how I hit upon the idea of deriving Planck's law in this way. Then he challenged me. He wanted to find out whether my hypothesis, this particular kind of statistics, did really mean something novel about the interaction of quanta, and whether I could work out the details of this business. . . . I tried to work out certain things along the lines Einstein had suggested. After I had been in Berlin for some time Heisenberg's paper [(1925c)] came out' (see Mehra, 1975b, p. 141).

Bose did not make any progress along these lines, although he had the opportunity of discussing matters with Walter Gordon (of the Klein–Gordon equation). The problem of illuminating the statistics of light-quanta and their interaction was eventually treated by Paul Dirac (Dirac, 1926f, 1927b).

Jordan, in the three-man paper, had written a section on cavity radiation from the point of view of matrix mechanics and had been able to calculate the expression for the energy fluctuations in agreement with Planck's law (Born, Heisenberg and Jordan, 1926, Chapter 4, Section 3). Thus, Jordan, on the basis of quantum mechanics, had been able to explain certain consequences which Einstein had ascribed to the corpuscular properties of radiation. However, in doing so he had assumed the existence of zero-point energy of $\frac{1}{2}h\nu$ per radiation frequency ν. Einstein argued that from this assumption an infinite zero-point energy arises for each volume which is filled with radiation in equilibrium, and that one would also not find with it the correct expression for the probability that all radiation energy is contained in a small fraction of the cavity's volume (Einstein to Jordan, 6 April 1926). Both objections remained unanswered for the moment, but they stimulated Jordan to reinvestigate the wave-particle problem in quantum mechanics. About two weeks after receiving Einstein's letter, Jordan submitted a paper to *Zeitschrift für Physik* dealing with the connection between Duane's quantum theory of interference and de Broglie's matter waves (Jordan, 1926). There, with the help of Born and Wiener's operator calculus and the quantum-mechanical action-angle formalism, he proved in particular that material particles were reflected from gratings only in definite directions; these directions could be explained—similar to the case of classical radiation scattering —by assuming waves to be associated with the particles. Jordan continued to work on the wave-particle paradoxes in quantum mechanics and became, together with Paul Dirac, the founder of the wave-quantization method or quantum field theory.[52]

As in the question of properly explaining the observed quantum features of radiation and matter, Einstein participated in the discussion of another problem of even greater fundamental importance. When Heisenberg went to Berlin in spring 1926 to address the physics colloquium there, he had occasion to talk privately to Einstein about the guiding principles of quantum mechanics. Einstein questioned in particular the assumption that only observable quantities should enter into the structure of the theory, although he had himself employed it in his formulation of relativity theory in 1905. He now argued that ultimately it was the theory which decided what quantities were observable. In this context he noted the great difficulty that occurred in quantum mechanics: while one declared that the orbits of electrons in atoms were unobservable and thus excluded from the theory, one could also not describe the macroscopic tracks of electrons observed in cloud chambers. Heisenberg frankly admitted the existence

[52]In a review of the light-quantum hypothesis, Jordan answered Einstein's questions in the following way (Jordan, 1928, pp. 195 and 201): First, an infinite zero-point energy of cavity radiation does not cause any difficulties because in every experiment only energy differences will show up. Second, the probability of an arbitrary energy distribution within partial volumes of the cavity can be obtained in principle by calculating the time averages of the energy, the square of the energy, and of all higher powers of the energy. Since this procedure is impossible in practice, one has to turn to a method in which the full equivalence of wave and particle theories is guaranteed. This is the method of so-called field quantization.

of this difficulty. Several months later, in his talk at Düsseldorf, he referred to this situation: 'The contradictions in the intuitive interpretations of phenomena occurring in the present scheme [of quantum mechanics] are totally unsatisfactory. In order to obtain an intuitive interpretation, free of contradictions, of experiments—which, by themselves, are admittedly without contradiction—some essential feature is still missing in our picture of the structure of matter' (Heisenberg, 1926d, p. 994). Several months later, in February 1927, Heisenberg would have the extra feature necessary in his hands: from the general mathematical scheme of quantum mechanics he would derive the uncertainty relations, which would allow one to interpret consistently such paradoxical quantum phenomena as the corpuscle-like properties of radiation and the wave-like properties of electrons, as well as to explain the reason why it made no sense to describe the motion of electrons in an atom but one could sufficiently accurately describe the tracks of electrons in a cloud chamber (Heisenberg, 1927b). In these matters Einstein's persistent questioning contributed essential hints to find the way.

In spite of all attempts by the pioneers to propagate quantum mechanics, the new theory appeared to most physicists in the first half of 1926 as a very difficult mathematical scheme. For example, Paul Ehrenfest, an experienced quantum physicist and one quite sympathetic towards the scientific developments of Göttingen and Copenhagen, had great trouble in keeping up with the new theory. In a letter dated 26 August 1926 he reported to Einstein: 'First I was in Göttingen, then in Oxford (British Association [Meeting]) and Cambridge. Now I am suffering from indigestion caused by the endless Heisenberg–Born–Dirac–Schrödinger sausage-machine–physics-mill ("*unendlicher Heisenberg–Born–Dirac–Schrödinger Wurstmachinen-Physik-Betrieb*").' While Ehrenfest found the entire state of quantum theory of that time indigestible, Erwin Schrödinger believed that it was only matrix mechanics that was so. Writing to Sommerfeld on 28 April 1926 he characterized matrix mechanics as follows:

> As for Heisenberg's mechanics, it appears to me now to present something like the following picture: Think of a person who has arrived accidentally at the theory of functions along *such* a path that he conceives a *function* as an enumerable sequence of constants, i.e., as that which *we* call the coefficients of its [the function's] power series. In principle, it must be possible to develop all of function theory on this basis, without taking any notice of the *conception* that *we* have of an analytic function. Perhaps, [in this way] one would obtain even interesting, new aspects, for one need not then restrict the concept of a function to those sequences which, when interpreted as coefficients of a power series, have a non-vanishing radius of convergence. However, along this ascetic path, the largest part of *our* function theory would be accessible only with the greatest labour, for along it [i.e., the ascetic path] one would use the abilities to visualize things—which we bring with us automatically from our daily life—to an entirely insufficient extent.'

In his paper establishing the equivalence of matrix and wave mechanics, Schrödinger remarked: '... I was discouraged ("*abgeschreckt*"), if not repelled ("*abgestoßen*"), by what appeared to me a rather difficult method of transcendental algebra, defying any visualization ("*Anschaulichkeit*")' (Schrödinger, 1926e, p. 735, Footnote 2). His wave mechanics, Schrödinger claimed, gave a conceptually simpler and more adequate understanding of quantum phenomena.

However, the events of 1926, which we have described in some detail, show that quantum physicists initially followed the more ascetic methods of Heisenberg, Born and Jordan, and Dirac, for they had been the first to change the situation in quantum theory completely. Quantum physics, in general, became a fashionable subject which began to be taught to students at an increasing number of universities in Europe and America. For example, in fall 1926 the number of lecture hours devoted to quantum theory in German universities suddenly doubled over what it had been in the preceding several years. Of all the places, where one could learn quantum mechanics during the early period, Göttingen was perhaps the most famous. After all, the new theory had been invented there and, in the beginning, Born, Heisenberg and Jordan were all in Göttingen. In particular Max Born, who was already a distinguished teacher of quantum theory before summer 1925 and had attracted research students from all over Germany, now became famous in the entire scientific world. In the following years young physicists from various countries showed up in his Institute in large numbers. Walter Heitler joined as an assistant in 1927, and between 1927 and 1931, J. Robert Oppenheimer from the United States, Victor Weisskopf from Vienna, and Max Delbrück from Berlin, among others, took their doctorates under Born. Many physicists, like the American Edward Uhler Condon and the Russian Vladimir Fock, wanted to improve their knowledge of quantum mechanics at the very source. As Born noted later with some pride about those days, 'Göttingen represented the most successful school, with which only Copenhagen competed' (Born, 1978, p. 237).

Werner Heisenberg, whose work in summer 1925 had led to the theory of quantum mechanics, received even greater fame than did Born. In the months following his discovery, Wolfgang Pauli, in particular, emphasized its crucial importance; he referred to the new quantum theory simply as 'Heisenberg's quantum mechanics.'[53] After the appearance of the papers of Dirac and Schrödinger, people frequently distinguished between various mathematical formulations of the theory and spoke about matrix mechanics as Heisenberg's quantum mechanics. Born, to some extent, was unhappy about this, for he and Jordan and not Heisenberg had first introduced the matrix scheme. Born could claim correctly that before this step was taken, 'Heisenberg did not even know what a matrix was' (Born to Einstein, 31 March 1948). On the other hand, the later events of 1926 justified to some extent the fact that Heisenberg's name became so

[53] Pauli first talked about 'Heisenberg's mechanics' in a letter to Ralph Kronig on 9 October 1925. A little later, in a letter to Niels Bohr on 17 November 1925, he used the phrase 'Heisenberg's quantum mechanics.' This phrase was taken over by Niels Bohr in his paper on 'Atomic Theory and Mechanics,' which appeared in the supplemental issue of *Nature* dated 5 December 1925 (Bohr, 1925b, p. 852); there it applied to the entire new theory, i.e., the formulation of Born and Jordan (1925b) as well as of Born, Heisenberg and Jordan (1926).

strongly attached to matrix mechanics; for at that time it was Heisenberg who adhered most rigidly to matrix methods against Schrödinger's claim that his wave mechanics offered a more appropriate physical interpretation of atomic phenomena.

Heisenberg emerged from these events and accomplishments as a man with an international scientific reputation, a hero of the new physics. He had completed his doctorate in Munich in 1923 at the age of 22, had become *Privatdozent* in Göttingen a year later, and was invited by Niels Bohr in fall 1925 to succeed Hendrik Kramers as his closest collaborator and lecturer in Copenhagen. When he was about to assume his duties there, Heisenberg received his first call from a German university: Leipzig invited him as *Extraordinarius* (extraordinary or associate professor) in theoretical physics. In connection with this call, Richard Courant wrote to Bohr:

> Don't be angry with me that I meddle in the matter of Heisenberg's call [to the associate professorship at Leipzig], and that too with an express letter. As you know well, I have always been and still am in favour of the idea that Heisenberg, in spite of all existing hindrances, should go to you in Copenhagen and that he should give preference to it above all positions of a similar kind. Now Heisenberg has received a very honourable call to [the University of] Leipzig, for the moment as an extraordinary [i.e., associate] professor, but with almost certain prospect of obtaining a full professorship there in just a few years. This is, of course, a situation, which—considered from the superficial, material viewpoint of career—may suggest urgently to accept the call from Leipzig.
>
> Nevertheless, in a conversation which I had with Heisenberg yesterday, I urged him to go to you now *under all circumstances* and not to sacrifice the scientific and human advantages of his stay in Copenhagen to the superficial advantages of the call from Leipzig. It is my opinion that Heisenberg can calmly forego this first opportunity [of a position] in Germany. All of us will certainly see to it that within the very near future he will obtain another similar position. My greatest wish is to create in Göttingen a second full professorship, beside Born's, and I do not consider this matter to be hopeless; in fact, I believe that as soon as the Prussian State will get out of its present financial crisis, the prospects for this plan will be very good. (Courant to Bohr, 24 April 1926)

Courant, one of the most influential members of the mathematical faculty of Göttingen, was very concerned about the future career of the young physicist. Heisenberg did indeed turn down the call in 1926 and went to Copenhagen. A year later, however, he accepted a renewed call to Leipzig and became, before he was quite twenty-six, a full professor of theoretical physics there.

Together with Peter Debye, who was also appointed in 1927 as professor of experimental physics, Heisenberg would turn Leipzig into one of the important centres of research in quantum physics, They would attract many talented students to work on the explanation of the properties of matter, on problems of atoms and molecules and later of nuclei and elementary particles. Both would receive the Nobel Prize, Debye for Chemistry in 1936 and Heisenberg the 1932 Physics Prize (for his discovery of quantum mechanics). In a letter to Max Born, his former teacher, thanking him for his congratulations, Heisenberg said:

> The fact that I am to receive the Nobel Prize alone for work done in Göttingen in collaboration—you, Jordan and I—this fact depresses me and I hardly know what to write to you. I am, of course, glad that our common efforts are now appreciated, and I enjoy the recollection of the beautiful time of collaboration. I also believe that all good physicists know how great was your and Jordan's contribution to the structure of quantum mechanics—and this remains unchanged by a wrong decision from outside. Yet I myself can do nothing but thank you again for all the fine collaboration, and feel a little ashamed. (Heisenberg to Born, 25 November 1933)[54]

At the time, when Heisenberg wrote this letter, Born had already been dismissed from his professorship and had left Germany; the Göttingen centre where quantum mechanics had been created did not exist anymore. What remained from those days, however, were the memories of many physicists from all over the world, who had learned the new theory from the pioneers, and a book entitled *Elementare Quantenmechanik* (Born and Jordan, 1930).

In 1926 Max Born began to think about writing a sequel to his earlier *Vorlesungen über Atommechanik* (Born, 1925), that is, a book which would contain the definitive quantum theory in deductive form as he had promised in fall 1924. While Born and Jordan worked on the project for several years, the equivalence of quantum-mechanical methods—that is, the matrix mechanics of the Göttingen group and the algebraic formulation of Dirac, on the one hand, and Schrödinger's wave mechanics, on the other hand—had been demonstrated and the two mathematical schemes had merged into a unified theory. A number of review articles and books presenting quantum mechanics essentially in terms of undulatory methods had appeared before Born and Jordan's book was completed.[55] They decided, however, 'once to attempt how far one can proceed with elementary, i.e., principally algebraic methods' (Born and Jordan, 1930, p. vi). Therefore, they presented the abundant riches of atomic physics in a deductive manner in the language of matrix mechanics.

[54]Until 1954, when he was awarded the Nobel Prize in Physics for his statistical interpretation of quantum mechanics, Max Born remained envious of and unhappy about the adulation Heisenberg had received for being regarded as the 'discoverer' of quantum mechanics. Several of Born's letters to Einstein testify to this unhappiness. In numerous conversations that I had with Born during 1953–1968 he expressed this unhappiness every time he recalled the period of the creation of quantum mechanics and its acceptance; he believed that Heisenberg had been singled out for a more favoured treatment by fortune. In my last conversation with Heisenberg, which took place in his office at the *Max-Planck-Institut für Physik* in Munich on 25 February 1975, we discussed this matter. Heisenberg remarked: 'I was *so relieved* when Born was awarded the Nobel Prize.' (J. Mehra)

[55]For example, Hans Thirring and Otto Halpern's reviews of the new quantum theory in *Ergebnisse der exakten Naturwissenschaften* (Thirring, 1928; Halpern and Thirring, 1929); Arnold Sommerfeld's wave mechanical supplement to his *Atombau und Spektrallinien* (Sommerfeld, 1929); Jakow Frenkel's introduction to wave mechanics (*Einführung in die Wellenmechanik*, Frenkel, 1929), and Alfred Landé's lectures on wave mechanics (*Vorlesungen über Wellenmechanik*, Landé, 1930). In the United States there existed the book of E. U. Condon and P. M. Morse (1929) and the review articles of E. C. Kemble (1929) and Kemble and E. L. Hill (1930), which treated wave mechanics.

Born and Jordan's book appeared eventually in 1930. It was difficult to read and did not achieve any success. Wolfgang Pauli, in a caustic review of the book wrote:

> This book is the second volume of a series, in which the goal and the significance of the nth volume is clarified by the virtual existence of an $(n + 1)$th volume. The first volume consists of the lectures on atomic mechanics by M. Born [*Vorlesungen über Atommechanik* (Born, 1925)], which treats the older quantum theory based on mechanical orbits, while wave mechanics is supposed to be the subject of the third volume, which will be written "as soon as time and strength will permit." What is then the "elementary" quantum mechanics of the present volume? The preface of the book provides information about it: Elementary is that quantum mechanics which makes use of elementary tools, and elementary tools are purely algebraic ones; the use of differential equations, for example, is therefore consistently avoided as much as possible.
>
> As much as this plan of the book is interesting as a mathematical attempt, it does have the consequence—due to the necessarily one-sided picture of the theory —that the book can be of any use only to a very small circle of readers (to which the reviewer also belongs). Many results of quantum theory (as, for instance, angular distribution of the photoelectrons, the intensity of continuous spectra, the Compton effect, the theory of metal electrons as well as Dirac's theory of the spinning electron) can indeed not be derived at all with the elementary methods defined above, while the others [can be derived] only by inconvenient and indirect methods. (Among the latter results belong, for instance, the derivation of the Balmer terms, which is carried out in matrix theory according to an earlier paper of Pauli's dealing with it. In this regard, one will not be able to accuse the reviewer that he finds the grapes to be sour because they hang too high for him.) The restriction to algebraic methods also often inhibits insight into the range and the inner logic of the theory, as may be seen especially from the preliminary exclusion of quantities with continuous eigenvalue spectra (see Chapter 2: Mathematical Foundations, and Chapter 6: Statistical Interpretation of Matrix Mechanics). . . . The getup of the book, as far as printing and paper are concerned, is splendid. (Pauli, 1930, p. 602)

In short, the same man who, in 1925, had scored perhaps the greatest triumph of the matrix scheme now considered the plan of the book and the presentation of quantum mechanics in the language of purely matrix methods as a complete failure, although he did not find any logical mistakes or errors of interpretation in the treatment.[56] The book in fact contained everything that was needed to explain quantum mechanics, including the Schrödinger equation. Thus, when his student Maria Goeppert (-Mayer) once asked Born, 'Why don't you have the Schrödinger equation in there?,' he picked out a formula which was not a differential equation at all and said, 'This is the Schrödinger equation' (Goeppert-Mayer, AHQP Interview, p. 1). She also recalled that she never heard Born treating wave mechanics in his lectures at Göttingen; this was done by Walter

[56] In his own review article on quantum mechanics, which appeared three years after the book of Born and Jordan, Pauli exclusively used wave mechanical methods (Pauli, 1933c).

Heitler when he taught the course on quantum mechanics in Born's absence. Max Born's preference for matrix methods was mainly a matter of habit, of being accustomed to them. But around 1930 most theoreticians tended to agree with Paul Ehrenfest's remark in this context: 'There are also bad habits' ('*Es gibt auch schlechte Gewohnheiten.*').

The algebraic methods with the help of which quantum mechanics was first formulated by Born, Heisenberg, Jordan, and Dirac, were certainly very difficult to handle for most physicists, although they did not necessarily have to be as clumsy as might appear from a comparison of Pauli's matrix calculation of the hydrogen spectrum with Schrödinger's wave mechanical one. With some justification Max von Laue remarked about the former: 'Pauli's work on the Balmer formula, for example, is really terrible' (Laue to Schrödinger, 12 October 1926). However, in Dirac's formulation, which he still retained in his book, quantum mechanics was presented as an elegant and beautiful algebraic scheme (Dirac, 1930). It was still another question whether matrix mechanics, as Heisenberg and Jordan thought for some time, expressed the theory in a more fundamental or elementary manner. One may argue in favour of such a view that the use of not-so-readily visualizable concepts, as are connected with the algebraic methods, reflects perfectly the strange and paradoxical features which separate the quantum from the classical description of phenomena. On the other hand, even the most ardent partisan of the matrix theory had to admit the difficulties of handling that this scheme presented. Most of the successes of quantum mechanics that were obtained after 1926 would not have occurred, certainly not as rapidly as they did, had Schrödinger's wave mechanics not been available. And yet, as Eugene Wigner remarked: 'You know, that is an important event in your life when you read the Born–Heisenberg–Jordan paper [which ushered the new era of quantum mechanics]' (Wigner, AHQP Interview, First Session, p. 15).

References

Note: We have used the following abbreviations for journals and periodicals. Journals and periodicals not mentioned here are cited by their full titles. In the list of references we have given the dates, if available, of the particular issues in which the papers were published—except the *Sitzungsberichte* or *Comptes rendus* of the European academies. The references are arranged in alphabetical order according to the author, and in time-order according to the date of publication of the papers. Books or contributions to books (though not the yearbooks, which are treated as journals) have always been placed at the end of the annual list of a given author. The time-ordering letters (a, b, c, . . .) refer to all the papers of an author cited in Volumes 1 to 4.

American Journal of Physics	*Amer. J. Phys.*
Annalen der Physik (fourth series)	*Ann. d. Phys. (4)*
Archives for the History of Quantum Physics (Interviews)	AHQP Interview
Comptes rendus hebdomadaires des séances de l'Académie des Sciences (Paris)	*Comptes rendus (Paris)*
Encyklopädie der mathematischen Wissenschaften mit Einschluß ihrer Anwendungen	*Encykl. d. math. Wiss.*
Ergebnisse der exakten Naturwissenschaften	*Ergeb. d. exakt. Naturwiss.*
Journal of the London Mathematical Society	*J. London Math. Soc.*
Journal of Mathematics and Physics of the M.I.T.	*J. Math. & Phys. M.I.T.*
Journal für die reine und angewandte Mathematik	*J. reine & angew. Math.*
Journal of the Optical Society of America and Review of Scientific Instruments	*J. Opt. Soc. America & Rev. Sci. Instr.*
Journal de physique théorique et appliquée (Paris)	*J. phys. (Paris)*
Monthly Notices of the Royal Astronomical Society of London	*Monthly Notices Roy. Astr. Soc. (London)*
Naturwissenschaften	*Naturwiss.*
Philosophical Magazine (fifth and sixth series)	*Phil. Mag. (5), (6)*

Philosophical Transactions of the Royal Society of London, series A	*Phil. Trans. Roy. Soc. (London)* A
The Physical Review (second series)	*Phys. Rev. (2)*
Physikalische Zeitschrift	*Phys. Zs.*
Proceedings of the Cambridge Philosophical Society	*Proc. Camb. Phil. Soc.*
Proceedings, Koninklijke Akademie van Wetenschappen te Amsterdam	*Proc. Kon. Akad. Wetensch. (Amsterdam)*
Proceedings of the National Academy of Sciences (U.S.A.)	*Proc. Nat. Acad. Sci. (U.S.A.)*
Proceedings of the Royal Society of London, later series A	*Proc. Roy. Soc. (London)* A
Sitzungsberichte der Preussischen Akademie der Wissenschaften (Berlin) (from 1922: Physikalisch-mathematische Klasse)	*Sitz.ber. Preuss. Akad. Wiss. (Berlin)*
Verhandlungen der Deutschen Physikalischen Gesellschaft (second series)	*Verh. d. Deutsch. Phys. Ges. (2)*
Verslag van de gewone Vergadering der wis- en natuurkundige Afdeeling, Koninklijke Akademie van Wetenschappente Amsterdam	*Versl. Kon. Akad. Wetensch. (Amsterdam)*
Zeitschrift für Elektrochemie und angewandte physikalische Chemie	*Z. Elektrochem.*
Zeitschrift für Physik	*Z. Phys.*

Adams, Walter S.

1925 The relativity displacement of the spectral lines in the companion of Sirius, *Proc. Nat. Acad. Sci. (U.S.A.)* **11**, 382–387 (communicated 18 May 1925, published in issue No. 7 of 15 July 1925).

Anderson, Phillip Warren

1968 Van Vleck and magnetism, *Physics Today* **21**, No. 10, 23–26 (published in the issue of October 1968).

Andrade, Edward Neville da Costa

1964 *Rutherford and the Nature of the Atom*, Garden City, N.Y.: Anchor Books, Doubleday & Company.

Aston, Francis William, and Ralph Howard Fowler

1922 Some problems of the mass-spectrograph, *Phil. Mag. (6)* **43**, 514–528 (dated November 1921, published in issue No. 255 of 1 March 1922).

Baker, Henry Frederick

1922 *Principles of Geometry, Volume I: Foundations*, Cambridge: Cambridge University Press.

BECKER, RICHARD

1925 M. v. Laue. Zum Prinzip der mechanischen Transformierbarkeit (Adiabatenhypothese). *Ann. d. Phys. (4)* **76**, 619–628, 1925, Nr. 6. P. A. M. Dirac. The Adiabatic Invariance of the Quantum Integrals. *Proc. Roy. Soc. London* **A107**, 725–734, 1925, Nr. 744.; *Physikalische Berichte* **6**, 1479–1480 (published in issue No. 22 of 15 November 1925).

BLACKETT, PATRICK MAYNARD STUART

1924 Angular momentum and electron impact, *Proc. Camb. Phil. Soc.* **22**, 56–66 (received 17 December 1923, published in Part I of 28 February 1924).

1925 The ejection of protons from nitrogen nuclei, photographed by the Wilson method, *Proc. Roy. Soc. (London)* **A107**, 349–360 (communicated by E. Rutherford, received 17 December 1924, published in issue No. A 742 of 2 February 1925).

BLACKETT, PATRICK MAYNARD STUART, AND GIUSEPPE P. S. OCCHIALINI

1933 Some photographs of the tracks of penetrating radiation, *Proc. Roy. Soc. (London)* **A139**, 699–726 (communicated by E. Rutherford, received 7 February 1933, published in issue No. A 839 of 3 March 1933).

BOHR, NIELS

1923a Über die Anwendung der Quantentheorie auf den Atombau. I. Die Grundpostulate der Quantentheorie, *Z. Phys.* **13**, 117–165 (received 15 November 1922, published in issue No. 3 of 31 January 1923). English translation (by L. F. Curtiss): On the application of the quantum theory to atomic structure. Part I. The fundamental postulates, *Proc. Camb. Phil. Soc. Supplement* **22**, 1–44 (1924); reprinted in *Collected Works 3*, pp. 455–499.

1923c Linienspektren und Atombau, *Ann. d. Phys. (4)* **71**, 228–288 (received 15 March 1923, published in issue No. 9–12 of 23 May 1923, honouring H. Kayser's seventieth birthday); reprinted in *Collected Works 4*, pp. 549–610. English translation: Line-spectra and atomic structure, in *Collected Works 4*, pp. 611–656.

1925a Über die Wirkung von Atomen bei Stößen, *Z. Phys.* **34**, 142–157 (received 30 March 1925, with a postscript of July 1925, published in issue No. 2/3 of 28 September 1925).

1925b Atomic theory and mechanics, *Nature* **116**, 845–852 (elaborated text of a lecture, presented on 30 August 1925 at the sixth Scandinavian Mathematical Congress, published in the Supplement to the issue of 5 December 1925). German translation: Atomtheorie und Mechanik, *Naturwiss.* **14**, 1–10 (published in the issue of 1 January 1926).

1976 *Collected Works. Volume 3: The Correspondence Principle (1918–1923)*, (J. Rud Nielsen, ed.), Amsterdam–New York–Oxford: North-Holland Publishing Company.

1977 *Collected Works. Volume 4: The Periodic System (1920–1923)*, (J. Rud Nielsen, ed.), Amsterdam–New York–Oxford: North-Holland Publishing Company.

BOHR, NIELS, HENDRIK ANTHONY KRAMERS AND JOHN CLARKE SLATER

1924 The quantum theory of radiation, *Phil. Mag. (6)* **47**, 785–822 (dated January 1924, published in issue No. 281 of April 1924); reprinted in *Sources of Quantum Mechanics* (van der Waerden, 1967), pp. 159–176. German publication: Über die Quantentheorie der Strahlung, *Z. Phys.* **24**, 69–87 (received 22 February 1924, published in issue No. 2 of 22 May 1924); reprinted in *Collected Scientific Papers* (Kramers, 1956), pp. 271–289.

BORN, MAX

1923b Atomtheorie des festen Zustandes (Dynamik der Kristallgitter), *Encykl. d. math. Wiss. V/3*, 527–781 (dated 7 September 1922, published in issue No. 4 of 24 October 1923).

1924b Über Quantenmechanik, *Z. Phys.* **26**, 379–395 (received 13 June 1924, published in issue No. 6 of 20 August 1924); reprinted in *Ausgewählte Abhandlungen 2*, pp. 61–77, and in *Begründung der Matrizenmechanik* (Born, Heisenberg and Jordan 1962), pp. 13–29. English translation: Quantum mechanics, in *Sources of Quantum Mechanics* (van der Waerden, 1967), pp. 181–198.

1925 *Vorlesungen über Atommechanik*, Berlin: J. Springer Verlag. English translation (by J. W. Fisher): *Mechanics of the Atom*, London: Bell, 1927.

1926d *Problems of Atomic Dynamics*, Cambridge, Mass.: M.I.T. Press.

1926e *Probleme der Atomdynamik (Dreißig Vorlesungen gehalten im Wintersemester 1925/26 am Massachusetts Institute of Technology)*, Berlin: J. Springer Verlag.

1955b Die statistische Deutung der Quantenmechanik, (Nobel lecture delivered on 11 December 1954 at Stockholm), in *Les Prix Nobel en 1954*, Stockholm: Nobel Foundation, pp. 79–90; reprinted in *Ausgewählte Abhandlungen 2*, pp. 430–441. English translation: The statistical interpretation of quantum mechanics, in *Nobel Lectures: Physics 1942–1962* (Nobel Foundation, ed.), Amsterdam–New York: Elsevier Publishing Company, 1964, pp. 256–267.

1962 *Ausgewählte Abhandlungen*, 2 volumes, Göttingen: Vandenhoeck & Ruprecht.

1978 *My Life: Recollections of a Nobel Laureate*, New York: Charles Scribner's Sons. German translation: *Mein Leben: Die Erinnerungen eines Nobelpreisträgers*, Munich: Nymphenburger Verlagshandlung GmbH, 1975.

BORN, MAX, AND WERNER HEISENBERG

1923b Die Elektronenbahnen im angeregten Heliumatom, *Z. Phys.* **16**, 229–243 (received 11 May 1923, published in issue No. 4 of 9 July 1923); reprinted in *Ausgewählte Abhandlungen* (Born, 1962), pp. 23–37.

BORN, MAX, WERNER HEISENBERG AND PASCUAL JORDAN

1926 Zur Quantenmechanik. II., *Z. Phys.* **35**, 557–615 (received 16 November 1925, published in issue No. 8/9 of 4 February 1926); reprinted in *Ausgewählte Abhandlungen 2* (Born, 1962), pp. 155–213, and in *Begründung der Matrizenmechanik* (Born, Heisenberg and Jordan, 1962). pp. 77–135. English translation: On quantum mechanics II, in *Sources of Quantum Mechanics* (van der Waerden, 1967), pp. 321–385.

1962 Zur Begründung der Quantenmechanik (*Dokumente der Naturwissenschaften—Abteilung Physik*, Vol. 2; A. Hermann, ed.), Stuttgart: E. Battenberg Verlag.

BORN, MAX, AND PASCUAL JORDAN

1925a Zur Quantentheorie aperiodischer Vorgänge, *Z. Phys.* **33**, 479–505 (received 11 June 1925, published in issue No. 7 of 15 August 1925); reprinted in *Ausgewählte Abhandlungen 2* (Born, 1962), pp. 97–123.

1925b Zur Quantenmechanik, *Z. Phys.* **34**, 858–888 (received 27 September 1925, published in issue No. 11/12 of 28 November 1925); reprinted in *Ausgewählte Abhandlungen* (Born, 1962), pp. 124–154, and in *Begründung der Matrizenmechanik* (Born, Heisenberg and Jordan, 1962), pp. 47–76. English translation (except Chapter 4): On quantum mechanics, in *Sources of Quantum Mechanics* (van der Waerden, 1967), pp. 277–306.

1930 *Elementare Quantenmechanik (Zweiter Band der Vorlesungen über Atommechanik)*, Berlin: J. Springer Verlag.

BORN, MAX, AND NORBERT WIENER

1926a A new formulation of the laws of quantization of periodic and aperiodic phenomena, *J. Math. & Phys. M.I.T.* **5**, 84–98.

1926b Eine neue Formulierug der Quantengesetze für periodische und nichtperiodische Vorgänge, *Z. Phys.* **36**, 174–187 (received 5 January 1926, published in issue No. 3 of 12 March 1926); reprinted in *Ausgewählte Abhandlungen 2*, pp. 214–227, and in M. Born: *Zur statistischen Deutung der Quantentheorie* (Dokumente der Naturwissenschaft, Abteilung Physik, Vol. 1; A. Hermann, ed.), Stuttgart: E. Battenberg Verlag, 1962, pp. 34–47.

BOSE, SATYENDRA NATH

1924a Plancks Gesetz und Lichtquantenhypothese, *Z. Phys.* **26**, 178–181 (received 2 July 1924, published in issue No. 3 of 11 August 1924).

1924b Wärmegleichgewicht im Strahlungsfeld bei Anwesenheit von Materie, *Z. Phys.* **27**, 384–393 (dated 14 June 1924, received 14 July 1924, published with an addendum of A. Einstein, pp. 392–393, in issue No. 5/6 of 17 September 1924).

BREIT, GREGORY

1926 A correspondence principle in the Compton effect, *Phys. Rev. (2)* **27**, 362–372 (presented at the Kansas meeting of the American Physical Society, 28–30 December 1925; paper dated 26 January 1926, published in issue No. 4 of April 1926).

BROGLIE, LOUIS DE

1924a A tentative theory of light quanta, *Phil. Mag. (6)* **47**, 446–458 (communicated by R. H. Fowler, dated 1 October 1923; published in issue No. 278 of February 1924).

BRYAN, GEORGE HARTLEY

1905 Three Cambridge mathematical works. *The Algebra of Invariants*. By J. H. Grace, M.A., and A. Young, M.A., *The Dynamical Theory of Gases.* By J. H. Jeans, M.A., *A Treatise on the Analytical Dynamics of Particles and Rigid Bodies.* By E. T. Whittaker, M.A., *Nature* **71**, 601–603 (published in the issue of 27 April 1905).

BURGERS, JOHANNES MARTINUS

1916a Adiabatische invarianten bij mechanische systemen. I., *Versl. Kon. Akad. Wetensch. (Amsterdam)* **25**, 849–857 (Supplement No. 41c to the *Communications from the Physical Laboratory at Leiden*; communicated by H. A. Lorentz to the meeting of 25 November 1916). English translation: Adiabatic invariants of mechanical systems. I., *Proc. Kon. Akad. Wetensch. (Amsterdam)* **20**, 149–157 (published in issue No. 2 of August 1918). German translation: Die adiabatischen Invarianten bedingt periodischer Systeme, *Ann. d. Phys. (4)* **52**, 195–203 (received 20 December 1916, published in issue No. 2 of 6 March 1917).

1917 Adiabatische invarianten bij mechanische systemen. III., *Versl. Kon. Akad. Wetensch. (Amsterdam)* **25**, 1055–1061 (Supplement No. 41e to the *Communications from the Physical Laboratory at Leiden*; communicated by H. A. Lorentz to the meeting of 27 January 1917). English translation: Adiabatic invariants of mechanical systems. III., *Proc. Kon. Akad. Wetensch. (Amsterdam)* **20**, 163–169.

1918 *Het Atoommodel van Rutherford–Bohr*, doctoral dissertation, University of Leyden (*Archives du Musée Teyler*, third series, Vol. IV), Haarlem: De Erven Loosjes.

CHANDRASEKHAR, SUBRAHMANYAN

1931 The stellar coefficients of absorption and opacity, *Proc. Roy. Soc. (London)* **A133**, 241–254 (communicated by P. A. M. Dirac, received 14 May 1931, published in issue No. A 821 of 1 September 1931).

CHRISTIANSEN, JENS ANTON, AND HENDRIK ANTHONY KRAMERS

1923 Über die Geschwindigkeit chemischer Reaktionen, *Zeitschrift für physikalische Chemie* **104**, 451–471 (received 12 February 1923, published in issue No. 5/6 of 9 May 1923); reprinted in *Collected Scientific Papers* (Kramers, 1956), pp. 249–267.

CHWOLSON, ORESTES

1921 Zur Frage über die Struktur des Atomkernes, *Z. Phys.* **7**, 268–284 (received 8 October 1921, published in issue No. 4/5 of 12 November 1921).

CLARK, RONALD W.

1971 *Einstein: The Life and Times*, New York: World Publishing Company; paperback edition, New York: Avon Books, 1972.

COCKCROFT, JOHN, AND ERNEST THOMAS SINTON WALTON

1932a Production of high velocity positive ions I. Further developments in the method of obtaining high velocity positive ions, *Proc. Roy. Soc. (London)* **A136**, 619–630 (communicated by E. Rutherford; received 23 February 1932, published in issue No. A 830 of 1 June 1932).

1932b Experiments with high velocity positive ions. II. The disintegration of elements by high velocity protons, *Proc. Roy. Soc. (London)* **A137**, 229–242 (communicated by E. Rutherford; received 15 June 1932, published in issue No. A 831 of 1 July 1932).

COMPTON, ARTHUR HOLLY

1923b A quantum theory of the scattering of X-rays by light elements, *Phys. Rev. (2)* **21**, 483–502 (dated 13 December 1922, published in issue No. 5 of May 1923).

1923e Absorption measurements of the change of wave-length accompanying the scattering of X-rays, *Phil. Mag. (6)* **46**, 897–911 (dated 23 June 1923, published in issue No. 275 of November 1923).

CONDON, EDWARD UHLER, AND PHILIP MCCORD MORSE

1929 *Quantum Mechanics*, New York: McGraw-Hill Book Company.

COURANT, RICHARD, AND DAVID HILBERT

1924 *Methoden der mathematischen Physik (Erster Band)*, Berlin: J. Springer Verlag.

CROWTHER, J. G.

1952 *British Scientists of the Twentieth Century*, London: Routledge & Kegan Paul, Limited.

1974 *The Cavendish Laboratory, 1874–1974*, London and Basingstoke: Macmillan.

CUNNINGHAM, EBENEZER

1907 On the electromagnetic mass of a moving electron, *Phil. Mag. (6)* **14**, 538–547 (published in issue No. 82 of October 1907).

1908 On the principle of relativity and the electromagnetic mass of the electron. A reply to Dr. A. H. Bucherer, *Phil. Mag. (6)* **16**, 423–428 (published in issue No. 93 of September 1908).

1909 The principle of relativity in electrodynamics and an extension thereof, *Proceedings of the London Mathematical Society (2)* **8**, 77–98 (received 1 May 1909, published in issue No. 1041).

1914 *The Principle of Relativity*, Cambridge: Cambridge University Press.

1915 *Relativity and the Electron Theory*, London: Longmans, Green & Company.

1921 Relativity: The growth of an idea, *Nature* **106**, 784–786 (published in the issue of 17 February 1921).

CUNNINGHAM, EBENEZER, AND ARTHUR STANLEY EDDINGTON

1916 Discussion on 'Gravitation,' in: Mathematics and physics as the British Association, *Nature* **98**, 120 (report on a discussion at the Newcastle meeting, September 1916, published in the issue of 12 October 1916).

DARWIN, CHARLES GALTON, AND RALPH HOWARD FOWLER

1922a On the partition of energy, *Phil. Mag. (6)* **44**, 450–479 (dated May 1922, published in issue No. 261 of September 1922).

1922b On the partition of energy—Part II. Statistical principles and thermodynamics, *Phil. Mag. (6)* **44**, 823–842 (published in issue No. 263 of November 1922).

1923c Fluctuations in an assembly in statistical equilibrium, *Proc. Camb. Phil. Soc.* **21**, 391–404 (received 31 October 1923, read 13 November 1923, published in Part IV).

DEBYE, PETER

1912a Einige Resultate einer kinetischen Theorie der Isolatoren, *Phys. Zs.* **13**, 97–100 (received 12 December 1911, published in issue No. 3 of 1 February 1912); Nachtrag zur Notiz über eine kinetische Theorie der Isolatoren, *Phys. Zs.* **13**, 295 (received 2 March 1912, published in issue No. 7 of 1 April 1912). English translation: Some results of a kinetic theory of insulators, in *Collected Papers of Peter J. W. Debye*, New York and London: Interscience Publishers, 1954, pp. 173–179.

1925 Note on the scattering of X-rays, *J. Math. & Phys. M.I.T.* **4**, 133–147 (dated March 1925); reprinted in *Collected Papers*, pp. 89–103.

1926 Die Grundgesetze der elektrischen und magnetischen Erregung vom Standpunkte der Quantentheorie, *Phys. Zs.* **27**, 67–74 (received 15 December 1925, published in issue No. 3 of 1 February 1926). English translation: The basic laws of electric and magnetic excitation from the point of view of quantum theory, in *Collected Papers*, pp. 193–206.

DENNISON, DAVID MATHIAS

1926 The rotation of molecules, *Phys. Rev. (2)* **28**, 318–333 (dated 27 April 1926, published in issue No. 2 of August 1926).

1927 A note on the specific heat of the hydrogen molecule, *Proc. Roy. Soc. (London)* **A115**, 483–486 (communicated by R. H. Fowler; received 3 June 1927, published in issue No. A 771 of 1 July 1927).

DEWEY, JANE M.

1926 Intensities in the Stark effect of helium, *Phys. Rev. (2)* **28**, 1108–1124 (dated 28 August 1926, published in issue No. 6 of December 1926).

1927 Intensities in the Stark effect of helium: II, *Phys. Rev. (2)* **30**, 770–780 (dated July 1927, published in issue No. 6 of December 1927).

DIRAC, PAUL ADRIEN MAURICE

1924a Dissociation under a temperature gradient, *Proc. Camb. Phil. Soc.* **22**, 132–137 (communicated by R. H. Fowler, received 3 March 1924).

1924b Note on the relativity dynamics of a particle, *Phil. Mag. (6)* **47**, 1158–1159 (communicated by A. S. Eddington, published in No. 282 of June 1924).

1924c Note on the Doppler principle and Bohr's frequency condition, *Proc. Camb. Phil. Soc.* **22**, 432–433 (communicated by R. H. Fowler; received 19 May, read 14 July 1924).

1924d The conditions for statistical equilibrium between atoms, electrons and radiation, *Proc. Roy. Soc. (London)* **A106**, 581–596 (communicated by E. Rutherford, received 8 July 1924, published in issue No. A 739 of 1 November 1924).

1925a The adiabatic invariance of quantum integrals, *Proc. Roy. Soc. (London)* **A107**, 725–734 (communicated by E. Rutherford; received 19 December 1924, read 19 February 1925 and published in issue No. A 744 of 1 April 1925).

1925b The effect of Compton scattering by free electrons in a stellar atmosphere, *Monthly Notices Roy. Astr. Soc. (London)* **85**, 825–832 (communicated by E. A. Milne, published in issue No. 8 of June 1925).

1925c The adiabatic hypotheses for magnetic fields, *Proc. Camb. Phil. Soc.* **23**, 69–72 (communicated by R. H. Fowler; received 5 November and read 7 December 1925).

1925d The fundamental equations of quantum mechanics, *Proc. Roy. Soc. (London)* **A109**, 642–653 (communicated by R. H. Fowler; received 7 November 1925, published in issue No. A 752 of 1 December 1925); reprinted in *Sources of Quantum Mechanics* (van der Waerden, 1967), pp. 307–320.

1926a Quantum mechanics and a preliminary investigation of the hydrogen atom, *Proc. Roy. Soc. (London)* **A110**, 561–579 (communicated by R. H. Fowler; received 22 January 1926, published in issue No. A 755 of 1 March 1926); reprinted, except Sections 5–7, in *Sources of Quantum Mechanics* (van der Waerden, 1967), pp. 417–427.

1926b The elimination of nodes in quantum mechanics, *Proc. Roy. Soc. (London)* **A111**, 281–305 (communicated by R. H. Fowler; received 27 March 1926, published in issue No. A 757 of 1 May 1926).

1926c Relativity quantum mechanics with an application to Compton scattering, *Proc. Roy. Soc. (London)* **A111**, 405–423 (communicated by R. H. Fowler; received 29 April 1926, published in issue No. A 758 or 2 June 1926).

1926d *Quantum Mechanics*, doctoral dissertation, Cambridge University.

1926e On quantum algebra, *Proc. Camb. Phil. Soc.* **23**, 412–418 (communicated by R. H. Fowler; received 17 July 1926, read 26 July 1926).

1926f On the theory of quantum mechanics, *Proc. Roy. Soc. (London)* **A112**, 661–677 (communicated by R. H. Fowler; received 26 August 1926, published in issue No. A 762 of 1 October 1926).

1927b The quantum theory of the emission and absorption of radiation, *Proc. Roy. Soc. (London)* **A114**, 243–265 (communicated by N. Bohr; received 2 February 1927, published in issue No. A 767 of 1 March 1927).

1930 *The Principles of Quantum Mechanics*, Oxford: The Clarendon Press.

1950 Generalized Hamiltonian dynamics, *Canadian Journal of Mathematics* **2**, 129–148 (based on lectures given at the Canadian Mathematical Seminar in Vancouver in August 1949).

1958a Generalized Hamiltonian dynamics, *Proc. Roy. Soc. (London)* **A246**, 326–332 (received 13 March 1958, published in issue No. A 1246 of 19 August 1958).

1958b The theory of gravitation in Hamiltonian form, *Proc. Roy. Soc. (London)* **A246**, 333–343 (received 21 April 1958, published in issue No. A 1246 of 19 August 1958).

1971 *The Development of Quantum Theory (J. Robert Oppenheimer Memorial Prize Acceptance Speech)*, New York: Gordon & Breach Publishers.

1972 Relativity and quantum mechanics, *Fields and Quanta* **3**, 139–164 (presented at the Symposium on the 'Past Decade in Particle Theory,' Austin, Texas, 14–17 April 1970).

1977 Recollections of an exciting era, in *History of Twentieth Century Physics* (Varenna Summer School, 31 July–12 August 1972), New York: Academic Press Inc.

DOUGLAS, ALLIE VIBERT

1956 *The Life of Arthur Stanley Eddington*, London and New York: Thomas Nelson & Sons, Ltd.

DREYER, JOHN LOUIS EMIL

1924 Address delivered by the President, Dr. J. L. E. Dreyer, on presenting the Gold Medal of the Society to Professor A. S. Eddington, F.R.S., *Monthly Notices Roy. Astr. Soc. (London)* **84**, 548–558 (published in issue No. 8 of 13 June 1924).

DUANE, WILLIAM

1923 The transfer in quanta of radiation momentum to matter, *Proc. Nat. Acad. Sci. (U.S.A.)* **9**, 158–164 (communicated 2 March 1923, published in issue No. 5 of 15 May 1923).

DYSON, FRANK WATSON

1917 On the opportunity afforded by the eclipse of 1919 May 29 of verifying Einstein's theory of gravitation, *Monthly Notices Roy. Astr. Soc. (London)* **77**, 445–447 (letter dated 2 March 1917, published in issue No. 5 of 9 March 1917).

ECKART, CARL

1926a The solution of the problem of the simple oscillator by a combination of the Schrödinger and the Lanczos theories, *Proc. Nat. Acad. Sci. (U.S.A.)* **12**, 473–476 (communicated 31 May 1926, published in issue No. 7 of 15 July 1926).

1926b Operator calculus and the solution of the equations of quantum dynamics, *Phys. Rev. (2)* **28**, 711–726 (dated 7 June 1926, note added with proof, 2 September 1926, published in issue No. 4 of October 1926).

1926c The hydrogen spectrum in the new quantum theory, *Phys. Rev. (2)* **28**, 927–935 (dated 17 July 1926, published in issue No. 5 of November 1926).

EDDINGTON, ARTHUR STANLEY

1914 *Stellar Movements and the Structure of the Universe*, London: Macmillan.

1918 *Report on the Relativity Theory of Gravitation*, London: The Physical Society.

1920 *Space, Time and Gravitation*, Cambridge: Cambridge University Press.

1922a On the absorption of radiation inside a star, *Monthly Notices Roy. Astr. Soc. (London)* **83**, 32–46 (published in issue No. 1 of 10 November 1922).

1922b Applications of the theory of the stellar absorption coefficient, *Monthly Notices Roy. Astr. Soc. (London)* **83**, 98–109 (published in issue No. 2 of December 1922).

1923a The problem of electron-capture in the stars, *Monthly Notices Roy. Astr. Soc. (London)* **83**, 431–436 (published in issue No. 8 of 8 June 1923).

1923b *The Mathematical Theory of Relativity*, Cambridge: Cambridge University Press.

1924 On the relation between the masses and the luminosities of stars, *Monthly Notices Roy. Astr. Soc. (London)* **84**, 308–332 (published in issue No. 5 of 14 March 1924).

1932 The hydrogen content of stars, *Monthly Notices Roy. Astr. Soc. (London)* **92**, 471–481 (presented at the meeting of 8 April 1932, published in issue No. 6 of April 1932).

1941 On the cause of Cepheid pulsation, *Monthly Notices Roy. Astr. Soc. (London)* **101**, 182–194 (received 2 April 1941, published in issue No. 4 of April 1941).

1943 Joseph Larmor, *Obituary Notices of the Fellows of the Royal Society* **4**, 197–207.

EDEN, RICHARD JOHN, AND JOHN CHARLTON POLKINGHORNE

1972 Dirac in Cambridge, in *Aspects of Quantum Theory* (A. Salam and E. P. Wigner, eds.), Cambridge: Cambridge University Press, pp. 1–5.

EHRENFEST, PAUL

1913c Een mechanische theorema van Boltzmann en zijne betrekking tot de quanta theorie, *Versl. Kon. Akad. Wetensch. (Amsterdam)* **22**, 586–593 (communicated by H. A. Lorentz to the meeting of 29 November 1913). English translation: A mechanical theorem of Boltzmann and its relation to the theory of quanta, *Proc. Kon. Akad. Wetensch. (Amsterdam)* **16**, 591–597; reprinted in Paul Ehrenfest: *Collected Scientific Papers* (M. J. Klein, ed.), Amsterdam: North-Holland Publishing Company, 1959, pp. 340–346.

EHRENFEST, PAUL, AND RICHARD CHACE TOLMAN

1924 Weak quantization, *Phys. Rev. (2)* **24**, 287–295 (dated April 1924, published in issue No. 3 of September 1924); reprinted in P. Ehrenfest: *Collected Scientific Papers*, pp. 498–506.

EHRENFEST, PAUL, AND VIKTOR TRKAL

1920 Deduction of the dissociation-equilibrium from the theory of quanta and a calculation of the chemical constants based on this, *Proc. Kon. Akad. Wetensch. (Amsterdam)* **23**, 162–183 (communicated to the meeting of 28 February 1920); reprinted in P. Ehrenfest: *Collected Scientific Papers*, pp. 414–435.

EINSTEIN, ALBERT

1905b Über einen die Erzeugung und Verwandlung des Lichtes betreffenden heuristischen Gesichtspunkt, *Ann. d. Phys. (4)* **17**, 132–148 (received 18 March 1905, published in issue No. 6 of 9 June 1905); reprinted in Albert Einstein: *Die Hypothese der Lichtquanten* (*Dokumente der Naturwissenschaften—Abteilung Physik*, Vol 7; A. Hermann, ed.), Stuttgart: E. Battenberg Verlag, 1966, pp. 26–42. English translation: On a heuristic point of view about the creation and conversion of light, in D. ter Haar: *The Old Quantum Theory*, Oxford–London–Edinburgh–New York–Toronto–Sydney–Paris–Braunschweig: Pergamon Press, pp. 91–107.

1906g Die Plancksche Theorie der Strahlung und die Theorie der spezifischen Wärme, *Ann d. Phys. (4)* **22**, 180–190 (received 9 November 1906, published in issue No. 1 of 28 December 1906); reprinted in A. Einstein, P. Debye, M. Born and T. von Kármán: *Die Quantentheorie der spezifischen Wärme* (*Dokumente der Naturwissenschaften—Abteilung Physik*, Vol. 8; A. Hermann, ed.), München, 1967, pp. 20–31.

1916c Die Grundlage der allgemeinen Relativisitätstheorie, *Ann. d. Phys. (4)* **49**, 769–822 (received 20 March 1916, published in issue No. 7 of 11 May 1916); reprinted in H. A. Lorentz, A. Einstein, H. Minkowski and H. Weyl: *Das Relativitätsprinzip*, third edition (A. Sommerfeld, ed.), Leipzig: B. G. Teubner, 1920. English translation: The foundation of the general theory of relativity, in H. A. Lorentz, A. Einstein, H. Minkowski and H. Weyl: *The Principle of Relativity* (translation of the fourth German edition of *Das Relativitätsprinzip*, 1922, by W. Perret and G. B. Jeffery), London: Methuen & Company, Ltd.; reprinted by Dover Publications, Inc., New York, pp. 111–164.

1916d Strahlungs-Emission und Absorption nach der Quantentheorie, *Verh. d. Deutsch. Phys. Ges. (2)* **18**, 318–323 (received 17 July 1916; presented at the meeting of 23 July 1916, published in issue No. 13/14 of 30 July 1916).

1924c Quantentheorie des einatomigen idealen Gases, *Sitz.ber. Preuss. Akad. Wiss. (Berlin)*, pp. 261–267 (presented at the meeting of 10 July 1924).

1925a Quantentheorie des einatomigen idealen Gases. 2. Abhandlung, *Sitz.ber. Preuss. Akad. Wiss. (Berlin)*, pp. 3–14 (presented at the meeting of 8 January 1925).

1925b Quantentheorie des idealen Gases, *Sitz.ber. Preuss. Akad. Wiss. (Berlin)*, pp. 18–25 (presented at the meeting of 29 January 1925).

EINSTEIN, ALBERT, AND PAUL EHRENFEST

1922 Quantentheoretische Bemerkungen zum Experiment von Stern und Gerlach, *Z. Phys.* **11**, 31–34 (received 21 August 1922, published in issue No. 1 of 16 September 1922); reprinted in P. Ehrenfest: *Collected Scientific Papers*, pp. 452–455.

1923 Zur Quantentheorie des Strahlungsgleichgewichts, *Z. Phys.* **19**, 301–306 (received 16 October 1923, published in issue No. 5/6 of 8 December 1923); reprinted in P. Ehrenfest: *Collected Scientific Papers*, pp. 485–490.

EPSTEIN, PAUL SOPHUS

1922a Die Störungsrechnung im Dienste der Quantentheorie. I. Eine Methode der Störungsrechnung, *Z. Phys.* **8**, 211–228 (received 9 November 1921, published in issue No. 4 of 31 January 1922).

1922b Die Störungsrechnung im Dienste der Quantentheorie. II. Die numerische Durchführung der Methode, *Z. Phys.* **8**, 305–320 (received 9 November 1921, published in issue No. 5 of 15 February 1922).

1922c Die Störungsrechnung im Dienste der Quantentheorie. III. Kritische Bemerkungen zur Dispersionstheorie, *Z. Phys.* **9**, 92–110 (received 18 December 1921, published in issue No. 1/2 of 15 March 1922).

1922d Problems of quantum theory in the light of the theory of perturbations, *Phys. Rev. (2)* **19**, 578–608 (dated September 1921, published in issue No. 6 of June 1922).

1926 The Stark effect from the point of view of Schrödinger's quantum theory, *Phys. Rev. (2)* **28**, 695–710 (dated 29 July 1926, published in issue No. 4 of October 1926).

1927 The magnetic dipole in Schrödinger's theory, *Phys. Rev. (2)* **29**, 750 (abstract of a paper presented at the Los Angeles meeting of the American Physical Society, 5 March 1927, published in issue No. 5 of May 1927).

EPSTEIN, PAUL SOPHUS, AND PAUL EHRENFEST

1924 The quantum theory of the Fraunhofer diffraction, *Proc. Nat. Acad. Sci. (U.S.A.)* **10**, 133–139 (communicated 14 February 1924, published in issue No. 4 of 15 April 1924); reprinted in P. Ehrenfest: *Collected Scientific Papers*, pp. 491–497.

EUCKEN, ARNOLD

1914 *Die Theorie der Strahlung und der Quanten. Verhandlungen einer von E. Solvay einberufenen Zusammenkunft (30. Oktober bis 3. November 1911). Mit einem Anhange über die Entwicklung der Quantentheorie vom Herbst 1911 bis Sommer 1913* (reports and discussions of the first Solvay Conference, edited in German by A. Eucken; appendix by A. Eucken), Halle an der Saale: Wilhelm Knapp Verlag.

FOSTER, JOHN STUART

1927a Stark patterns observed in helium, *Proc. Roy. Soc. (London)* **A144**, 47–66 (communicated by A. S. Eve; received 1 November 1926, published in issue No. A 766 of 1 February 1927).

1927b Theory of the Stark-effect in the arc spectra of helium, *Phys. Rev. (2)* **29**, 916 (abstract of a paper submitted to the Washington meeting of the American Physical Society, 22–23 April 1927, published in issue No. 6 of June 1927).

1927c Application of quantum mechanics to the Stark effect in helium, *Proc. Roy. Soc. (London)* **A117**, 137–163 (communicated by N. Bohr; received 8 August 1927, published in issue No. A 776 of 1 December 1927).

FOWLER, RALPH HOWARD

1920 *The Elementary Differential Geometry of Plane Curves* (*Cambridge Tracts on Mathematics*, No. 20), Cambridge (second edition 1929).

1921 A simple extension of Fourier's integral theorem and some physical applications, in particular to the theory of quanta, *Proc. Roy. Soc. (London)* **A99**, 462–471 (communicated by E. Rutherford; received 8 June 1921, published in issue No. A 701 of 1 September 1921).

1922 Notes on the kinetic theory of gases. Sutherland's constant S and van der Waals' a and their relation to the intermolecular field, *Phil. Mag. (6)* **43**, 785–800 (communicated by C. G. Darwin, published in issue No. 257 of May 1922).

1923a Dissociation-equilibria by the method of partitions, *Phil. Mag. (6)* **45**, 1–33 (dated August 1922, published in issue No. 265 of January 1923).

1923b Contributions to the theory of motion of α-particles through matter. Part I, Ranges., *Proc. Camb. Phil. Soc.* **21**, 521–540 (read 5 February 1923, published in Part IV).

1923c Bohr's atom in relation to the problem of covalency, *Transactions of the Faraday Society* **19**, 459–468 (contribution to the General Discussion on 'The Electronic Theory of Valency,' held by the Faraday Society, Cambridge, 13–14 July 1923; manuscript received 9 July 1923).

1924a Statistical equilibrium with special reference to the mechanism of ionization by electronic impacts, *Phil. Mag. (6)* **47**, 257–277 (published in issue No. 278 of February 1924).

1924b A tentative theory of the capture and loss of electrons by swift nuclei, *Phil. Mag. (6)* **47**, 416–430 (published in issue No. 278 of February 1924).

1924c The statistical theory of dissociation and ionization by collision, with applications to the capture and loss of electrons by α-particles, *Proc. Camb. Phil. Soc.* **22**, 253–272 (read 19 May 1924, addendum of 28 June 1924).

1925a Applications of the correspondence principle to the theory of line-intensities in band-spectra, *Phil. Mag. (6)* **49**, 1272–1288 (published in issue No. 294 of June 1925).

1925b A theoretical study of the stopping power of hydrogen atoms for α-particles, *Proc. Camb. Phil. Soc.* **22**, 793–803 (received 6 April 1925, read 4 May 1925).

1925c Notes on the theory of absorption lines in stellar spectra, *Monthly Notices Roy. Astr. Soc. (London)* **85**, 970–977 (published in issue No. 9 of October 1925).

1925d A note on the summation rules for the intensities of spectral lines, *Phil. Mag. (6)* **50**, 1079–1083 (published in issue No. 299 of November 1925).

1929 *Statistical Mechanics: The Theory of the Properties of Matter in Equilibrium* (expanded Adams Prize Essay for 1923–24), Cambridge: Cambridge University Press.

FOWLER, RALPH HOWARD, AND EDWARD ARMAND GUGGENHEIM

1925 Applications of statistical mechanics to determine the properties of matter in stellar interiors, *Monthly Notices Roy. Astr. Soc. (London)* **85**, 939–970 (published in issue No. 9 of October 1925).

FOWLER, RALPH HOWARD, AND EDWARD ARTHUR MILNE

1923 The intensities of absorption lines in stellar spectra, and the temperatures and pressures in the reversing layers of stars, *Monthly Notices Roy. Astr. Soc. (London)* **83**, 403–424 (published in issue No. 7 of 11 May 1923).

1924 The maxima of absorption lines in stellar spectra (Second Paper), *Monthly Notices Roy. Astr. Soc. (London)* **84**, 499–515 (published in issue No. 7 of 9 May 1924).

1925 A note on the principle of detailed balancing, *Proc. Nat. Acad. Sci. (U.S.A.)* **11**, 400–402 (communicated 25 May 1925, published in issue No. 7 of 15 July 1925).

FRENKEL, JAKOW (JAKOB)

1926 Die Elektrodynamik des rotierenden Elektrons, *Z. Phys.* **37**, 243–262 (received 2 May 1926, published in issue No. 4/5 of 5 June 1926).

1929 *Einführung in die Wellenmechanik*, Berlin: J. Springer Verlag.

GAMOW, GEORGE

1966 *Thirty Years that Shook Physics: The Story of Quantum Theory*, New York: Doubleday.

GEEL, WILLEM CHRISTIAAN VAN

1926 Intensitäten der Zeemankomponenten im partiellen Paschen-Back-Effekt, *Z. Phys.* **39**, 877–878 (received 30 September 1926, published in issue No. 10/11 of 16 November 1926).

GOLDBERG, STANLEY

1970 In defense of ether: The British response to Einstein's special theory of relativity, 1905–1911, *Historical Studies in the Physical Sciences 2*, 89–125.

GRAßMANN, HERMANN

1844 *Die Wissenschaft der extensiven Größen oder die Ausdehnungslehre. Eine neue mathematische Disziplin, dargestellt und durch Anwendungen erläutert. 1. Theil*, Leipzig: O. Wiegand.

1855 Sur les différents genres de multiplication, *J. reine & angew. Math.* **49**, 123–141.

1862 *Die Ausdehnungslehre. Vollständig und in strenger Form bearbeitet*, Berlin: Th. Enslin.

GREBE, LEONHARD

1920 Über die Gravitationsverschiebung der Fraunhoferschen Linien, *Phys. Zs.* **21**, 662–666 (presented at the 86th *Naturforscherversammlung*, Bad Nauheim, 19–25 September 1920, published in issue No. 23/24 of 1/15 December 1920).

HALPERN, OTTO

1924 Über das Wärmegleichgewicht zwischen Hohlraumstrahlung und Quantenatomen, *Z. Phys.* **21**, 151–158 (received 4 December 1923, published in issue No. 3 of 8 February 1924).

1926a Über die Quantelung des Rotators und die Koordinatenwahl in der neuen Quantenmechanik, *Naturwiss.* **14**, 488–489 (letter dated March 1926, published in the issue of 21 May 1926).

1926b Notiz über die Quantelung des Rotators und die Koordinatenwahl in der neuen Quantenmechanik, *Z. Phys.* **38**, 8–11 (received 5 June 1926, published in issue No. 1/2 of 24 July 1926).

HALPERN, OTTO, AND HANS THIRRING

1929 Die Grundgedanken der neueren Quantentheorie. Zweiter Teil: Die Weiterentwicklung seit 1926, *Ergeb. d. exakt. Naturwiss.* **8**, 367–508.

HARDY, GODFREY HAROLD

1921 Srinivasa Ramanujan, 1887–1920, *Proc. Roy. Soc. (London)* **A99**, XIII–XXIX (published in the issue of September 1921).

1967 *A Mathematician's Apology*, reprint, Cambridge: Cambridge University Press.

HEAVISIDE, OLIVER

1893a On operators in physical mathematics. Part I. *Proc. Roy. Soc. (London)* **52**, 504–529 (received 15 December 1892, read 2 February 1893, published in issue No. 320 of February 1893).

1893b On operators in physical mathematics. Part II., *Proc. Roy. Soc. (London)* **54**, 105–143 (received 8 June 1893, read 15 June 1893, published in issue No. 326 of June 1893).

HEISENBERG, WERNER

1924e Über eine Abänderung der formalen Regeln der Quantentheorie beim Problem der anomalen Zeemaneffekte, *Z. Phys.* **26**, 291–307 (received 13 June 1924, published in issue No. 4/5 of 14 August 1924).

1925a Über eine Anwendung des Korrespondenzprinzips auf die Frage nach der Polarisation des Fluoreszenzlichtes, *Z. Phys.* **31**, 617–626 (received 30 November 1924, published in issue No. 7/8 of 7 March 1925).

1925b Zur Quantentheorie der Multiplettstruktur und der anomalen Zeemaneffekte, *Z. Phys.* **32**, 841–860 (received 10 April 1925, published in issue No. 11/12 of 30 June 1925).

1925c Über die quantentheoretische Umdeutung kinematischer und mechanischer Beziehungen, *Z. Phys.* **33**, 879–893 (received 29 July 1925, published in issue No. 12 of 18 September 1925); reprinted in *Begründung der Quantenmechanik* (Born, Heisenberg and Jordan, 1962), pp. 30–46. English translation: Quantum theoretical reinterpretation of kinematic and mechanical relations, in *Sources of Quantum Mechanics* (van der Waerden, 1967), pp. 261–276.

1926a Über quantentheoretische Kinematik und Mechanik, *Mathematische Annalen* **95**, 683–705 (received 21 December 1925).

1926c Über die Spektra von Atomsystemen mit zwei Elektronen, *Z. Phys.* **39**, 499–518 (received 24 July 1926, published in issue No. 7/8 of 26 October 1926).

1926d Quantenmechanik, *Naturwiss.* **14**, 989–994 (presented on 26 September 1926 at the 89th *Naturforscherversammlung*, Düsseldorf, published in the issue of 5 November 1926).

1927b Über den anschaulichen Inhalt der quantentheoretischen Kinematik und Mechanik, *Z. Phys.* **43**, 172–198 (received 23 March 1927, published in issue No. 3/4 of 31 May 1927).

1930 *The Physical Principles of Quantum Theory* (lectures delivered in spring 1929 at the University of Chicago, translated by C. Eckart and F. C. Hoyt), Chicago: The University of Chicago Press. German edition: *Die physikalischen Prinzipien der Quantentheorie*, Leipzig: S. Hirzel Verlag, 1930.

1969a Theory, criticism and a philosophy, in *From a Life in Physics: Evening Lectures at the International Centre for Theoretical Physics, Trieste, Italy*, Vienna: International Atomic Energy Agency, pp. 31–46.

1971 *Physics and Beyond: Encounters and Conversations* (translation of *Der Teil und das Ganze*, Munich: Piper Verlag, 1969, by A. J. Pomerans), New York–Evanston–London: Harper & Row, Publishers.

HEISENBERG, WERNER, AND PASCUAL JORDAN

1926 Anwendung der Quantenmechanik auf das Problem der anomalen Zeemaneffekte, *Z. Phys.* **37**, 263–277 (received 16 March 1926, published in issue No. 4/5 of 5 June 1926).

HENDERSON, J. B., AND HENRY RONALD HASSÉ

1922 A contribution to the thermodynamical theory of explosions. Parts I. and II. *Proc. Roy. Soc. (London)* **A100**, 461–482 (received 6 May 1921, published in issue No. A 706 of 1 February 1922).

HILBERT, DAVID

1899 *Grundlagen der Geometrie*, Leipzig: B. G. Teubner.

HILBERT, DAVID, JOHANNES VON NEUMANN, AND LOTHAR NORDHEIM

1928 Über die Grundlagen der Quantemechanik, *Mathematische Annalen* **98**, 1–30 (received 6 April 1927).

HILL, EDWARD, AND JOHN HASBROUCK VAN VLECK

1928 On the quantum mechanics of the rotational distortion of multiplets in molecular spectra, *Phys. Rev (2)* **32**, 250–272 (dated 1 June 1928, published in issue No. 2 of August 1928).

HODGE, WILLIAM VALLANCE DOUGLAS

1956 Henry Frederick Baker, 1866–1956, *Biographical Memoirs of the Fellows of the Royal Society* **2**, 49–68.

1959 Peter Fraser, *J. London Math. Soc.* **34**, 111–112 (published in Part 1, No. 133 of January 1959).

HÖNL, HELMUT

1925 Die Intensitäten der Zeemankomponenten, *Z. Phys.* **31**, 340–354 (received 26 November 1924, published in issue No. 1–4 of 11 February 1925).

HOYT, FRANK C.

1925 The harmonic analysis of electron orbits, *Phys. Rev. (2)* **25**, 174–186 (dated 3 October 1924, published in issue No. 2 of February 1925).

1926a Transition probabilities and principal quantum numbers, *Phys. Rev. (2)* **27**, 105 (abstract of a paper presented at the Chicago meeting of the American Physical Society, 27–28 November 1925, published in issue No. 1 of January 1926).

1926b Intensities in spectra and the new quantum mechanics, *Phys. Rev. (2)* **27**, 805 (abstract of a paper presented at the Washington meeting of the American Physical Society, 23–24 April 1926, published in issue No. 6 of June 1926).

HUND, FRIEDRICH

1925f Isotropes Ion und Bau der Molekeln und Kristallgitter, *Phys. Zs.* **26**, 682–685 (presented at the Danzig meeting of the German Physical Society, 11–17 September 1925, published in issue No. 19 of 1 October 1925).

1925g Versuch einer Ableitung der Gittertypen aus der Vorstellung des isotropen polarisierbaren Ions, *Z. Phys.* **34**, 833–857 (received 16 October 1925, published in issue No. 11/12 of 28 November 1925).

JACOBI, CARL GUSTAV JACOB

1838 Ein neues Theorem der analytischen Mechanik, *Monatsberichte der Kgl. Preussischen Akademie der Wissenschaften (Berlin)*, pp. 178–182; reprinted in *Gesammelte Werke 4*, (K. Weierstrass, ed.), Berlin: G. Reimer, 1886, pp. 137–142.

1841 Sur un théorème de Poisson, *Comptes rendus (Paris)* **11**, 529 (letter to the President of the Academy); reprinted in *Gesammelte Werke 4*, pp. 143–146.

1843 L' élimination des noeds dans le problème des trois corps, *Comptes rendus (Paris)* **15**, 236–255; reprinted in *J. reine & angew. Math.* **26**, 115–131 (1843), and in *Gesammelte Werke 4*, pp. 297–314.

1866 *Vorlesungen über Dynamik, gehalten an der Universität zu Königsberg im Wintersemester 1842–1843 und nach einem von C. W. Borchardt ausgearbeiteten Hefte* (A. Clebsch, ed.), Berlin: G. Reimer; reprinted in *Gesammelte Werke, Supplementband* (E. Lottner, ed.), Berlin: G. Reimer.

JAMMER, MAX

1966 *The Conceptual Development of Quantum Mechanics*, New York–St. Louis–San Francisco–Toronto–London–Sydney: McGraw-Hill Book Company.

JAUNCEY, GEORGE ERIC MACDONNELL

1924c A corpuscular quantum theory of the scattering of polarized X-rays, *Phys. Rev. (2)* **23**, 313–317 (dated 15 October 1923, published in issue No. 3 of March 1924).

JEANS, JAMES HOPWOOD

1904 *The Dynamical Theory of Gases*, Cambridge: Cambridge University Press (second edition 1916, third edition 1920, fourth edition 1925, reprinted New York: Dover Publications, Inc.).

1905a The dynamical theory of gases and of radiation, *Nature* **72**, 101–102 (letter dated 20 May 1905, published in the issue of 1 June 1905).

1905b On the partition of energy between matter and aether, *Phil. Mag. (6)* **10**, 91–98 (communicated April 1905, with a postscript added 7 June published in issue No. 55 of July 1905).

1910 On non-Newtonian mechanical systems, and Planck's theory of radiation, *Phil. Mag. (6)* **20**, 943–954 (dated 17 August 1910, published in issue No. 120 of December 1910).

1914a Discussion on radiation, *Report on the 83rd Meeting of the British Association, Birmingham 1913* (presented on 12 September 1913) London: John Murray.

1914b *Report on Radiation and the Quantum-Theory*, London: The Physical Society of London ("The Electrician" Printing & Publishing Co.).

1916 *The Dynamical Theory of Gases*, second edition, Cambridge: Cambridge University Press.

1919 *Problems of Cosmogony and Stellar Dynamics* (Adams Prize Essay 1917), Cambridge: Cambridge University Press.

1924 *Report on Radiation and the Quantum-Theory*, second edition, London: The Physical Society of London (Fleetway Press, Ltd.).

1925a Address, delivered by the President, Dr. J. H. Jeans, on presenting the Gold Medal of the Society to Sir Frank Watson Dyson, Astronomer Royal, for his contributions to astronomy in general, and, in particular, for his work on the proper motions of the stars, *Monthly Notices Roy. Astr. Soc. (London)* **85**, 672–677 (published in No. 8 of 12 June 1925).

1925b *The Dynamical Theory of Gases*, fourth edition, Cambridge: Cambridge University Press.

1928 *Astronomy and Cosmogony*, Cambridge: Cambridge University Press.

JORDAN, PASCUAL

1926 Bemerkung über einen Zusammenhang zwischen Duanes Quantentheorie der Interferenz und den de Broglieschen Wellen, *Z. Phys.* **37**, 376–382 (received 22 April 1926, published in issue No. 4/5 of 5 June 1926).

1928 Die Lichtquantenhypothese. Entwicklung und gegenwärtiger Stand, *Ergeb. d. exakt. Naturwiss.* **7**, 158–208.

KAPITZA, PETER L., AND PAUL ADRIEN MAURICE DIRAC

1933 The reflection of electrons from standing light waves, *Proc. Cambr. Phil. Soc.* **29**, 297–300 (received 24 March 1933, read 1 May 1933).

KELLAND, PHILIP, AND PETER GUTHRIE TAIT

1873 *Introduction to Quanternions*, London: Macmillan & Company (later editions 1891 and 1904, the last edited by C. G. Knott).

KEMBLE, EDWIN CRAWFORD

1916a The distribution of angular velocities among diatomic molecules, *Phys. Rev. (2)* **8**, 689–700 (dated 7 August 1916, published in issue No. 6 of December 1916).

1916b On the occurrence of harmonics in the infra-red absorption spectra of gases, *Phys. Rev. (2)* **8**, 701–714 (dated 9 August 1916, published in issue No. 6 of December 1916).

1920 The Bohr theory and the approximate harmonics in the infra-red spectra of diatomic gases, *Phys. Rev. (2)* **15**, 95–109 (published in issue No. 2 of February 1920).

1921 The probable normal state of the helium atom, *Phil. Mag. (6)* **42**, 123–133 (communicated by T. Lyman; dated 11 February 1921, published in issue No. 247 of July 1921).

1926b *Problems of Atomic Dynamics. (Lectures delivered at the Massachusetts Institute of Technology). Max Born, Phys. Rev. (2)* **28**, 423–424 (book review, published in issue No. 2 of August 1926).

1929 General principles of quantum mechanics. Part I., *Reviews of Modern Physics* **1**, 157–215 (published in issue No. 2 of October 1929).

Kemble, Edwin Crawford, and Edward Lee Hill

1930 General principles of quantum mechanics. Part II., *Reviews of Modern Physics* **2**, 1–58 (published in issue No. 1 of January 1930).

Kemble, Edwin Crawford, and John Hasbrouck Van Vleck

1923 On the theory of the temperature variation of the specific heat of hydrogen, *Phys. Rev. (2)* **21**, 653–661 (dated 13 December 1922, published in issue No. 6 of June 1923).

Kirchner, Fritz

1926 Experimentelle Untersuchungen über die Richtungsverteilung der von Röntgenstrahlen ausgelösten Elektronen, *Phys. Zs.* **27**, 799–801 (presented on 24th September 1926 at the 89th *Naturforscherversammlung*, Düsseldorf, published in issue No. 23 of 1 December 1926).

Kittel, Charles

1974 Larmor and the prehistory of the Lorentz transformations, *Amer. J. Phys.* **42**, 726–729 (received 27 December 1973, published in the issue of September 1974).

Klein, Oskar

1924a The simultaneous action on a hydrogen atom of crossed homogeneous electric and magnetic fields, *Phys. Rev. (2)* **23**, 308 (abstract of a paper presented at the Cincinnati meeting of the American Physical Society, 27–29 December 1923, published in issue No. 2 of February 1924).

1924b Über die gleichzeitige Wirkung von gekreuzten homogenen elektrischen und magnetischen Feldern auf das Wasserstoffatom. I., *Z. Phys.* **22**, 109–118 (dated 31 December 1923, received 23 January 1924, published in issue No. 1/2 of 6 March 1924).

Klein, Oskar, and Svein Rosseland

1921 Über Zusammenstöße zwischen Atomen und freien Elektronen, *Z. Phys.* **4**, 46–51 (received 20 November 1920, published in issue No. 1 of January 1921).

Korn, Arthur

1926 *Die Konstitution der chemischen Atome*, Berlin: G. Siemens.

Kramers, Hendrik Anthony

1920b Über den Einfluß eines elektrischen Feldes auf die Feinstruktur der Wasserstofflinien, *Z. Phys.* **3**, 199–223 (received 1 October 1920, published in issue No. 4 of November 1920); reprinted in *Collected Scientific Papers*, pp. 109–133.

1923a Über das Modell des Heliumatoms, *Z. Phys.* **13**, 312–341 (received 31 December 1922, published in issue No. 5 of 19 February 1923); reprinted in *Collected Scientific Papers*, pp. 192–221.

1923d Theory of X-ray absorption and of the continuous X-ray spectrum, *Phil. Mag. (6)* **46**, 836–871 (published in issue No. 275 of November 1923); reprinted in *Collected Scientific Papers*, pp. 156–191.

1924a The law of dispersion and Bohr's theory of spectra, *Nature* **113**, 673–674 (letter dated 25 March 1924, published in the issue of 10 May 1924); reprinted in *Collected Scientific Papers*, pp. 290–291, and in *Sources of Quantum Mechanics* (van der Waerden, 1967), pp. 177–180.

1924b The quantum theory of dispersion, *Nature* **114**, 310–311 (letter dated 22 July 1924, published in the issue of 30 August 1924); reprinted in *Collected Scientific Papers*, p. 292, and in *Sources of Quantum Mechanics* (van der Waerden, 1967), pp. 199–201.

1925c Eenige Opmerkingen over de Quantenmechanica van Heisenberg, *Physica* **5**, 369–376 (dated November 1925, published in issue No. 11/12 of November–December 1925).

1956 *Collected Scientific Papers*, Amsterdam: North-Holland Publishing Company.

KRAMERS, HENDRIK ANTHONY, AND WERNER HEISENBERG

1925 Über die Streuung von Strahlung durch Atome, *Z. Phys.* **31**, 681–708 (received 5 January 1925, published in issue No. 9 of 17 March 1925); reprinted in *Collected Scientific Papers* (Kramers, 1956), pp. 293–320. English translation: On the dispersion of radiation by atoms, in *Sources of Quantum Mechanics* (van der Waerden, 1967), pp. 223–251.

KRONIG, RALPH DE LAER

1925a Über die Intensität der Mehrfachlinien und ihre Zeemankomponenten, *Z. Phys.* **31**, 885–897 (received 18 February 1925, published in issue No. 12 of 14 April 1925).

1925b Über die Intensität der Mehrfachlinien und ihre Zeemankomponenten. II., *Z. Phys.* **33**, 261–272 (received 2 June 1925, published in issue No. 4 of 1 August 1925).

1926b The theory of X-ray dispersion, *Phys. Rev.* **27**, 797 (abstract of a paper presented at the Washington meeting of the American Physical Society, 23–24 April 1926, published in issue No. 6 of June 1926).

1926c On the theory of dispersion of X-rays, *J. Opt. Soc. Amer. & Rev. Sci. Instr.* **12**, 547–557 (dated 29 January 1926, published in issue No. 6 of June 1926).

1926d The dielectric constant of diatomic dipole-gases on the new quantum mechanics, *Proc. Nat. Acad. Sci. (U.S.A)* **12**, 488–493 (communicated 3 July 1926, published in issue No. 8 of 15 August 1926).

KRONIG, RALPH DE LAER, AND ISIDOR ISAAC RABI

1927 The symmetrical top in the undulatory mechanics, *Phys. Rev. (2)* **29**, 262–269 (dated 4 November 1926, published in issue No. 2 of February 1927).

LANCZOS, KORNEL (CORNELIUS)

1926a Über eine feldmäßige Darstellung der neuen Quantenmechanik, *Z. Phys.* **35**, 812–830 (received 22 December 1925, published in issue No. 11/12 of 26 February 1926).

LANDAU, LEV DAVIDOVICH

1926 Zur Theorie der Spektren der zweiatomigen Moleküle, *Z. Phys.* **40**, 621–627 (received 13 November 1926, published in issue No. 8 of 22 December 1926).

Landé, Alfred

1923a Termstruktur und Zeemaneffekt der Multipletts, *Z. Phys.* **15**, 189–205 (received 5 March 1923, published in issue No. 4/5 of 26 May 1923).

1926c Neue Wege in der Quantenmechanik, *Naturwiss.* **14**, 455–458 (published in the issue of 14 May 1926).

1926d *Die neuere Entwicklung der Quantentheorie,* (*Wissenschaftliche Forschungsberichte —Naturwissenschaftliche Reihe*, Vol. 5, R. Liesegang, ed.), Dresden-Leipzig: T. Steinkopff (second edition of *Fortschritte der Quantentheorie*, 1922).

1930 *Vorlesungen über Wellenmechanik (gehalten an der Staatsuniversität zu Columbus, U.S.A.),* Leipzig: Akademische Verlagsgesellschaft.

Langevin, Paul, and Maurice de Broglie (Editors)

1912 *La Théorie de Rayonnement et les Quanta—Rapports et Discussions de la Réunion tenue à Bruxelles, du 30 Octobre au 3 Novembre 1911*, Paris: Gauthier-Villars.

Laporte, Otto

1926 Interpretation of complex spectra, *Phys. Rev. (2)* **27**, 512–513 (abstract of a paper presented at the Montreal meeting of the American Physical Society, 26–27 February 1926, published in issue No. 4 of April 1926).

Larmor, Joseph

1897b A dynamical theory of the electric and luminiferous medium.—Part III. Relations with material media, *Phil. Trans. Roy. Soc. (London)* **A190**, 205–300 (received 21 April 1897, read 13 May 1897).

1900b *Aether and Matter: A Development of the Dynamical Relations of the Aether to Material Systems on the Basis of the Atomic Constitution of Matter, Including a Discussion of the Earth's Motion on Optical Phenomena, Being an Adams Prize Essay in the University of Cambridge*, Cambridge: Cambridge University Press.

1909 On the statistical and thermodynamical relations of radiant energy, *Proc. Roy. Soc. (London)* **A83**, 82–95 (Bakerian lecture delivered on 18 November 1909, published in issue No. A 560 of December 1909).

1919 On generalized relativity in connection with Mr. W. J. Johnston's symbolic calculus, *Proc. Roy. Soc. (London)* **A96**, 334–363 (received 28 August 1919, published in issue No. A 678 of 15 December 1919).

1921 On electro-crystalline properties as conditioned by atomic lattices, *Proc. Roy. Soc. (London)* **A99**, 1–10 (received 6 January 1921, published in issue No. A 696 of 6 April 1921).

Larsen, Egon

1962 *The Cavendish Laboratory: Nursery of Genius*, London: E. Ward Ltd., New York: F. Watts, Inc.

Laue, Max von

1925 Zum Prinzip der mechanischen Transformierbarkeit. (Adiabatenhypothese.), *Ann. d. Phys. (4)* **76**, 619–628 (received 3 February 1925, published in issue No. 6 of April 1925).

LEHRER, ERWIN

1926 Über die Druckabhängigkeit der Suszeptibilität diamagnetischer Gase (Vorläufige Mitteilung.), *Z. Phys.* **37**, 155–156 (received 14 April 1926, published in issue No. 1/2 of 22 May 1926).

LENNARD-JONES, JOHN EDWARD

1924 On the determination of molecular fields.—I. From the variation of the viscosity of a gas with temperature, *Proc. Roy. Soc. (London)* **A106**, 441–462 (communicated by S. Chapman; received 22 April 1924, published in issue No. A 738 of 1 October 1924).

1929 The electronic structure of some diatomic molecules, *Transactions of the Faraday Society* **25**, 668–686 (presented at the General Discussion on 'Molecular Spectra and Molecular Structure,' held by the Faraday Society at Bristol, 24–25 September 1929; manuscript received 5 September 1929).

LENNARD-JONES, JOHN EDWARD, AND BERYL M. DENT

1926 The forces between atoms and ions. II., *Proc. Roy. Soc. (London)* **A112**, 230–234 (communicated by S. Chapman; received 8 June 1926, published in issue A 760 of 3 August 1926).

1927a Some theoretical determinations of the structure of carbonate crystals. I., *Proc. Roy. Soc. (London)* **A113**, 673–689 (communicated by S. Chapman; received 3 November 1926, published in issue No. A 765 of 1 January 1927).

1927b Some theoretical determinations of the structure of carbonate crystals. II., *Proc. Roy. Soc. (London)* **A113**, 690–696 (communicated by S. Chapman; received 17 November 1926, published in issue No A 765 of 1 January 1927).

1928 The change in lattice spacing at a crystal boundary, *Proc. Roy. Soc. (London)* **A121**, 247–259 (communicated by S. Chapman; received 27 April 1928, published in issue No. 787 of 1 November 1928).

LEWIS, GILBERT NEWTON

1925 A new principle of equilibrium, *Proc. Nat. Acad. Sci. (U.S.A.)* **11**, 179–183 (communicated 21 January 1925, published in issue No. 3 of 15 March 1925).

LIE, SOPHUS

1875 Begründung einer Invarianten-Theorie der Berührungs-Transformationen, *Mathematische Annalen* **8**, 215–303 (dated 5 July 1874).

1878 *Theorie der Transformationsgruppen. Erster Abschnitt,* Leipzig: B. G. Teubner.

LINDEMANN, FREDERICK ALEXANDER

1910 Über die Berechnung molekularer Eigenfrequenzen, *Phys. Zs.* **11**, 609–612 (received 25 June 1910, published in issue No. 14 of 15 July 1910).

LODGE, OLIVER

1921a The geometrization of physics, and its supposed basis on the Michelson–Morley experiment, *Nature* **106**, 795–800 (published in the issue of 17 February 1921).

1921b Remarks on simple relativity and the relative velocity of light, *Nature* **107**, 716–719, 748–751, 784–785, 814–818 (published in the issues of 4, 11, 18 and 25 August 1921).

LONDON, FRITZ

1926a Energiesatz und Rydbergprinzip in der Quantenmechanik, *Z. Phys.* **36**, 775–777 (received 17 March 1926, published in issue No. 7 of 13 April 1926).

1926b Über die Jacobischen Transformationen der Quantenmechanik, *Z. Phys.* **37**, 915–925 (received 22 May 1926, published in issue No. 12 of 10 July 1926).

LORENTZ, HENDRIK ANTOON

1903 On the emission and absorption by metals of rays of heat of great wave-lengths, *Proc. Kon. Akad. Wetensch. (Amsterdam)* **5**, 666–685 (presented at the meeting of 24 April 1903); reprinted in H. A. Lorentz: *Collected Papers III* (P. Zeeman and A. D. Fokker, eds.), The Hague: Martinus Nijhoff, 1936, pp. 155–175.

1908a Le partage de l'énergie entre la matière pondérable et l'éther, *Il Nuovo Cimento* **16**, 5–34 (1908), and *Atti del IV Congresso Internazionale dei Matematici (Roma, 6–11 Aprile 1908) I,* Rome: R. Accademia dei Lincei, 1909, pp. 145–165; reprinted in *Revue génerale des sciences* **20**, 14–26 (1909), and in H. A. Lorentz: *Collected Papers VII* (P. Zeeman and A. D. Fokker, eds.), The Hague: Martinus Nijhoff, 1934, pp. 317–341.

1909b *The Theory of Electrons and its Application to the Phenomena of Light and Radiant Heat. (A Course of Lectures Delivered in Columbia University, New York, in March and April 1906),* Leipzig: B. G. Teubner.

1925b L'ancienne et la nouvelle mécanique, in *Le Livre du Cinquantenaire de la Société Française de Physique,* Paris: Editions de la Revue d'Optique Théorique et Instrumentale, pp. 99–114 (lecture delivered at the Sorbonne on 10 December 1923); reprinted in H. A. Lorentz: *Collected Papers VII,* pp. 274–289.

MANNEBACK, CHARLES

1926 Die Dielektrizitätskonstante der zweiatomigen Dipolgase nach der Wellenmechanik, *Phys. Zs.* **27**, 563–569 (received 20 July 1926, published in issue No. 17 of 1 September 1926).

MEHRA, JAGDISH

1972 The golden age of theoretical physics: P. A. M. Dirac's scientific work from 1924–1933, in *Aspects of Quantum Theory* (A. Salam and E. P. Wigner, eds.), Cambridge: Cambridge University Press, pp. 17–59.

1975b Satyendra Nath Bose, *Biographical Memoirs of Fellows of the Royal Society (London)* **21**, 117–154.

1975c *The Solvay Conferences on Physics: Aspects of the Development of Physics since 1911*, Dordrecht–Boston: D. Reidel Publishing Company.

MENSING, LUCIE (LUCY)

1926a Die Rotations-Schwingungsbanden nach der Quantenmechanik, *Z. Phys.* **36**, 814–823 (received 29 March 1926, published in issue No. 11/12 of 11 May 1926).

MENSING, LUCIE, AND WOLFGANG PAULI

1926 Über die Dielektrizitätskonstante von Dipolgasen nach der Quantenmechanik, *Phys. Zs.* **27**, 509–512 (received 25 June 1926, published in issue No. 15 of 1 August 1926).

MIE, GUSTAV

1925a Bremsstrahlung und Comptonsche Streustrahlung, *Z. Phys.* **33**, 33–41 (received 6 June 1925, published in issue No. 1/2 of 18 July 1925).

MILNE, EDWARD ARTHUR

1924a Statistical equilibrium in relation to the photo-electric effect, and its application to the determination of absorption coefficients, *Phil. Mag. (6)* **47**, 209–241 (dated 12 November 1923, published in issue No. 277 of January 1924).

1924b Recent work in stellar physics, *Proceedings of the Physical Society (London)* **36**, 94–113 (presented on 25 January 1924).

1925 Note on Saha's ionization formula, and on the theoretical value of the photo-electric absorption coefficient, *Phil. Mag. (6)* **50**, 547–550 (dated May 1925, published in issue No. 296 of August 1925).

1929 The structure and opacity of a stellar atmosphere, *Phil. Trans. Roy. Soc. (London)* **A228**, 421–461 (Bakerian lecture presented on 6 June 1929, manuscript received 6 June 1929, published in issue No. A 668 of 12 November 1929).

1945 Ralph Howard Fowler, *Obituary Notices of Fellows of the Royal Society (London)* **5**, 61–78.

MILNE, EDWARD ARTHUR, AND RALPH HOWARD FOWLER

1921 Siren harmonics and a pure tone siren, *Proc. Roy. Soc. (London)* **A98**, 414–427 (communicated by A. V. Hill; received 9 November 1920, published in issue No. A 694 of 3 March 1921).

NERNST, WALTHER, AND FREDERICK ALEXANDER LINDEMANN

1911a Untersuchungen über die spezifische Wärme bei tiefen Temperaturen. V., *Sitz. ber. Preuss. Akad. Wiss. (Berlin)*, pp. 494–501 (presented at the meeting of 6 April 1911).

1911b Spezifische Wärme und Quantentheorie, *Z. Elektrochem.* **17**, 817–827 (published in issue No. 18 of September 1911).

NEWALL, HUGH FRANK

1912 Stellar spectroscopy in 1911, *Monthly Notices Roy. Astr. Soc. (London)* **72**, 327–336 (report at the 92nd Annual General Meeting of the Royal Astronomical Society, published in issue No. 4 of 9 February 1912).

NEWMAN, MAXWELL HERMAN ALEXANDER

1925 On the theorem of Pappus, *Proc. Camb. Phil. Soc.* **22**, 919–923 (received 24 June 1925, read 20 July 1925).

NICHOLSON, JOHN WILLIAM

1912b The constitution of the solar corona. II., *Monthly Notices Roy. Astr. Soc. (London)* **72**, 677–692 (dated 28 April 1912, published in issue No. 8 of June 1912).

OLIPHANT, MARCUS

1972 *Rutherford: Recollections of the Cambridge Days*, Amsterdam–London–New York: Elsevier Publishing Company.

Parkinson, E. M.

1976 George Gabriel Stokes, in *Dictionary of Scientific Biography XIII*, New York: Charles Scribner's Sons, pp. 74–79.

Pauli, Wolfgang

1921a Zur Theorie der Dielektrizitätskonstante zweiatomiger Dipolgase, *Z. Phys.* **6**, 319–327 (received 30 July 1921, published in issue No. 6 of 3 October 1921); reprinted in *Collected Scientific Papers* **2**, pp. 39–47.

1923b Über das thermische Gleichgewicht zwischen Strahlung und freien Elektronen, *Z. Phys.* **18**, 272–286 (received 9 August 1923, published in issue No. 5 of 18 October 1923); reprinted in *Collected Scientific Papers 2*, pp. 161–175.

1926a Über das Wasserstoffspektrum vom Standpunkt der neuen Quantenmechanik, *Z. Phys.* **36**, 336–363 (received 17 January 1926, published in issue No. 5 of 27 March 1927); reprinted in *Collected Scientific Papers 2*, pp. 252–279. English translation: On the hydrogen spectrum from the standpoint of the new quantum mechanics, in *Sources of Quantum Mechanics* (van der Waerden, 1967), pp. 387–415.

1930 Born, M., and P. Jordan, *Elementare Quantenmechanik, Naturwiss.* **18**, 602 (book review published in the issue of 27 June 1930); reprinted in *Collected Scientific Papers* **2**, p. 1397.

1933c Die allgemeinen Prinzipien der Wellenmechanik, in *Handbuch der Physik 24/1* (H. Geiger and K. Scheel, eds.), second edition, Berlin: J. Springer Verlag, pp. 83–272; reprinted (except the last 30 pages) in *Handbuch der Physik V/1* (S. Flügge, ed.), Berlin: Springer-Verlag, 1958, pp. 1–168, and in *Collected Scientific Papers 1*, pp. 771–938.

1964 *Collected Scientific Papers*, 2 volumes (R. Kronig and V. G. Weisskopf, eds.), New York–London–Sydney: Interscience Publishers.

Pauling, Linus

1926a The quantum theory of the dielectric constant of hydrogen chloride and similar gases, *Proc. Nat. Acad. Sci. (U.S.A.)* **12**, 32–35 (communicated 14 December 1925, published in issue No. 1 of 15 January 1926).

1926b The quantum theory of the dielectric constant of hydrogen chloride and similar gases, *Phys. Rev. (2)* **27**, 568–577 (dated February 1926, published in issue No. 5 of May 1926).

1926c Die Abschirmkonstanten der relativistischen oder magnetischen Röntgendubletts, *Z. Phys.* **40**, 344–350 (received 27 October 1926, published in issue No. 5 of 8 December 1926).

1927 The influence of a magnetic field on the dielectric constant of a dipole gas, *Phys. Rev. (2)* **29**, 145–160 (dated 10 September 1926, published in issue No. 1 of January 1927).

Planck, Max

1906 *Vorlesungen über die Theorie der Wärmestrahlung*, Leipzig: J. A. Barth.

1910b *Acht Vorlesungen über theoretische Physik, gehalten 1909 an der Columbia Universität*, Leipzig: S. Hirzel Verlag.

1911a Eine neue Strahlungshypothese, *Verh. d. Deutsch. Phys. Ges. (2)* **13**, 138–148 (presented at the meeting of 3 February 1911, published in issue No. 3 of 15 February 1911); reprinted in *Physikalische Abhandlungen und Vorträge II*, pp. 249–259.

1926 Physikalische Gesetzlichkeit im Lichte neuerer Forschung, *Naturwiss.* **14**, 249–261 (presented on 14 February 1926 at the *Akademische Kurse*, Düsseldorf, published in the issue of 26 March 1926); reprinted in *Physikalische Abhandlungen und Vorträge III*, pp. 159–171.

1958 *Physikalische Abhandlungen und Vorträge,* (*Verband Deutscher Physikalischer Gesellschaften and Max-Planck-Gesellschaft zur Förderung der Wissenschaften e.V.*, eds.), 3 volumes, Braunschweig: Fr. Vieweg & Sohn.

Poincaré, Henri

1893 *Les Méthodes Nouvelles de la Mécanique Céleste. Tome II: Méthodes de MM. Newcomb, Lindstedt & Bohlin*, Paris: Gauthier-Villars; paperback reprint New York: Dover Publications, Inc., 1957.

1912a Sur la théorie des quanta, *J. Phys. (Paris) (5)* **2**, 5–34 (published in issue No. 1 of 12 January 1912).

Poisson, Siméon Denis

1809 Mémoire sur la variation des constantes arbitraire dans les questions de mécanique, *Journal de l'École Polytechnique* **8**, 266–344.

Rayleigh, John William Strutt, Lord

1900b Remarks upon the law of complete radiation, *Phil. Mag. (5)* **49**, 539–540 (published in issue No. 301 of June 1900).

1905b The dynamical theory of gases and radiation, *Nature* **72**, 54–55 (letter published in the issue of 18 May 1905).

1905c The constant of radiation as calculated from molecular data, *Nature* **72**, 243–244 (letter published in the issue of 13 July 1905).

Reiche, Fritz

1926 Landé, A., *Die neuere Entwicklung der Quantentheorie, Naturwiss.* **14**, 793–794 (book review published in the issue of 20 August 1926).

Reid, Constance

1970 *Hilbert*, New York–Heidelberg–Berlin: Springer-Verlag.

Rosenthal-Schneider, Ilse

1957 *Erinnerung an Gespräche mit Einstein*, unpublished manuscript, dated 23 July 1957, *Einstein Archives*, Princeton, N.J.

Russell, Bertrand

1903 *The Principles of Mathematics*, London: George Allen & Unwin.

Rutherford, Ernest

1919 Collision of α-particles with light atoms IV. An anomalous effect in nitrogen, *Phil. Mag. (6)* **37**, 581–587 (dated April 1919, published in issue No. 222 of June

1919); reprinted in *The Collected Papers of Lord Rutherford. Volume Two: Manchester* (J. Chadwick, ed.), London: George Allen & Unwin, Ltd., pp. 585–590.

1920 Nuclear constitution of atoms, *Proc. Roy. Soc. (London)* **A97**, 374–400 (Bakerian lecture delivered 3 June 1920, manuscript received 3 June 1920, published in issue No. A 686 of 1 July 1920); reprinted in *The Collected Papers of Lord Rutherford. Volume Three: Cambridge* (J. Chadwick, ed.), London: George Allen & Unwin, Ltd., pp. 14–38.

1928 Address of the President at the anniversary meeting, November 30, 1927, *Proc. Roy. Soc. (London)* **A117**, 300–316 (published in issue No. A 777 of 2 January 1928).

SAHA, MEGH NAD

1920 Ionization in the solar chromosphere, *Phil. Mag. (6)* **40**, 472–488 (dated 4 March 1920, published in issue No. 238 of October 1920).

SÄNGER, R.

1926 Temperaturempfindlichkeit der Dielektrizitätskonstanten von CH_4, CH_3Cl, CH_2Cl_2, $CHCl_3$, CCl_4 im dampfförmigen Zustand, *Phys. Zs.* **27**, 556–563 (received 17 July 1926, published in issue No. 17 of 1 September 1926).

SCHLAPP, R.

1928 The Stark effect of the fine-structure of hydrogen, *Proc. Roy. Soc. (London)* **A119**, 313–334 (communicated by C. G. Darwin; received 2 April 1928, published in issue No. A 782 of 1 June 1928).

SCHRÖDINGER, ERWIN

1922a Dopplerprinzip und Bohrsche Frequenzbedingung, *Phys. Zs.* **23**, 301–303 (received 7 June 1922, published in issue No. 15 of 1 August 1922).

1922c Eine bemerkenswerte Eigenschaft der Quantenbahnen des einzelnen Elektrons, *Z. Phys.* **12**, 13–23 (received 5 October 1922, published in issue No. 1/2 of 9 December 1922).

1926c Quantisierung als Eigenwertproblem. (Erste Mitteilung.), *Ann. d. Phys. (4)* **79**, 361–376 (received 27 January 1926, published in issue No. 4 of 13 March 1926). English translation: Quantization as a problem of proper values. (Part I), in *Collected Papers on Wave Mechanics* (translated by J. F. Shearer and W. M. Deans), Glasgow: Blackie & Son, 1928, pp. 1–12.

1926d Quantisierung als Eigenwertproblem. (Zweite Mitteilung.), *Ann. d. Phys. (4)* **79**, 489–527 (received 23 February 1926, published in issue No. 6 of 6 April 1925). English translation: Quantization as a problem of proper values. (Part II), in *Collected Papers on Wave Mechanics*, pp. 13–40.

1926e Über das Verhältnis der Heisenberg-Born-Jordanschen Quantenmechanik zu der meinen, *Ann. d. Phys. (4)* **79**, 734–756 (received 18 March 1926, published in issue No. 8 of 4 May 1926). English translation: On the relation between the quantum mechanics of Heisenberg, Born, and Jordan and that of Schrödinger, in *Collected Papers on Wave Mechanics*, pp. 45–61.

1926f Quantisierung als Eigenwertproblem. (Dritte Mitteilung: Störungstheorie, mit Anwendung auf den Starkeffekt der Balmerlinien.), *Ann. d. Phys. (4)* **80**, 437–490 (received 16 May 1926, published in issue No. 13 of 13 July 1926). English translation: Quantization as a problem of proper values. (Part III) Perturbation theory, with application to the Stark effect of the Balmer lines, in *Collected Papers on Wave Mechanics*, pp 62–101.

SITTER, WILLEM DE

1916a On Einstein's theory of gravitation and its astronomical consequences, *Monthly Notices Roy. Astr. Soc. (London)* **76**, 699–728 (dated 16 August 1916, published in the Supplementary Number, issue No. 9, of October 1916).

1916b On Einstein's theory of gravitation and its astronomical consequences. Second Paper, *Monthly Notices Roy. Astr. Soc. (London)* **77**, 155–184 (dated September–October 1916, published in issue No. 2 of 8 December 1916).

SLATER, JOHN CLARKE

1924a Radiation and atoms, *Nature* **113**, 307–308 (letter dated 28 January 1924, published in the issue of 1 March 1924).

1924b Compressibility of the alkali halides, *Phys. Rev. (2)* **23**, 488–500 (dated 21 July 1923, published in issue No. 4 of April 1924).

1925a A quantum theory of optical phenomena, *Phys. Rev. (2)* **25**, 395–428 (dated 1 December 1924, published in issue No. 4 of April 1925).

1975 *Solid State and Molecular Theory: A Scientific Biography*, New York–Sydney–Toronto: J. Wiley & Sons.

SOMMERFELD, ARNOLD

1922d *Atombau und Spektrallinien*, third edition, Braunschweig: Fr. Vieweg & Sohn. English translation (by H. L. Brose): *Atomic Structure and Spectral Lines*, London: Methuen, and New York: Dutton, 1923.

1924d *Atombau und Spektrallinien*, fourth edition, Braunschweig: Fr. Vieweg & Sohn.

1929 *Atombau und Spektrallinien. Wellenmechanisher Ergänzungsband,* Braunschweig: Fr. Vieweg & Sohn.

SOMMERFELD, ARNOLD, AND ALBRECHT UNSÖLD

1926 Über das Spektrum des Wasserstoffs, *Z. Phys.* **36**, 259–275 (received 5 February 1926, published in issue No. 4 of 22 March 1926); reprinted in A. Sommerfeld: *Gesammelte Schriften III*, Braunschweig: Fr. Vieweg & Sohn, 1968, pp. 770–786.

STONER, EDMUND CLIFTON

1924 The distribution of electrons among atomic levels, *Phil. Mag. (6)* **48**, 719–736 (dated July 1926, published in issue No. 286 of October 1924).

TAIT, PETER GUTHRIE

1867 *An Elementary Treatise on Quanternions*, Oxford: The Clarendon Press. (Later enlarged editions: 1873, 1890).

TAMM, IGOR

1925 Versuch einer quantitativen Fassung des Korrespondenzprinzips und die Berechnung der Spektrallinien. I., *Z. Phys.* **34**, 59–80 (received 15 July 1925, published in issue No. 1 of 21 September 1925).

1926 Zur Quantenmechanik des Rotators, *Z. Phys.* **37**, 685–698 (received 23 April 1926, published in issue No. 9 of 21 June 1926).

THIRRING, HANS

1928 Die Grundgedanken der neueren Quantenmechanik. Erster Teil: Die Entwicklung bis 1926, *Erg. d. exakt. Naturwiss.* **7**, 384–431.

THOMAS, LLEWELLYN HILLETH

1925 An extended form of Kronecker's theorem with an application which shows that Burgers's theorem on adiabatic invariants is statistically true for an assembly, *Proc. Camb. Phil. Soc.* **22**, 886–903 (received 28 July 1925, read 26 October 1925).

1926 The motion of the spinning electron, *Nature* **117**, 514 (letter dated 20 February 1926, published in the issue of 10 April 1926).

THOMSON, GEORGE PAGET

1964 *J. J. Thomson and the Cavendish Laboratory in his Day*, London: Thomas Nelson & Sons, Ltd.

THOMSON, JOSEPH JOHN

1897a Cathode rays, *Notices on the Proceedings at the Meetings of the Members of the Royal Institution of Great Britain* **15**, 419–432 (presented on 30 April 1897, published in issue No. 91 of 1897).

1897b Cathode-rays, *Phil. Mag. (5)* **44**, 293–316 (dated 7 August 1897, published in issue No. 269 of October 1897).

1904c *Electricity and Matter* (1903 Silliman lectures at Yale University), New Haven: Yale University Press.

1912b The unit theory of light, *Proc. Camb. Phil. Soc.* **16**, 643–652 (read 6 May 1912, published in Part VIII).

1913a Rays of positive electricity, *Proc. Roy. Soc. (London)* **A89**, 1–20 (Bakerian lecture delivered on 22 May 1913, manuscript received 4 June 1913, published in issue No. A 607 of 1 August 1913).

TOLMAN, RICHARD CHACE

1914a The principle of similitude, *Phys. Rev. (2)* **3**, 244–255 (dated 18 January 1914, published in issue No. 4 of April 1914).

1914b The specific heat of solids and the principle of similitude, *Phys. Rev. (2)* **4**, 145–153 (dated 12 May 1914, published in issue No. 2 of August 1914).

1918 A general theory of energy partition with application to quantum theory, *Phys. Rev. (2)* **11**, 261–275 (published in issue No. 4 of April 1918).

1923 Rotational specific heat and half quantum numbers, *Phys. Rev. (2)* **22**, 470–478 (dated 16 May 1923, published in issue No. 5 of November 1923).

1924 Duration of molecules in upper quantum states, *Phys. Rev. (2)* **23**, 693–709 (dated 10 January 1924, published in issue No. 6 of June 1924).

1925a The principle of microscopic reversibility, *Proc. Nat. Acad. Sci. (U.S.A.)* **11**, 436–439 (communicated 19 May 1925, published in issue No. 7 of 15 July 1925).

1925b On the estimation of maximum coefficients of absorption, *Phys. Rev. (2)* **26**, 431–432 (dated 24 June 1925, published in issue No. 4 of October 1925).

TOLMAN, RICHARD CHACE, AND RICHARD M. BADGER

1926 A new kind of test of the correspondence principle based on the prediction of the absolute intensities of spectral lines, *Phys. Rev. (2)* **27**, 383–396 (dated 19 January 1926, published in issue No. 4 of April 1926).

UHLENBECK, GEORGE EUGENE, AND SAMUEL GOUDSMIT

1925 Ersetzung der Hypothese vom unmechanischen Zwang durch eine Forderung bezüglich des inneren Verhaltens jedes einzelnen Elektrons, *Naturwiss.* **13**, 953–954 (letter dated 17 October 1925, published in the issue of 20 November 1925).

VAN DER WAERDEN, BARTEL LEENDERT

1967 *Sources of Quantum Mechanics*, Amsterdam: North-Holland Publishing Company; paperback reprint in the *Classics of Science Series, Vol. V*, New York: Dover Publications, Inc., 1968.

VAN VLECK, JOHN HASBROUCK

1922a The dilemma of the helium atom, *Phys. Rev. (2)* **19**, 419–420 (abstract of a paper presented at the Toronto meeting of the American Physical Society, 28–30 December 1921, published in issue No. 4 of April 1922).

1922b The normal helium atom and its relation to the quantum theory, *Phil. Mag. (6)* **44**, 842–869 (dated 13 March 1922, published in issue No. 263 of November 1922).

1924a The absorption of radiation by multiply periodic orbits, and its relation to the correspondence principle and the Rayleigh–Jeans law. Part I. Some extensions of the correspondence principle, *Phys. Rev. (2)* **24**, 330–346 (dated 19 June 1924, published in issue No. 4 of October 1924); reprinted in *Sources of Quantum Mechanics* (van der Waerden, 1967), pp. 203–222.

1924b The absorption of radiation by multiply periodic orbits, and its relation to the correspondence principle and the Rayleigh–Jeans law. Part II. Calculation of the absorption by multiply periodic orbits, *Phys. Rev. (2)* **24**, 347–365 (dated 19 June 1924, published in issue No. 4 of October 1924).

1925 On the quantum theory of the polarization of the resonance radiation in magnetic fields, *Proc. Nat. Acad. Sci. (U.S.A.)* **11**, 612–618 (communicated 20 August 1925, published in issue No. 10 of 15 October 1925).

1926a Magnetic susceptibilities and dielectric constants in the new quantum mechanics, *Nature* **118**, 226–227 (letter published in the issue of 14 August 1926).

1926b On the quantum theory of the specific heat of hydrogen. Part I. Relation to the new mechanics, band spectra, and chemical constants, *Phys. Rev. (2)* **28**, 980–1021 (dated August 1926, published in issue No. 5 of November 1926).

1926c *Quantum Principles and Line Spectra*, Washington, D.C.: National Research Council (*Bull. Nat. Res. Council* **10**, *Part 4*, published in March 1926).

1927a A general proof of the Langevin-Debye formula for susceptibilities of O_2 and NO, *Phys. Rev. (2)* **29**, 613 (abstract of a paper presented at the New York meeting of the American Physical Society, 25–26 February 1927, published in issue No. 4 of April 1927).

1927b On dielectric constants and magnetic susceptibilities in the new quantum mechanics. Part I. A general proof of the Langevin–Debye formula, *Phys. Rev. (2)* **29**, 727–744 (dated 28 February 1927, published in issue No. 5 of May 1927).

1927c On dielectric constants and magnetic susceptibilities in the new quantum mechanics. Part II. Application to dielectric constants, *Phys. Rev. (2)* **30**, 31–54 (dated 5 May 1927, published in issue No. 1 of July 1927).

1928 On the dielectric constants and magnetic susceptibilities in the new quantum mechanics. Part III. Application to dia- and paramagnetism, *Phys. Rev. (2)* **31**, 587–613 (dated 1 February 1928, published in issue No. 4 of April 1928).

1934 A new method of calculating the mean value of $1/r^s$ for Keplerian systems in quantum mechanics, *Proc. Roy. Soc. (London)* **A143**, 679–681 (communicated by P. A. M. Dirac; received 3 November 1933, published in issue No. A 850 of 1 February 1934).

1968 My Swiss visits of 1906, 1926, and 1930, *Helvetica Physica Acta* **41**, 1234–1237 (dated 22 April 1968).

WALLER, IVAR

1926 Der Starkeffekt zweiter Ordnung bei Wasserstoff und die Rydbergkorrektion der Spektra von He and Li^+, *Z. Phys.* **38**, 635–646 (received 21 June 1926, published in issue No. 8 of 21 August 1926).

WENTZEL, GREGOR

1924a Zur Quantenoptik, *Z. Phys.* **22**, 193–199 (received 2 February 1924, published in issue No. 3 of 10 March 1924).

1924d Zur Quantentheorie des Röntgenbremsspektrums, *Z. Phys.* **27**, 257–284 (received 24 July 1924, published in issue No. 4 of 12 September 1924).

1926a Die mehrfach periodischen Systeme in der Quantenmechanik, *Z. Phys.* **37**, 80–94 (received 27 March 1926, published in issue No. 1/2 of 22 May 1926).

1926b Eine Schwierigkeit für die Theorie des Kreiselelektrons, *Z. Phys.* **37**, 911–914 (received 22 May 1926, published in issue No. 12 of 10 July 1926).

WEYL, HERMANN

1908 *Singuläre Integralgleichungen mit besonderer Berücksichtigung des Fourierschen Integraltheorems*, doctoral dissertation, University of Göttingen. Revised as: Singuläre Integralgleichungen, *Mathematische Annalen* **66**, 273–324 (published in issue No. 3 of 15 December 1908).

WHITEHEAD, ALFRED NORTH, AND BERTRAND RUSSELL

1910, 1912, 1913 *Principia Mathematica*, Volumes I, II and III, Cambridge: Cambridge University Press.

WHITTAKER, EDMUND TAYLOR

1902 *A Course on Modern Analysis*, Cambridge: Cambridge University Press.

1904 *A Treatise on the Analytical Dynamics of Particles and Rigid Bodies,* Cambridge: Cambridge University Press.

1917 *A Treatise on the Analytical Dynamics of Particles and Rigid Bodies,* second edition, Cambridge: Cambridge University Press.

1951 *A History of the Theories of Aether and Electricity. Vol. 1: The Classical Theories*, London–Edinburgh–Paris–Melbourne–Toronto–New York: Th. Nelson & Sons (revised and enlarged edition of the book published first in 1910).

1953 *A History of the Theories of Aether and Electricity. Vol. 2: The Modern Theories 1900–1926*, London–Edinburgh–Paris–Melbourne–Toronto–New York: Th. Nelson & Sons.

WILSON, WILLIAM

1915 The quantum theory of radiation and line spectra, *Phil. Mag. (6)* **29**, 795–802 (dated 15 March 1915, published in issue No. 174 of June 1915).

1923 The quantum theory and electromagnetic phenomena, *Proc. Roy. Soc. (London)* **A102**, 478–483 (communicated by J. W. Nicholson; received 21 August 1922, published in issue A 717 of 1 January 1923).

ZAHN, C. T.

1924 The electric moment of gaseous molecules of halogen hydrides, *Phys. Rev. (2)* **24**, 400–417 (dated 8 March 1924, published in issue No. 4 of October 1924).

Author Index

Note: Pages where biographical data are given are indicated in italics.